Smart Cities: Big Data Prediction Methods and Applications

Hui Liu

Smart Cities: Big Data Prediction Methods and Applications

Science Press
Beijing

Springer

Hui Liu
School of Traffic and Transportation Engineering
Central South University
Changsha, Hunan, China

ISBN 978-981-15-2836-1 ISBN 978-981-15-2837-8 (eBook)
https://doi.org/10.1007/978-981-15-2837-8

This Springer imprint is published by the registered company Springer Nature Singapore Pte Ltd.
The registered company address is: 152 Beach Road, #21-01/04 Gateway East, Singapore 189721, Singapore

Preface

Smart city is to make full use of the new generation of information technology in various industries in the city based on the next generation of innovation in the knowledge society of the advanced form of urban information to achieve the deep integration of information, industrialization, and urbanization. The smart city in the twenty-first century can make the most of big data processing technology to analyze and mine the key information of the core system of urban operation, to make an intelligent response to a variety of urban needs, including transportation, environmental protection, public safety, and industrial and commercial activities, and to achieve a better urban life.

A smart city often intersects with big data city, intelligent city, ecological city, low-carbon city, and other regional development concepts, and even mixes with intelligent environment, intelligent traffic, smart grid, and other industry information concepts. At present, research on smart cities has different emphasis, mostly focused on technology application discussion, network construction analysis, wisdom effect research, etc. The research of smart cities is still in the new stage of vigorous development. Therefore, from the point of view of data science, the author has completed the work based on his theoretical achievements in the field of time series and the results of a large number of engineering experiments in smart grid, smart traffic, and smart environment.

Under the background of smart city, the book carries on big data prediction from three aspects of the power grid and building energy, road network traffic flow, environmental index data, which is of great practical significance to the construction and planning of smart city. With the help of big data prediction of power grid and building energy, it can further improve the power grid system and building energy management, the safety of power consumption, and the efficiency of energy utilization. With the help of big data prediction of road network traffic flow, the road network structure can be further improved, alleviate traffic congestion, and enhance the traffic capacity of the road network. With the help of big data prediction of the environment, it can further improve the quality of resident life and provide a healthy and green living environment for residents. The prediction methods proposed of smart cities based on big data in the book have a certain reference significance for the planning, construction, and development of green cities and smart cities.

Changsha, China
August 2019

Hui Liu

Acknowledgements

The studies in the book are supported by the National Natural Science Foundation of China, the National Key R&D Program of China, the Innovation Drive of Central South University, China. In the process of writing the book, Huipeng Shi, Zhihao Long, Hengxin Yin, Guangxi Yan, Zeyu Liu, and other team members have done a lot of model verification and other works; the author would like to express heartfelt appreciations.

About the Book

This book proposes theoretical methods and predictive models for large-scale data prediction and analysis in the smart cities using artificial intelligence and big data. At the same time, this book uses the method of case analysis to verify and analyze the proposed model. All content is divided into four parts, totaling 10 chapters. Part 1 summarizes the research background and development status of several key issues in smart city involved in the book and summarizes the related technologies. This part uses the statistical analysis of the literature to expound the research hotspots and research progress of smart cities with big data technology. It provides a theoretical basis for the study of theoretical algorithms in subsequent chapters. Part 2 uses the cross-correlation analysis to mine and analyze the regional correlation and time-domain variation in urban power load data. At the same time, this part uses ARIMA, SARIMA, ARCH, and SVM to predict the urban electricity consumption and household load consumption and constructs the framework of power load forecasting under the background of big data. Part 3 analyzes the traffic flow data and the vehicle trajectory data and establishes a traffic flow prediction model and a vehicle trajectory prediction model, respectively. Firstly, this part starts with single data-driven traffic flow data and establishes a deterministic prediction model and interval prediction model for traffic flow. Then, the multi-dimensional traffic flow data including the spatiotemporal mapping relationship is analyzed to realize the traffic flow prediction driven by multi-dimensional data. Part 4 starts with the three aspects of urban air quality, urban hydrological status, and urban noise and introduces the intelligent environment prediction technology in the context of big data. The ELM model, Bayesian model, RF algorithm, BFGS algorithm, and GRU model are used to predict the urban environmental time series.

This book can help scientists, engineers, managers, and students engaged in artificial intelligence, big data analysis and forecasting, computational intelligence, smart city, smart grids and energy, intelligent traffic systems, smart environment, air pollutant controlling, and other related research fields.

Contents

Part I Exordium

1 Key Issues of Smart Cities 3
1.1 Smart Grid and Buildings 3
1.1.1 Overview of Smart Grid and Building 4
1.1.2 The Importance of Smart Grid and Buildings in Smart City 5
1.1.3 Framework of Smart Grid and Buildings 6
1.2 Smart Traffic Systems 6
1.2.1 Overview of Smart Traffic Systems 6
1.2.2 The Importance of Smart Traffic Systems for Smart City 6
1.2.3 Framework of Smart Traffic Systems 8
1.3 Smart Environment 8
1.3.1 Overview of Smart Environment for Smart City 8
1.3.2 The Importance of Smart Environment for Smart City 10
1.3.3 Framework of Smart Environment 11
1.4 Framework of Smart Cities 11
1.4.1 Key Points of Smart City in the Era of Big Data 11
1.4.2 Big Data Time-series Forecasting Methods in Smart Cities 12
1.4.3 Overall Framework of Big Data Forecasting in Smart Cities 13
1.5 The Importance Analysis of Big Data Forecasting Architecture for Smart Cities 14
1.5.1 Overview and Necessity of Research 14
1.5.2 Review on Big Data Forecasting in Smart Cities 15
1.5.3 Review on Big Data Forecasting in Smart Gird and Buildings 18

1.5.4 Review on Big Data Forecasting in Smart Traffic Systems ... 21
1.5.5 Review on Big Data Forecasting in Smart Environment ... 22
References ... 23

Part II Smart Grid and Buildings

2 Electrical Characteristics and Correlation Analysis in Smart Grid ... 27
2.1 Introduction ... 27
2.2 Extraction of Building Electrical Features ... 28
2.2.1 Analysis of Meteorological Elements ... 29
2.2.2 Analysis of System Load ... 30
2.2.3 Analysis of Thermal Perturbation ... 31
2.3 Cross-Correlation Analysis of Electrical Characteristics ... 33
2.3.1 Cross-Correlation Analysis Based on MI ... 33
2.3.2 Cross-Correlation Analysis Based on Pearson Coefficient ... 35
2.3.3 Cross-Correlation Analysis Based on Kendall Coefficient ... 37
2.4 Selection of Electrical Characteristics ... 40
2.4.1 Electrical Characteristics of Construction Power Grid ... 40
2.4.2 Feature Selection Based on Spearman Correlation Coefficient ... 41
2.4.3 Feature Selection Based on CFS ... 43
2.4.4 Feature Selection Based on Global Search-ELM ... 45
2.5 Conclusion ... 46
References ... 48

3 Prediction Model of City Electricity Consumption ... 51
3.1 Introduction ... 51
3.2 Original Electricity Consumption Series ... 54
3.2.1 Regional Correlation Analysis of Electricity Consumption Series ... 54
3.2.2 Original Sequences for Modeling ... 55
3.2.3 Separation of Sample ... 56
3.3 Short-Term Deterministic Prediction of Electricity Consumption Based on ARIMA Model ... 58
3.3.1 Model Framework of ARIMA ... 58
3.3.2 Theoretical Basis of ARIMA ... 59
3.3.3 Modeling Steps of ARIMA Predictive Model ... 60
3.3.4 Forecasting Results ... 64

3.4 Power Consumption Interval Prediction Based on ARIMA-ARCH Model 69
3.4.1 Model Framework of ARCH 69
3.4.2 The Theoretical Basis of the ARCH 69
3.4.3 Modeling Steps of ARIMA-ARCH Interval Predictive Model 70
3.4.4 Forecasting Results 71
3.5 Long-Term Electricity Consumption Prediction Based on the SARIMA Model 76
3.5.1 Model Framework of the SARIMA 76
3.5.2 The Theoretical Basis of the SARIMA 77
3.5.3 Modeling Steps of the SARIMA Predictive Model 78
3.5.4 Forecasting Results 79
3.6 Big Data Prediction Architecture of Household Electric Power 81
3.7 Comparative Analysis of Forecasting Performance 84
3.8 Conclusion 86
References 88

4 Prediction Models of Energy Consumption in Smart Urban Buildings 89
4.1 Introduction 89
4.2 Establishment of Building Simulating Model 91
4.2.1 Description and Analysis of the BEMPs 91
4.2.2 Main Characters of DeST Software 94
4.2.3 Process of DeST Modeling 95
4.3 Analysis and Comparison of Different Parameters 101
4.3.1 Introduction of the Research 101
4.3.2 Meteorological Parameters 102
4.3.3 Indoor Thermal Perturbation 103
4.3.4 Enclosure Structure and Material Performance 105
4.3.5 Indoor Design Parameters 106
4.4 Data Acquisition of Building Model 108
4.4.1 Data After Modeling 108
4.4.2 Calculation of Room Temperature and Load 108
4.4.3 Calculation of Shadow and Light 108
4.4.4 Calculation of Natural Ventilation 109
4.4.5 Simulation of the Air-Conditioning System 110
4.5 SVM Prediction Model for Urban Building Energy Consumption 110
4.5.1 The Theoretical Basis of the SVM 110
4.5.2 Steps of Modeling 112
4.5.3 Forecasting Results 114
4.6 Big Data Prediction of Energy Consumption in Urban Building 115

4.6.1 Big Data Framework for Energy Consumption 117
4.6.2 Big Data Storage and Analysis for Energy Consumption 117
4.6.3 Big Data Mining for Energy Consumption 117
4.7 Conclusion 119
References 120

Part III Smart Traffic Systems

5 Characteristics and Analysis of Urban Traffic Flow in Smart Traffic Systems 125
5.1 Introduction 125
5.1.1 Overview of Trajectory Prediction of Smart Vehicle 125
5.1.2 The Significance of Trajectory Prediction for Smart City 126
5.1.3 Overall Framework of Model 127
5.2 Traffic Flow Time Distribution Characteristics and Analysis 129
5.2.1 Original Vehicle Trajectory Series 129
5.2.2 Separation of Sample 131
5.3 The Spatial Distribution Characteristics and Analysis of Traffic Flow 132
5.3.1 Trajectory Prediction of Urban Vehicles Based on Single Data 132
5.3.2 Trajectory Prediction of Urban Vehicles Based on Multiple Data 140
5.3.3 Trajectory Prediction of Urban Vehicles Under EWT Decomposition Framework 146
5.3.4 Comparative Analysis of Forecasting Performance 153
5.4 Conclusion 156
References 157

6 Prediction Model of Traffic Flow Driven Based on Single Data in Smart Traffic Systems 159
6.1 Introduction 159
6.2 Original Traffic Flow Series for Prediction 161
6.3 Traffic Flow Deterministic Prediction Driven by Single Data 162
6.3.1 Modeling Process 162
6.3.2 The Prediction Results 167
6.4 Traffic Flow Interval Prediction Model Driven by Single Data 167
6.4.1 The Framework of the Interval Prediction Model 167
6.4.2 Modeling Process 170
6.4.3 The Prediction Results 174

6.5 Traffic Flow Interval Prediction Under Decomposition Framework 175
6.5.1 The Framework of the WD-BP-GARCH Prediction Model 175
6.5.2 Modeling Process 184
6.5.3 The Prediction Results 187
6.6 Big Data Prediction Architecture of Traffic Flow 190
6.7 Comparative Analysis of Forecasting Performance 191
6.8 Conclusion 193
References 193

7 Prediction Models of Traffic Flow Driven Based on Multi-Dimensional Data in Smart Traffic Systems 195
7.1 Introduction 195
7.2 Analysis of Traffic Flow and Its Influencing Factors 196
7.3 Elman Prediction Model of Traffic Flow Based on Multiple Data 198
7.3.1 The Framework of the Elman Prediction Model 198
7.3.2 Modeling Process 198
7.3.3 The Prediction Results 201
7.4 LSTM Prediction Model of Traffic Flow Based on Multiple Data 202
7.4.1 The Framework of the LSTM Prediction Model 202
7.4.2 Modeling Process 205
7.4.3 The Prediction Results 206
7.5 Traffic Flow Prediction Under Wavelet Packet Decomposition 207
7.5.1 The Framework of the WPD-Prediction Model 207
7.5.2 Modeling Process 210
7.5.3 The Prediction Results 214
7.6 Comparative Analysis of Forecasting Performance 216
7.7 Conclusion 220
References 222

Part IV Smart Environment

8 Prediction Models of Urban Air Quality in Smart Environment 227
8.1 Introduction 227
8.2 Original Air Pollutant Concentrations Series for Prediction 228
8.2.1 Original Sequence for Modeling 228
8.2.2 Separation of Sample 231
8.3 Air Quality Prediction Model Driven by Single Data 232
8.3.1 Model Framework 232
8.3.2 Theoretical Basis of ELM 232

8.3.3 Steps of Modeling 233
8.3.4 Forecasting Results 233
8.4 Air Quality Mixture Prediction Model Driven by Multiple Data 234
8.4.1 Model Framework 234
8.4.2 Steps of Modeling 235
8.4.3 Forecasting Results 237
8.5 Air Quality Prediction Under Feature Extraction Framework 238
8.5.1 Model Framework 238
8.5.2 Theoretical Basis of Feature Extraction Method 238
8.5.3 Steps of Modeling 250
8.5.4 Forecasting Results 251
8.6 Big Data Prediction Architecture of Urban Air Quality 253
8.6.1 The Idea of Urban Air Quality Prediction Based on Hadoop 253
8.6.2 Parallelization Framework of the ELM 254
8.6.3 The Parallelized ELM Under the MapReduce Framework 254
8.7 Comparative Analysis of Forecasting Performance 256
8.8 Conclusion 258
References 259

9 Prediction Models of Urban Hydrological Status in Smart Environment 261
9.1 Introduction 261
9.2 Original Hydrological State Data for Prediction 262
9.2.1 Original Sequence for Modeling 262
9.2.2 Separation of Sample 265
9.3 Bayesian Classifier Prediction of Water Level Fluctuation 265
9.3.1 Model Framework 265
9.3.2 Theoretical Basis of the Bayesian Classifier 266
9.3.3 Steps of Modeling 267
9.3.4 Forecasting Results 268
9.4 The Elman Prediction of Urban Water Level 269
9.4.1 Model Framework 269
9.4.2 The Theoretical Basis of the Elman 270
9.4.3 Steps of Modeling 270
9.4.4 Forecasting Results 271
9.5 Urban River Water Level Decomposition Hybrid Prediction Model 272
9.5.1 Model Framework 272
9.5.2 The Theoretical Basis 272
9.5.3 Steps of Modeling 276
9.5.4 Forecasting Results 277

9.5.5 Influence and Analysis of Decomposition Parameters on Forecasting Performance of Hybrid Models 280
9.6 Comparative Analysis of Forecasting Performance 284
9.7 Conclusion 287
References 288

10 Prediction Model of Urban Environmental Noise in Smart Environment 289
10.1 Introduction 289
10.1.1 Hazard of Noise 289
10.1.2 The Significance of Noise Prediction for Smart City 290
10.1.3 Overall Framework of Model 291
10.2 Original Urban Environmental Noise Series 292
10.2.1 Original Sequence for Modeling 292
10.2.2 Separation of Sample 294
10.3 The RF Prediction Model for Urban Environmental Noise 295
10.3.1 The Theoretical Basis of the RF 295
10.3.2 Steps of Modeling 295
10.3.3 Forecasting Results 296
10.4 The BFGS Prediction Model for Urban Environmental Noise 298
10.4.1 The Theoretical Basis of the BFGS 298
10.4.2 Steps of Modeling 299
10.4.3 Forecasting Results 299
10.5 The GRU Prediction Model for Urban Environmental Noise 302
10.5.1 The Theoretical Basis of the GRU 302
10.5.2 Steps of Modeling 303
10.5.3 Forecasting Results 304
10.6 Big Data Prediction Architecture of Urban Environmental Noise 305
10.6.1 Big Data Framework for Urban Environmental Noise Prediction 307
10.6.2 Big Data Storage for Urban Environmental Noise Prediction 308
10.6.3 Big Data Processing of Urban Environmental Noise Prediction 308
10.7 Comparative Analysis of Forecasting Performance 310
10.8 Conclusion 312
References 313

List of Figures

Fig. 1.1 Framework of smart grid and buildings 7
Fig. 1.2 Framework of smart traffic systems 9
Fig. 1.3 Framework of smart environment. 12
Fig. 1.4 Overall framework of big data forecasting in smart cities 14
Fig. 1.5 Citation report of subject retrieval on “TS= (smart cities) AND TS= (big data forecasting OR big data prediction)”. 16
Fig. 1.6 The network diagram of documents based on the subject retrieval of “TS= (smart cities) AND TS= (big data forecasting OR big data prediction)”. 17
Fig. 1.7 Research direction map of document based on the subject retrieval of “TS= (smart cities) AND TS= (big data forecasting OR big data prediction)”. 18
Fig. 1.8 The overlay map of document based on the subject retrieval of “TS= (smart cities) AND TS= (big data forecasting OR big data prediction)” 19
Fig. 1.9 The item density map of document based on the subject retrieval of “TS= (smart cities) AND TS= (big data forecasting OR big data prediction)” 20
Fig. 1.10 The cluster density map of document based on the subject retrieval of “TS= (smart cities) AND TS= (big data forecasting OR big data prediction)”. 20
Fig. 1.11 Subject retrieval of annual publication volume of various types of literature on “TS= (smart grid OR smart buildings) AND TS= (big data forecasting OR big data prediction)”. 21
Fig. 1.12 Subject retrieval of annual publication volume of various types of literature on “TS= (smart traffic OR smart transportation) AND TS= (big data forecasting OR big data prediction)” 22
Fig. 1.13 Subject retrieval of annual publication volume of various types of literature on “TS= (smart environment) AND TS= (big data forecasting OR big data prediction)”. 23

Fig. 2.1 Original meteorological factors series 29
Fig. 2.2 Original system load factors series. 31
Fig. 2.3 Original thermal perturbation indoors series 32
Fig. 2.4 Heat map of cross-correlation result based on MI 34
Fig. 2.5 Heat map of cross-correlation result based on Pearson coefficient. 36
Fig. 2.6 Heat map of cross-correlation result based on Kendall coefficient. 38
Fig. 2.7 Original supply air volume series. 39
Fig. 2.8 Original loop pressure loss series. 39
Fig. 2.9 Original power consumption series 40
Fig. 2.10 Feature selection result based on Spearman correlation analysis . . 42
Fig. 2.11 Flowchart of CFS 43
Fig. 2.12 Flowchart of Global Search-ELM 46
Fig. 2.13 Feature selection result based on Global Search-ELM. 47
Fig. 3.1 Long-term electricity consumption series of different regions. . . . 54
Fig. 3.2 Short-term electricity consumption series 56
Fig. 3.3 Long-term electricity consumption series 57
Fig. 3.4 Separation of short-term electricity consumption series 57
Fig. 3.5 Separation of long-term electricity consumption series 58
Fig. 3.6 Modeling process of ARIMA predictive model 59
Fig. 3.7 Short-term electricity consumption first-order difference series 61
Fig. 3.8 Long-term electricity consumption first-order difference series 61
Fig. 3.9 Short-term series calculation results of autocorrelation coefficient and the partial autocorrelation coefficient 63
Fig. 3.10 Long-term series calculation results of autocorrelation coefficient and the partial autocorrelation coefficient 63
Fig. 3.11 Short-term series forecasting results of the ARIMA models in 1-step 65
Fig. 3.12 Short-term series forecasting results of the ARIMA models in 2-step 65
Fig. 3.13 Short-term series forecasting results of the ARIMA models in 3-step 66
Fig. 3.14 Long-term series forecasting results of the ARIMA models in 1-step 67
Fig. 3.15 Long-term series forecasting results of the ARIMA models in 2-step 67
Fig. 3.16 Long-term series forecasting results of the ARIMA models in 3-step 68
Fig. 3.17 Modeling process of the ARCH predictive model 69
Fig. 3.18 Short-term series predicted residuals series 71
Fig. 3.19 Long-term series predicted residuals series 72

Fig. 3.20 Short-term series interval forecasting results of the ARIMA models in 1-step . . . 74
Fig. 3.21 Short-term series interval forecasting results of the ARIMA models in 2-step . . . 74
Fig. 3.22 Short-term series interval forecasting results of the ARIMA models in 3-step . . . 75
Fig. 3.23 Long-term series interval forecasting results of the ARIMA models in 1-step . . . 75
Fig. 3.24 Long-term series interval forecasting results of the ARIMA models in 2-step . . . 76
Fig. 3.25 Long-term series interval forecasting results of the ARIMA models in 3-step . . . 76
Fig. 3.26 Modeling process of SARIMA predictive model . . . 77
Fig. 3.27 Short-term series relationship between minimum BIC detection and period . . . 79
Fig. 3.28 Long-term series relationship between minimum BIC detection and period . . . 80
Fig. 3.29 Short-term series forecasting results of the SARIMA models in 1-step . . . 81
Fig. 3.30 Short-term series forecasting results of the SARIMA models in 2-step . . . 82
Fig. 3.31 Short-term series forecasting results of the SARIMA models in 3-step . . . 82
Fig. 3.32 Long-term series forecasting results of the SARIMA models in 1-step . . . 83
Fig. 3.33 Long-term series forecasting results of the SARIMA models in 2-step . . . 83
Fig. 3.34 Long-term series forecasting results of the SARIMA models in 3-step . . . 84
Fig. 3.35 The MapReduce prediction architecture of electricity consumption . . . 85

Fig. 4.1 Connection diagram between smart city and smart building . . . 91
Fig. 4.2 Building energy modeling programs . . . 93
Fig. 4.3 Annual base temperature chart of 10 rooms . . . 94
Fig. 4.4 Daily dry-bulb temperature . . . 95
Fig. 4.5 2-D floor plan of 6 story building . . . 96
Fig. 4.6 3-D display of 6 story building . . . 97
Fig. 4.7 Building model plan . . . 99
Fig. 4.8 Shadow analysis of the building model from daily and annual solar path . . . 100
Fig. 4.9 The steps of modeling . . . 101
Fig. 4.10 Annual hourly load in the building . . . 102
Fig. 4.11 Annual hourly load per area in the building . . . 103
Fig. 4.12 The temperature of the hottest month . . . 103

Fig. 4.13 The temperature of the coldest month 104
Fig. 4.14 Annual direct solar radiation 104
Fig. 4.15 The original data of 60 days load 105
Fig. 4.16 The temperature of 60 days after adjustment of meteorological parameter 105
Fig. 4.17 Indoor thermal perturbation setting of a room in a week 106
Fig. 4.18 The temperature of 60 days after adjustment of indoor thermal perturbation. 106
Fig. 4.19 The material change of the exterior wall 107
Fig. 4.20 The temperature of 60 days after adjustment of enclosure structure and material 108
Fig. 4.21 The construction of a 3-floor building with elevator 109
Fig. 4.22 The temperature of 60 days after adjustment of building construction 110
Fig. 4.23 Forecasting results of 1-step strategy 115
Fig. 4.24 Forecasting results of 2-step strategy 115
Fig. 4.25 Forecasting results of 3-step strategy 116
Fig. 4.26 The framework of Hadoop-SVM model. 119

Fig. 5.1 The general framework of the chapter 128
Fig. 5.2 D1: (**a**) Original longitude series; (**b**) Original latitude series; (**c**) Original speed series; (**d**) Original angle series 129
Fig. 5.3 D2: (**a**) Original longitude series; (**b**) Original latitude series; (**c**) Original speed series; (**d**) Original angle series 130
Fig. 5.4 D3: (**a**) Original longitude series; (**b**) Original latitude series; (**c**) Original speed series; (**d**) Original angle series 130
Fig. 5.5 The prediction principle of BP network 133
Fig. 5.6 The general framework of this section 134
Fig. 5.7 Trajectory prediction results of the ELM model (single data) for D1 135
Fig. 5.8 Trajectory prediction results of the ELM model (single data) for D2 135
Fig. 5.9 Trajectory prediction results of the ELM model (single data) for D3 136
Fig. 5.10 Trajectory prediction results of the BPNN (single data) for D1. . 138
Fig. 5.11 Trajectory prediction results of the BPNN (single data) for D2. . 138
Fig. 5.12 Trajectory prediction results of the BPNN (single data) for D3. . 139
Fig. 5.13 The general framework of this section 140
Fig. 5.14 Trajectory prediction results of the ELM model (multiple data) for D1 141
Fig. 5.15 Trajectory prediction results of the ELM model (multiple data) for D2 142
Fig. 5.16 Trajectory prediction results of the ELM model (multiple data) for D3 142
Fig. 5.17 Trajectory prediction results of the BPNN (multiple data) for D1 144

Fig. 5.18 Trajectory prediction results of the BPNN (multiple data) for D2 . . . 144
Fig. 5.19 Trajectory prediction results of the BPNN (multiple data) for D3 . . . 145
Fig. 5.20 The general framework of this chapter . . . 148
Fig. 5.21 Trajectory prediction results of the EWT-ELM model for D1 . . . 149
Fig. 5.22 Trajectory prediction results of the EWT-ELM model for D2 . . . 150
Fig. 5.23 Trajectory prediction results of the EWT-ELM model for D3 . . . 150
Fig. 5.24 Trajectory prediction results of the EWT-BPNN for D1 . . . 152
Fig. 5.25 Trajectory prediction results of the EWT-BPNN for D2 . . . 152
Fig. 5.26 Trajectory prediction results of the EWT-BPNN for D3 . . . 153

Fig. 6.1 Distribution of traffic flow series . . . 161
Fig. 6.2 The input/output data structure of single data . . . 163
Fig. 6.3 The training set and testing set . . . 163
Fig. 6.4 The results of the 1-step prediction for the traffic flow series . . . 164
Fig. 6.5 The results of the 2-step prediction for the traffic flow series . . . 164
Fig. 6.6 The results of the 3-step prediction for the traffic flow series . . . 165
Fig. 6.7 The results of the 4-step prediction for the traffic flow series . . . 165
Fig. 6.8 The results of the 5-step prediction for the traffic flow series . . . 166
Fig. 6.9 Comparison of prediction results based on different prediction steps . . . 166
Fig. 6.10 Interval prediction model flow . . . 168
Fig. 6.11 The division of datasets . . . 169
Fig. 6.12 The results of the 1-step interval prediction for the traffic flow series . . . 171
Fig. 6.13 The results of the 2-step interval prediction for the traffic flow series . . . 172
Fig. 6.14 The results of the 3-step interval prediction for the traffic flow series . . . 172
Fig. 6.15 The results of the 4-step interval prediction for the traffic flow series . . . 173
Fig. 6.16 The results of the 5-step interval prediction for the traffic flow series . . . 173
Fig. 6.17 The structure of the WD . . . 175
Fig. 6.18 The prediction process of the WD-BP-GARCH model for traffic flow series . . . 176
Fig. 6.19 The eight sub-series of the traffic flow series based on the WD . . . 178
Fig. 6.20 The prediction results of 3-step for each sub-series based on the BP prediction model . . . 179
Fig. 6.21 The prediction results of 3-step for each sub-series based on the BP-GARCH interval model . . . 180
Fig. 6.22 The results of the 1-step prediction for the traffic flow series . . . 181
Fig. 6.23 The results of the 2-step prediction for the traffic flow series . . . 181
Fig. 6.24 The results of the 3-step prediction for the traffic flow series . . . 182

Fig. 6.25 The results of the 4-step prediction for the traffic flow series . . . 182
Fig. 6.26 The results of the 5-step prediction for the traffic flow series . . . 183
Fig. 6.27 Comparison of prediction results based on different prediction steps . . . 183
Fig. 6.28 The results of the 1-step interval prediction for the traffic flow series . . . 185
Fig. 6.29 The results of the 2-step interval prediction for the traffic flow series . . . 185
Fig. 6.30 The results of the 3-step interval prediction for the traffic flow series . . . 186
Fig. 6.31 The results of the 4-step interval prediction for the traffic flow series . . . 186
Fig. 6.32 The results of the 5-step interval prediction for the traffic flow series . . . 187
Fig. 6.33 The MapReduce prediction architecture of traffic flow . . . 189

Fig. 7.1 Traffic flow motion diagram . . . 197
Fig. 7.2 The principle of the Elman network . . . 199
Fig. 7.3 The map of Rd1, Rd2, and Rd3 . . . 200
Fig. 7.4 The sample distribution of the three traffic flow series . . . 200
Fig. 7.5 The input/output data structure of one-step prediction . . . 201
Fig. 7.6 The results of the 1-step prediction for the traffic flow series . . . 202
Fig. 7.7 The results of the 2-step prediction for the traffic flow series . . . 203
Fig. 7.8 The results of the 3-step prediction for the traffic flow series . . . 203
Fig. 7.9 The results of the 4-step prediction for the traffic flow series . . . 204
Fig. 7.10 The results of the 5-step prediction for the traffic flow series . . . 204
Fig. 7.11 The principle of LSTM network . . . 205
Fig. 7.12 The results of the 1-step prediction for the traffic flow series . . . 206
Fig. 7.13 The results of the 2-step prediction for the traffic flow series . . . 207
Fig. 7.14 The results of the 3-step prediction for the traffic flow series . . . 207
Fig. 7.15 The results of the 4-step prediction for the traffic flow series . . . 208
Fig. 7.16 The results of the 5-step prediction for the traffic flow series . . . 208
Fig. 7.17 The structure of the WPD . . . 209
Fig. 7.18 The principle of WPD-prediction model . . . 210
Fig. 7.19 The sample distribution of the three traffic flow series . . . 211
Fig. 7.20 The eight sub-series of Rd1 traffic flow series based on the WPD . . . 212
Fig. 7.21 The eight sub-series of Rd2 traffic flow series based on the WPD . . . 212
Fig. 7.22 The eight sub-series of Rd3 traffic flow series based on the WPD . . . 213
Fig. 7.23 The input/output data structure of 1-step prediction . . . 213
Fig. 7.24 The results of the 1-step prediction for the traffic flow series . . . 215
Fig. 7.25 The results of the 2-step prediction for the traffic flow series . . . 215
Fig. 7.26 The results of the 3-step prediction for the traffic flow series . . . 216

Fig. 7.27 The results of the 4-step prediction for the traffic flow series ... 216
Fig. 7.28 The results of the 5-step prediction for the traffic flow series ... 217
Fig. 7.29 The results of the 1-step prediction for the traffic flow series ... 218
Fig. 7.30 The results of the 2-step prediction for the traffic flow series ... 218
Fig. 7.31 The results of the 3-step prediction for the traffic flow series ... 219
Fig. 7.32 The results of the 4-step prediction for the traffic flow series ... 219
Fig. 7.33 The results of the 5-step prediction for the traffic flow series ... 220

Fig. 8.1 Schematic diagram of the prediction process of the proposed prediction models ... 229
Fig. 8.2 Original AQI series ... 230
Fig. 8.3 Original air pollutant concentrations series ... 230
Fig. 8.4 Modeling flowchart of ELM prediction model driven by single data ... 232
Fig. 8.5 The forecasting results of AQI time series by ELM (single data) in 1-step ... 234
Fig. 8.6 The forecasting results of AQI time series by ELM (single data) in 2-step ... 234
Fig. 8.7 The forecasting results of AQI time series by the ELM (single data) in 3-step ... 235
Fig. 8.8 Modeling flowchart of the ELM prediction model driven by multiple data ... 236
Fig. 8.9 The forecasting results of AQI series by the ELM (multiple data) in 1-step ... 236
Fig. 8.10 The forecasting results of AQI series by the ELM (multiple data) in 2-step ... 237
Fig. 8.11 The forecasting results of AQI series by the ELM (multiple data) in 3-step ... 238
Fig. 8.12 Modeling flowchart of the ELM prediction model under feature extraction framework ... 239
Fig. 8.13 Box diagram of original feature data ... 241
Fig. 8.14 Box diagram of standardized feature data ... 241
Fig. 8.15 Identification results of PCA ... 243
Fig. 8.16 Sample series of selected principal components ... 243
Fig. 8.17 The identification results of the Gaussian KPCA ... 247
Fig. 8.18 Identification results of the FA ... 250
Fig. 8.19 Sample series of selected common factors ... 251
Fig. 8.20 The forecasting results of AQI time series by the PCA-ELM in 1-step ... 252
Fig. 8.21 The forecasting results of AQI time series by the KPCA-ELM in 1-step ... 252
Fig. 8.22 The forecasting results of AQI time series by the FA-ELM in 1-step ... 253
Fig. 8.23 Structure-based parallelized ELM ... 254
Fig. 8.24 The data-based parallelized ELM ... 255

Fig. 8.25 The forecasting results of AQI time series by proposed models in 1-step 256
Fig. 8.26 The forecasting results of AQI time series by proposed models in 2-step 257
Fig. 8.27 The forecasting results of AQI time series by proposed models in 3-step 257

Fig. 9.1 Original water level height series $\{X_1\}$. 263
Fig. 9.2 Original water level height series $\{X_2\}$. 263
Fig. 9.3 Fluctuation state of water level series $\{X_1\}$ 264
Fig. 9.4 Fluctuation state of water level series $\{X_2\}$ 264
Fig. 9.5 Modeling flowchart of Naive Bayesian classification predictor model. 266
Fig. 9.6 Fluctuation trend prediction result of original water level series $\{X_1\}$. 268
Fig. 9.7 Fluctuation trend prediction result of original water level series $\{X_2\}$. 269
Fig. 9.8 Modeling flowchart of the Elman water level prediction model 269
Fig. 9.9 The forecasting results of water level time series $\{X_1\}$ by the Elman. 271
Fig. 9.10 The forecasting results of water level time series $\{X_2\}$ by the Elman. 272
Fig. 9.11 Modeling flowchart of the Elman water level prediction model under decomposition framework. 273
Fig. 9.12 MODWT results of the original water level series $\{X_1\}$. 274
Fig. 9.13 The EMD results of the original water level series $\{X_1\}$ 275
Fig. 9.14 The SSA results of the original water level series $\{X_1\}$ 276
Fig. 9.15 The forecasting results of water level series $\{X_1\}$ by the MODWT-Elman. 278
Fig. 9.16 The forecasting results of water level series $\{X_1\}$ by the EMD-Elman. 278
Fig. 9.17 The forecasting results of water level series $\{X_1\}$ by the SSA-Elman 279
Fig. 9.18 Forecasting performance indices of different decomposition layer of the MODWT 281
Fig. 9.19 Forecasting performance indices of different mother wavelet of the MODWT. 282
Fig. 9.20 Forecasting performance indices of different types of mother wavelet of the MODWT 282
Fig. 9.21 Forecasting performance indices of different window length of the SSA. 283
Fig. 9.22 Forecasting results of water level series $\{X_1\}$ by optimal models in 1-step 284

Fig. 9.23 Forecasting results of water level series $\{X_1\}$ by optimal models in 2-step ... 285
Fig. 9.24 Forecasting results of water level series $\{X_1\}$ by optimal models in 3-step ... 285

Fig. 10.1 The general framework of the chapter ... 292
Fig. 10.2 Original public noise series ... 293
Fig. 10.3 Original neighborhood noise series ... 293
Fig. 10.4 Original traffic noise series... 294
Fig. 10.5 Forecasting results of the RF model for D1 ... 296
Fig. 10.6 Forecasting results of the RF model for D2 ... 297
Fig. 10.7 Forecasting results of the RF model for D3 ... 297
Fig. 10.8 Forecasting results of the BFGS model for D1 ... 300
Fig. 10.9 Forecasting results of the BFGS model for D2 ... 301
Fig. 10.10 Forecasting results of the BFGS model for D3 ... 301
Fig. 10.11 The structure of the GRU ... 303
Fig. 10.12 The framework of the GRU in the section ... 303
Fig. 10.13 Forecasting results of the GRU model for D1 ... 305
Fig. 10.14 Forecasting results of the GRU model for D2 ... 306
Fig. 10.15 Forecasting results of the GRU model for D3 ... 306
Fig. 10.16 The framework of Spark-RF model ... 309
Fig. 10.17 Forecasting results of proposed models for D1 ... 311
Fig. 10.18 Forecasting results of proposed models for D2 ... 311
Fig. 10.19 Forecasting results of proposed models for D3 ... 312

List of Tables

Table 2.1 Statistical characteristics of the original meteorological factors series ... 29
Table 2.2 Statistical characteristics of the original system load factors series ... 31
Table 2.3 Statistical characteristics of the original thermal perturbation indoors series ... 32
Table 2.4 Cross-correlation coefficient based on MI ... 34
Table 2.5 Cross-correlation coefficient based on Pearson coefficient ... 36
Table 2.6 Cross-correlation coefficient based on Kendall coefficient ... 38
Table 2.7 Statistical characteristics of the original electrical characteristics series ... 40
Table 2.8 Spearman coefficient of feature selection ... 42
Table 2.9 Correlation coefficient of 8 groups of variables ... 43
Table 2.10 Calculation result based on forward selection search ... 44
Table 2.11 Calculation result based on backward elimination search ... 44

Table 3.1 Cross-correlation coefficient of different regions based on Pearson ... 55
Table 3.2 Cross-correlation coefficient of different regions based on Kendall ... 55
Table 3.3 Cross-correlation coefficient of different regions based on Spearman ... 55
Table 3.4 Correlation coefficient characteristics of ARIMA model ... 62
Table 3.5 The forecasting performance indices of the ARIMA model for series 1 ... 64
Table 3.6 The forecasting performance indices of the ARIMA model for series 2 ... 64
Table 3.7 The forecasting performance indices of the ARIMA-ARCH model in 1-step ... 72
Table 3.8 The forecasting performance indices of the ARIMA-ARCH model in 2-step ... 73

Table 3.9 The forecasting performance indices of the ARIMA-ARCH model in 3-step . . . 73
Table 3.10 The forecasting performance indices of the SARIMA model for series 1 . . . 80
Table 3.11 The forecasting performance indices of the SARIMA model for series 2 . . . 81
Table 3.12 The forecasting performance comparison of different model for series 1 . . . 86
Table 3.13 The forecasting performance comparison of different model for series 2 . . . 86

Table 4.1 The change of structure materials . . . 107
Table 4.2 Output after modeling . . . 110
Table 4.3 Output after room temperature calculation . . . 111
Table 4.4 Output after room load calculation . . . 111
Table 4.5 Output after calculation of building daily shadow . . . 112
Table 4.6 Output after calculation of building illumination . . . 112
Table 4.7 Output after calculation of natural ventilation. . . . 113
Table 4.8 Output after calculation of air-conditioning scheme. . . . 113
Table 4.9 Output after calculation of wind network . . . 113
Table 4.10 Output after calculation of AHC simulation . . . 113
Table 4.11 The forecasting performance indices of the SVM model . . . 117

Table 5.1 Statistical characteristics of the original series . . . 131
Table 5.2 The forecasting performance indices of the ELM (single data) model . . . 137
Table 5.3 The forecasting performance indices of the BPNN (single data) model . . . 139
Table 5.4 The forecasting performance indices of the ELM (multiple data) model . . . 143
Table 5.5 The forecasting performance indices of the BPNN (multiple data) model . . . 146
Table 5.6 The forecasting performance indices of EWT-ELM model . . . 151
Table 5.7 The forecasting performance indices of EWT-BPNN model . . . 153
Table 5.8 The comprehensive forecasting performance indices of the ELM models . . . 154
Table 5.9 The comprehensive forecasting performance indices of the BP models . . . 155

Table 6.1 The mathematical statistical information of original series . . . 162
Table 6.2 The performance estimating results of the BP prediction model . . . 167
Table 6.3 The performance estimating results of the interval prediction model . . . 174
Table 6.4 The performance estimating results of the WD-BP prediction model . . . 184

Table 6.5 The performance estimating results of the WD-BP-GARCH prediction model 188
Table 6.6 The performance estimating results of the deterministic prediction models 189
Table 6.7 The performance estimating results of the interval prediction model 190

Table 7.1 The specific meanings of Rd1, Rd2, and Rd3 199
Table 7.2 The performance estimating results of the Elman prediction model 205
Table 7.3 The performance estimating results of the LSTM prediction model 209
Table 7.4 The performance estimating results of the WPD-prediction model 217
Table 7.5 The performance estimating results of the involved prediction models 221

Table 8.1 Statistical characteristics of the original series 231
Table 8.2 The forecasting performance indices of the ELM (single data) model 235
Table 8.3 The forecasting performance indices of the ELM (multiple data) model 238
Table 8.4 Identification results of PCA 242
Table 8.5 The identification results of the Gaussian KPCA 246
Table 8.6 Identification results of the FA 250
Table 8.7 The forecasting performance indices under feature extraction framework 253
Table 8.8 The comprehensive forecasting performance indices of proposed models 258

Table 9.1 Statistical characteristics of the original water level series 265
Table 9.2 The forecasting performance indices of the Elman model 273
Table 9.3 The forecasting performance indices of three hybrid models under decomposition framework 279
Table 9.4 The forecasting performance indices of optimal hybrid models 286

Table 10.1 Statistical characteristics of the original series 294
Table 10.2 The forecasting performance indices of the RF model 298
Table 10.3 The forecasting performance indices of the BFGS model 302
Table 10.4 The forecasting performance indices of the GRU model 307
Table 10.5 The comprehensive forecasting performance indices of proposed models 312

Abbreviations

AHU	Air handling unit
AIC	Akaike information criterion
ANFIS	Adaptive network-based fuzzy inference system
ANN	Artificial neural network
AQI	Air quality index
AR	Autoregressive
ARCH	Autoregressive conditional heteroscedasticity
ARCH-LM	Autoregressive conditional heteroscedasticity-Lagrange multiplier
ARIMA	Autoregressive integrated moving average
ARMA	Autoregressive moving average
BEMPs	Building energy modeling programs
BFGS	which is made up of the initials C. G. Broyden, R. Fletcher, D. Goldfarb, and D. F. Shanno
BIC	Bayesian information criterion
BM	Bayesian model
BP	Back propagation
RFID	Radio frequency identification
BPNN	Back propagation neural network
CA	Cellular automata
CAD	Computer-aided design
CDH	Cloudera's distribution including apache hadoop
CFS	Correlation-based feature selection
CNN	Convolutional neural network
CWC	Coverage width-based criterion
CWT	Continuous wavelet transform
DAG	Directed acyclic graph
DOE	Department of energy
DWT	Discrete wavelet transform
ELM	Extreme learning machine
EMD	Empirical mode decomposition
ERM	Empirical risk minimization

ES	Exponential smoothing
ESP	Energy simulation program
EWT	Empirical wavelet transform
FA	Factor analysis
FITNETs	Function fitting neural networks
FTS	Fuzzy time series
GA	Genetic algorithm
GARCH	Generalized autoregressive conditional heteroscedasticity
GFS	Google file system
GIS	Geographic information system
GPS	Global positioning system
GRU	Gated recurrent units
GWO	Grey wolf optimization
HDFS	Hadoop distributed file system
HDP	Hortonworks data platform
HHT	Hilbert–Huang transform
HMM	Hidden Markov model
HVAC	Heating, ventilating, and air-conditioning
HVACSIM	HVAC simulation
IMF	Intrinsic mode function
IOT	Internet of things
ITS	Intelligent transportation system
KMO	Kaiser-Meyer-Olkin
KPCA	Kernel principal component analysis
LSTM	Long short-term memory
MA	Moving average
MAE	Mean absolute error
MAPE	Mean absolute percent error
MI	Mutual information
MODWT	Maximal overlap discrete wavelet transform
NARX	Nonlinear autoregressive with external input
PCA	Principal component analysis
PICP	Prediction interval coverage probability
PINAW	Prediction interval normalized average width
PJM	An US regional transmission organization named Pennsylvania–New Jersey–Maryland
PSO	Particle swarm optimization
RBF	Radial basis function
RF	Random forest
RMSE	Root mean square error
SD	Seasonal decomposition
SDE	Standard deviation of error
SPARK	Simulation problem analysis and research kernel
SQL	Structured query language
SRM	Structural risk minimization

SSA	Singular spectrum analysis
SVM	Support vector machine
SVR	Support vector regression
TRNSYS	Transient system simulation tool
WD	Wavelet decomposition
WPD	Wavelet packet decomposition
YARN	Yet another resource negotiator

Part I
Exordium

With the continuous updating and development of information technology, a new generation of information technology such as the Internet of Things, the Internet, and cloud computing has gradually entered the field of vision. Smart city is a new urban development and construction model developed on information technology support systems. It provides new ideas and methods for solving a series of problems brought by the rapid development of urbanization. This part briefly introduces several key issues and related technologies of the smart city involved in the book. It describes the origins, implications, and developments of smart cities around the world, and focuses on smart grid and buildings, smart traffic systems, and smart environment to analyze the important impact of smart cities on future urban development. For these key issues, this part elaborates the research background and developing status of each issue. The necessity of research on the relevant content is fully analyzed. At the same time, it also briefly introduces the relevant technical framework for implementing each issue.

On the one hand, this part has carried out a statistical analysis on relevant research literatures by the research hot spots of smart cities. At the same time, it discusses current research hot spots and future research trends. On the other hand, this book is based on the application of big data in the relevant fields of smart cities, and discusses in detail the theoretical background, core technologies, and developing trend in the developing process of big data in smart city. It systematically provides the theoretical framework of big data prediction technology in the key issues of smart cities.

This part combines the basic overview of time-series analysis provided by related references with typical algorithm theory. It provides guide to the following research of smart cities and it is the basis for the subsequent theoretical algorithm research in the book.

Chapter 1
Key Issues of Smart Cities

Abstract With the continuous development of various cutting-edge technologies, the concept of smart cities has become increasingly hot in recent years. It will be the future direction of the development of cities. This chapter is the general chapter of the book. In the chapter, the big data forecasting technology is used as the basic point to elaborate and analyze from the aspects of smart grid and buildings, smart traffic, and smart environment. In each part, the relevant research significance and technical characteristics are described. Then, from the perspective of bibliometrics, this book reviews the domestic and foreign research on big data forecasting technology in smart cities. It can be seen from the literature analysis that the big data forecasting technology of smart cities is still in its infancy, and the research work of this book has extremely high academic value.

1.1 Smart Grid and Buildings

With the rapid development and in-depth application of a new round of information technologies, such as the global Internet of things, the new generation of mobile broadband network, the next generation of Internet, cloud computing, etc., the development of information technology, information, and innovation is brewing major changes and new breakthrough for a higher stage of intelligent development. And it has become an inevitable trend. In this context, some countries, regions, and cities take the lead in putting forward development strategies for building smart countries and smart cities (Cocchia 2014; Winden and Buuse 2017). In recent years, various cities in the first front in China have carried out detailed smart city construction planning, and some cities in the second front also have started smart city development planning. So far, nearly 50 cities at or above the provincial level have developed smart city construction plan and the total investment in smart city construction has reached hundreds of billions yuan.

H. Liu, *Smart Cities: Big Data Prediction Methods and Applications*,
https://doi.org/10.1007/978-981-15-2837-8_1

Although the term "Smart City" has appeared since 1998 (Van Bastelaer 1998), it is still confusing regarding its meaning (Anthopoulos and Fitsilis 2013). Until 2010, IBM officially put forward the vision of "smart cities," hoping to contribute to the urban development over the world. According to the research of IBM, cities are composed of six core systems with the different types of networks, infrastructure, and environment that affect the main functions of cities: organization, business/government, transportation, communication, water, and energy. These systems are not separated but work together in a collaborative way. The city itself is a macrosystem composed of these systems (Arafah and Winarso 2017).

In twenty-first century, smart cities mean to make full use of information and communication technology of sensor, analysis, and integration of city core system running in all kinds of key information in order to cover the livelihood, environmental protection, public security, urban services, commercial, and industrial activities (Pablo et al. 2018). The response of the various requirements is to make intelligence and create a better city life for human beings (Kyriazopoulou 2015; Öberg et al. 2017).

1.1.1 Overview of Smart Grid and Building

Smart grid is highly informationalized, interactive, and automated. It focuses on the application of advanced data processing technology, sensing technology, and intelligent control technology in the power grid. Smart grid focuses on the improvement of the intelligent level of power grid, bidirectional interactive intelligent electricity service, etc., which aims to support the construction of smart energy system. With the boom in smart grid construction, China, the USA, Europe, Japan, South Korea, and other countries and regions have implemented a large number of pilot issues to test new technologies, standards, and equipment of smart grid to seize the power equipment market. The governments of these countries and regions have carried out a series of policies to promote the construction and the process of smart grid. Relevant enterprises, research institutions, and universities have carried out a lot of research work and practice work about smart grid, forming a relatively complete theoretical and practical system. In recent years, power grid companies in many countries have implemented a large number of smart grid pilot demonstration issues. The issue covers the whole process of power production, configuration, and consumption, which is closely related to the safety, reliability, and efficiency of electricity by users.

From the perspective of connotation, smart grid is a powerful, reliable, efficient, interactive, and environment-friendly supply network service. It supports the construction and operation of green, low-carbon, efficient, and sustainable modern cities. In order to realize the full load of power grid for urban operation, it must start from the development of the local economy and consider the multi-dimensional demands of urban development such as energy conservation, emission reduction, and livelihood construction. Finally, the comprehensive construction of smart grid issues covers the fields of power generation, power distribution, electricity consumption, dispatching, communication for information, and other fields to build a modern intelligent power service system (Khajenasiri et al. 2017).

From a technical perspective, smart grids should further optimize the production, distribution, and consumption of electricity by obtaining more information about electricity consumption. In the smart grid, advanced sensor technology, Internet of things technology, communication technology for information, and other modern technologies are highly integrated forming a new comprehensive network system and realizing information exchange between power grid equipment. Based on technology integration, smart grid can realize advanced functions such as real-time intelligent control, remote debugging, online analysis, decision-making, and multi-location collaborative work. Therefore, smart grid is essentially the application of big data technology in power system.

1.1.2 The Importance of Smart Grid and Buildings in Smart City

Smart city has become an important trend for modern city development, involving all aspects of urban operation, and relevant solutions are also involved in many fields. The efficient operation and rapid development of cities cannot be separated from the continuous supply of electricity. As a critical direction of the development of power infrastructure, smart grid is gradually integrating with smart cities. The construction of smart grid for smart urban areas is not only a cross-link systematic issue but also a connection for the construction of grid with the concept of city establishment to achieve the harmonious development of electric power and city.

Smart grid is an important basis and objective need for cities to achieve orderly operation, management, balance of energy supply and demand (Yan et al. 2017), improvement of public services, and optimization of city-related industries. With the implementation of the smart grid development strategy, many smart grid themed issues have been launched in countries around the world. It is an effective method to build smart grid by constructing a comprehensive index system of smart grid to support smart cities and proposing different smart grid construction strategies with different emphases for different cities.

Smart cities cannot go on without smart grid. In addition, with the increasing energy crisis and environmental problems, the large number of access to energy distributing and storage devices leads to higher requirements for the reliability and quality of power supply in smart cities.

Among the related technologies of the smart grid, power demand forecasting is an important task for the power industry. It is very important in the operation, plan, and control of power system. Power demand forecasting is an important method to ensure the safe operation of power systems, scientific management, unified dispatch of power grid, and to make rational development plan. The electricity demand forecast is to forecast the electricity market demand.

1.1.3 Framework of Smart Grid and Buildings

In the book, forecasting experiments will be conducted from two aspects: regional electricity consumption prediction and building energy consumption. The overview framework of the smart grid and building is shown in Fig. 1.1.

1.2 Smart Traffic Systems

1.2.1 Overview of Smart Traffic Systems

Intelligent transportation systems (ITS) originated in the USA in the 1980s. It is an efficient, accurate, and real-time comprehensive traffic management system based on the traditional road facilities and the research of the key basic theory model. ITS effectively integrates advanced information technology, data communication technology, sensor technology, electronic control technology, and system engineering technology. ITS is the development direction of modern transportation system and has become the best way to solve traffic problems, which covers all transportation modes. Through the collection, processing, analysis, release, and utilization of traffic information, the transformation from traditional transportation systems to intelligent transportation system is realized.

Smart traffic systems are developed based on the ITS. Smart traffic systems refer to the application of the Internet of things, cloud computing, big data, artificial intelligence, automatic control, mobile Internet, and other information technologies of new generation in the field of transportation on the basis of ITS (Aldegheishem et al. 2018). Nowadays, smart traffic is a new way of sustainable urban development. It utilizes the new generation of information technology to optimize and integrate traffic data resources and enhance effective cooperation among various departments of traffic management. Through such an intelligent way of urban traffic management, the intelligent innovation of urban traffic development, the improvement of public service level, and the continuous improvement of traffic environment can be achieved.

1.2.2 The Importance of Smart Traffic Systems for Smart City

The development of smart transportation is in line with the internal needs of smart city construction. On the one hand, traffic congestion and traffic pollution are the main resource of the urban disease. Smart traffic systems can reduce road congestion and traffic accidents, and indirectly reduce automobile exhaust emissions through effective integration of new and high technologies, becoming the breakthrough point for effective treatment of urban traffic diseases (Aldegheishem et al.

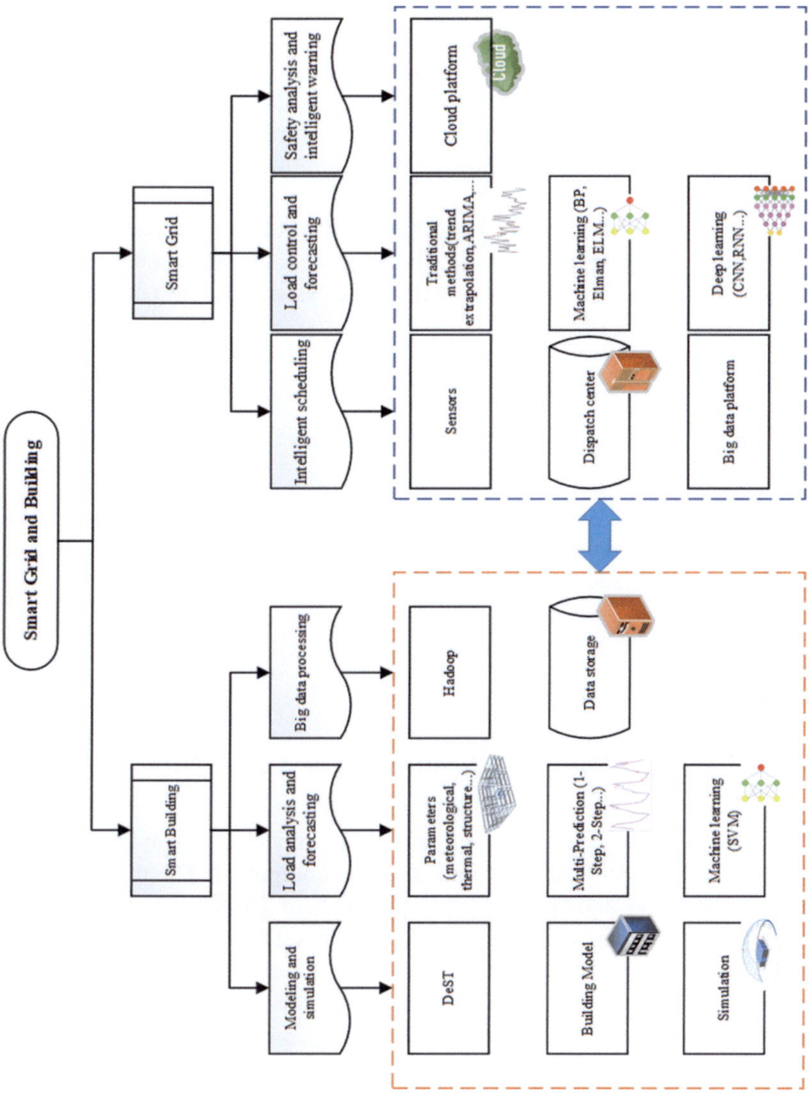

Fig. 1.1 Framework of smart grid and buildings

2018). On the other hand, based on the demand for intelligent transportation development, the government and transportation departments apply cloud computing, Internet of things, and other advanced technologies to daily traffic management. Therefore, the development of intelligent transportation is combined with emerging industries such as mobile Internet to better meet the needs of smart city construction.

Smart traffic system can speed up the construction of smart cities. The development of smart traffic system promotes the development of intelligent buses, intelligent parking, and traffic information service, which reflects the characteristics of smart cities to improve residential livelihood. Therefore, intelligent transportation plays an important role in the construction of smart cities and can promote the construction of smart cities.

Meanwhile, the construction of smart cities also provides great opportunities for the development of smart traffic system. In the process of continuously improving the urban intelligent construction, the smart traffic system also lays a good foundation for the development of itself.

1.2.3 Framework of Smart Traffic Systems

In the book, forecasting experiments will be conducted from two aspects: traffic trajectory prediction and traffic flow prediction. The overview framework of the smart traffic systems is shown in Fig. 1.2.

1.3 Smart Environment

1.3.1 Overview of Smart Environment for Smart City

Smart environment mainly refers to the traditional model of urban environmental governance on the basis of the big data and cloud computing. The thought and the idea integrate the ideas and concepts of the Internet of Things-based software and hardware system “Internet +” into urban environmental governance to improve the efficiency of urban environmental monitoring systems, change urban environmental pollution control models, and use information systems and intelligent systems to optimize urban environments (Liu et al. 2015).

The intellectualization of governance mode and means is an important feature of intelligent environmental protection. In the information environment, smart environment technology can collect and organize environmental sensor data, apply “Internet +” technology to environmental data analysis, and use cloud computing and big data technology for data analysis. Finally, on the basis of city distributed data, artificial intelligence aided environmental management is realized. The

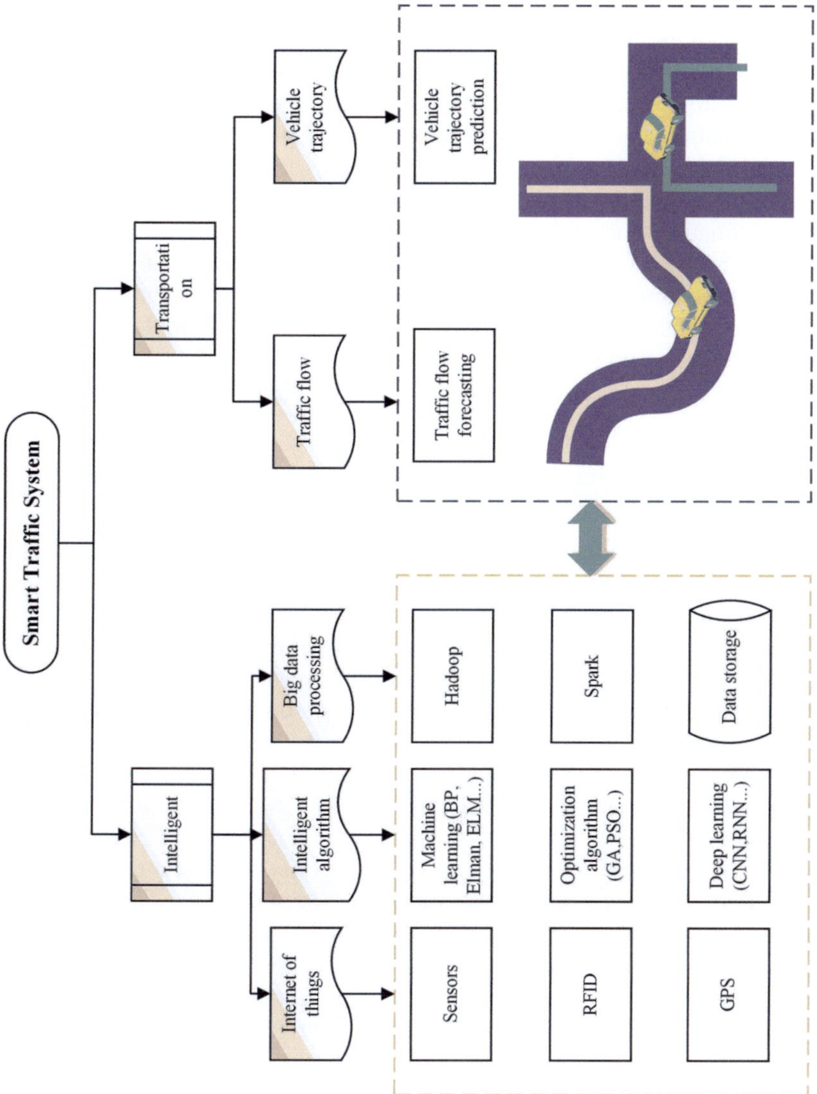

Fig. 1.2 Framework of smart traffic systems

computer-aided responses to various environmental problems and realizes the digital environmental protection to the wisdom of environmental change. At this time, the specific environmental data monitoring work does not require too much manual intervention, and the main tasks of environmental protection personnel have also completed the transition from inspection and supervision to monitoring and supervision (Alvear et al. 2018).

Interconnected and coordinated environmental governance is the key to smart environmental protection. In addition to the smart transformation and upgrading of external hardware, the core of smart environmental protection is to change the way of thinking about urban environmental protection and governance.

Various environmental problems are not related to each other, the relevant regulatory departments and different regions are organically integrated to achieve information coordination and linkage, and the integration of urban environmental governance is realized with the help of advanced hardware and intelligent software. The reform of the model for urban environmental governance can be realized only by responding to and dealing with various urban environmental problems with coordinated and integrated thinking.

1.3.2 The Importance of Smart Environment for Smart City

With the increasing development of economy, science, and technology, the urbanization process is constrained by the shortage of resources such as land, space, energy, and clean water. With the increase of urban population, environmental protection and other issues are facing increasing pressure. The problems of energy consumption and ecology in cities are becoming increasingly prominent. The ecological problems in urban development have become important factors affecting the development of residents' work and life as well as the sustainable development of cities, which have affected the healthy development of cities. Traditional technologies and management methods have become difficult to efficiently solve urban ecological problems. At present, many countries are studying the use of communication technology and intelligent solutions to redefine the city. This new urban management thinking will effectively solve the shortcomings of traditional urban management methods and improve the efficiency and ability of urban management. In particular, the integration of smart city construction into the urbanization construction will redefine the transportation, environment, energy, finance, management, and other systems of the city, which can especially effectively solve the ecological problems of the city and improve the efficiency of the whole city. With the rapid development of urbanization, cities are facing many challenges of sustainable development. To seize opportunities and develop harmoniously, cities need to become smarter.

Smart cities bring new opportunities for future urban development. Smart cities face the current urban ecological environment problems, and the construction of smart cities will greatly promote the construction of ecological civilization.

Considering the connotation of smart city in the construction of ecological civilization to improve the level of ecological construction, the Internet of things and sensors will bring a large amount of real-time data to urban operation. It will also effectively understand all aspects of urban operation information, which will help government agencies and enterprises to monitor, warn and control pollution, reduce pollutant emissions, and overcome the shortcomings of traditional regulatory departments of monitoring, management, and service. Real-time monitoring can sense natural disasters. Unified data center of smart city can effectively integrate the data of various systems running in the city and make the systems to share information, effectively reduce the problem of repeated construction in multiple departments and improve information efficiency.

1.3.3 Framework of Smart Environment

In the book, forecasting experiments of smart environment are conducted from three aspects: air quality prediction, hydrological forecasting, and noise prediction. The overview framework of the smart environment is shown in Fig. 1.3.

1.4 Framework of Smart Cities

1.4.1 Key Points of Smart City in the Era of Big Data

There are two forces driving the gradual formation of smart cities, one is the new generation of information technology represented by the Internet of things (Souza and Francisco 2019), cloud computing, and mobile Internet, and the other is the open urban innovation ecology gradually nurtured in the knowledge-based social environment (Gupta et al. 2019). The former is the technical factor of technological innovation, and the latter is the social and economic factor of social innovation. This shows the driving role of innovation in the development of smart cities. *Professor Meng* put forward the idea that the new generation of information technology and Innovation 2.0 are two indispensable components of smart city (Silvestre and Dyck 2017).

Smart cities need not only the support of the new generation of information technologies such as the Internet of things and cloud computing but also to cultivate the next generation of innovation oriented to the knowledge society (Innovation 2.0). The integration and development of information and communication technology breaks down the barrier of information and knowledge sharing, the boundary of innovation, promotes the formation of Innovation 2.0, and furthers the “melting” of various social organizations and activity boundaries.

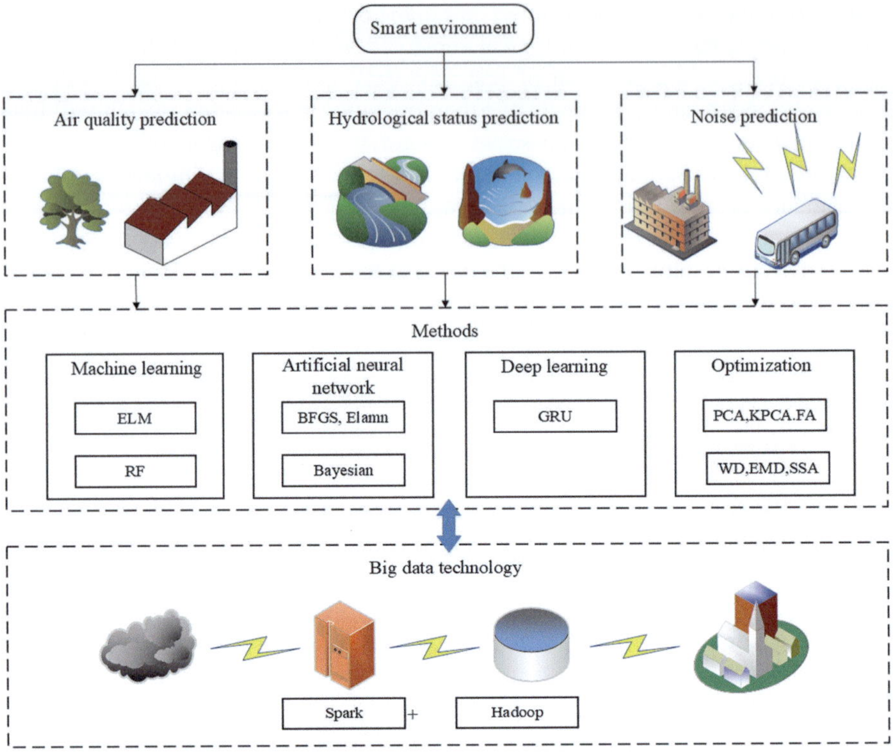

Fig. 1.3 Framework of smart environment

1.4.2 Big Data Time-series Forecasting Methods in Smart Cities

At present, there are many kinds of time-series prediction. According to the differences in forecasting time length, it can be divided into three categories: short-term prediction, medium-term prediction, and long-term prediction. According to the different prediction algorithm models, this section introduces three aspects of time-series forecasting model: traditional forecasting method, intelligent forecasting method, and intelligent hybrid prediction method.

The traditional forecasting method is based on mathematical statistics methods to analyze and forecast time series. Its main idea is to take the historical data of time series as observation value, use mathematical statistics method to study the trend law of historical data, and establish a regression model to fit the future data, so as to achieve the purpose of prediction. Common conventional forecasting methods include random forest model (RF) (Svetnik et al. 2003), autoregressive model (AR) (Akaike 1969), autoregressive moving average model (ARMA) (Neusser 2016), autoregressive integrated moving average model (ARIMA) (Akhtar and Rozi 2009), hidden Markov model (HMM) (Krogh et al. 2001), Bayesian model (BM) (Stephan and Wddaunizeau 2014), etc.

The intelligent forecasting method is based on machine learning and optimization algorithm. Machine learning is a hot research field in recent years, which can achieve nonlinear fitting of data by imitating the process of information transfer between neurons. Common machine learning algorithms include back propagation neural network (BPNN) (Huang et al. 2007), Elman neural network (Gao et al. 1996), long-short term memory (LSTM) (Graves 2012) neural network, convolutional neural network (CNN) (Krizhevsky et al. 2012), support vector machine (SVM) (Adankon and Cheriet 2002), extreme learning machine (ELM) (Huang et al. 2006), etc. Optimization algorithm, also known as evolutionary algorithm, can play a good role in parameter optimization and avoid machine learning algorithms falling into the state of over-fitting, under-fitting, or local optimal affecting. Common optimization algorithms include particle swarm optimization (PSO) (Kennedy and Eberhart 2002), genetic algorithm (GA) (Maulik and Bandyopadhyay 2000), grey wolf optimization algorithm (GWO) (Shakarami and Davoudkhani 2016), etc. The main idea of intelligent forecasting method is to use machine learning algorithm to build prediction model, and can use optimization algorithm to optimize model parameters and achieve the purpose of prediction.

The main idea of an intelligent hybrid forecasting method is to combine multiple algorithms, methods, and models to optimize each other. It can make full use of the advantages of each algorithm to promote the prediction performance of the model. Pre-processing of the data runs before forecasting model, such as wavelet decomposition (WD) (Pati et al. 2002), wavelet packet decomposition (WPD) (Eren and Devaney 2004), empirical mode decomposition (EMD) (Huang et al. 1998), principal component analysis (PCA) (Wold et al. 1987), factor analysis (FA) (Akaike 1987), etc. Using the method of pre-processing can decrease the data noise, reduce the complexity of data, or sufficiently extract the characteristics of the data. The process of building a forecasting model is based on machine learning algorithm, and the algorithm can be used to optimize the model parameters, such as the learning rate of the model, the number of hidden layers, and other parameters, to find better model parameters, which can improve the generalization ability and prediction accuracy of the model. When the forecasting model is completed, the forecasting results can be optimized, for example, the error analysis method is used to analyze the error to correct the prediction error, and the weighted algorithm is used to calculate the weight of the prediction results of multiple forecasting models. Intelligent hybrid model has been widely used in time-series prediction for its good forecasting performance.

1.4.3 Overall Framework of Big Data Forecasting in Smart Cities

Big data time-series forecasting technology has been widely used in all aspects of smart cities, such as urban electricity consumption forecast, traffic flow forecast, urban air quality forecast, etc. These applications can be summarized into three aspects: smart grid and buildings, smart transportation, and smart environment. This book will focus on the three aspects of time-series prediction modeling and big data prediction research. The overview framework of big data forecasting in smart cities is shown in Fig. 1.4.

1.5 The Importance Analysis of Big Data Forecasting Architecture for Smart Cities

1.5.1 Overview and Necessity of Research

Smart city aims at making full use of the new generation of information technologies in various industries in the city based on the innovation in the knowledge society of the advanced form of urban information to achieve the deep integration of information, industrialization, and urbanization. The smart city in the twenty-first century can make the most of big data technology to analyze and mine the key infor-

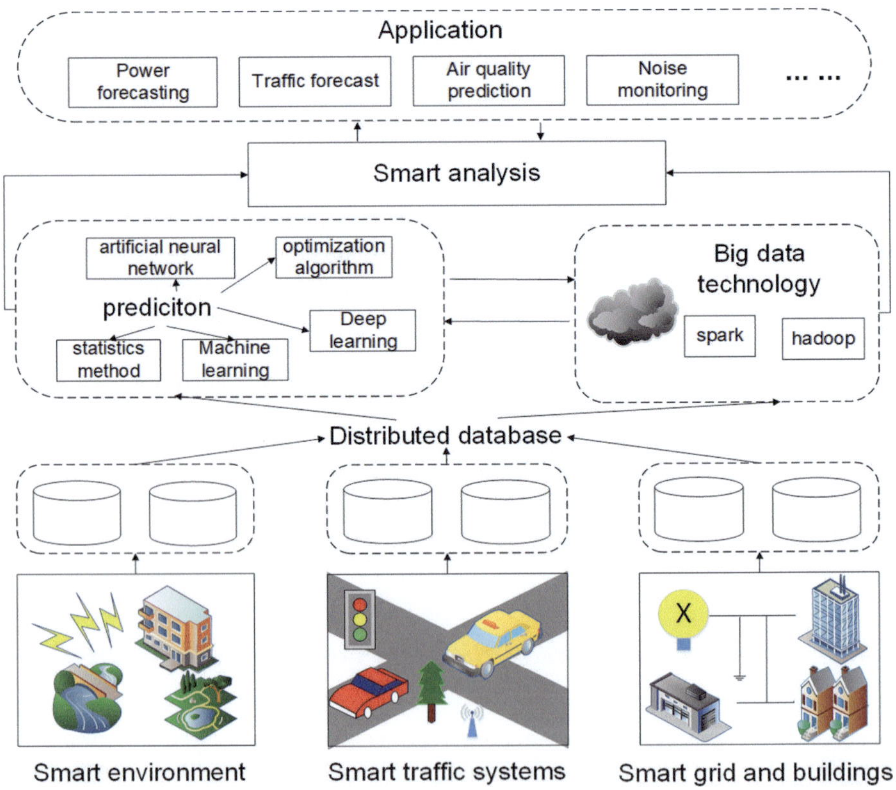

Fig. 1.4 Overall framework of big data forecasting in smart cities

mation of the core system of urban operation, thus make an intelligent response to a variety of urban needs including transportation, environmental protection, public safety, and industrial activities to achieve a better urban life.

Smart city often intersects with big data city, intelligent city, ecological city, low-carbon city and other regional development concepts, or even mixed with intelligent environment, intelligent transportation, smart grid, and other industry information concepts. At present, researches on smart cities have different emphasis, mostly focusing on technology application, network construction analysis, intelligent effect research, etc. The research of smart city is still in the new stage of vigorous development.

Under the background of smart cities, big data prediction from three aspects of power grid energy, traffic flow in road network, and environmental index data is of great practical significance to the construction and planning of smart cities. With the help of big data prediction of power grid energy, the power grid system and the safety of power consumption can be improved. With the help of big data prediction of traffic flow in road networks, it can further improve the road network structure, alleviate traffic congestion, and enhance the traffic capacity of the road network. With the help of big data's prediction of the environment, it can improve the quality of life of residents and provide a healthy and green living environment for residents. The prediction method of smart city big data proposed in the book has a certain reference significance for the planning, construction, and development of green cities and smart cities.

1.5.2 Review on Big Data Forecasting in Smart Cities

In recent years, smart cities have gradually become a hot topic worldwide, which is also the future development direction and new development path of cities. From the perspective of bibliometrics, the section analyzes the research status of smart cities in big data forecasting and several typical key issues of smart cities involved in the book.

On the *Web of Science* platform, taking "TS= (smart cities) AND TS= (big data forecasting OR big data prediction)" as the subject retrieval, 106 articles can be searched by July 2019. On this basis, 2 articles can be obtained by using high-level papers in the field as the refining basis. The citation report is shown in Fig. 1.5 and the literature network diagram, document overlay map, document item density map, and document cluster density map are shown in Figs. 1.6, 1.8, 1.9, and 1.10, respectively. If "TI= (smart cities) AND TI= (big data forecasting OR big data prediction)" is used as the title retrieval, 7 articles can be searched. Similarly, only 1 article can be obtained on the basis of high-level papers in the field. From the pictures, it can be seen that in recent years, the development of related research on the big data prediction of smart cities is relatively stable, showing an overall upward trend. However, on the whole, it is still in the initial stage, and the relevant research results are imperfect, one-sided, and unsystematic.

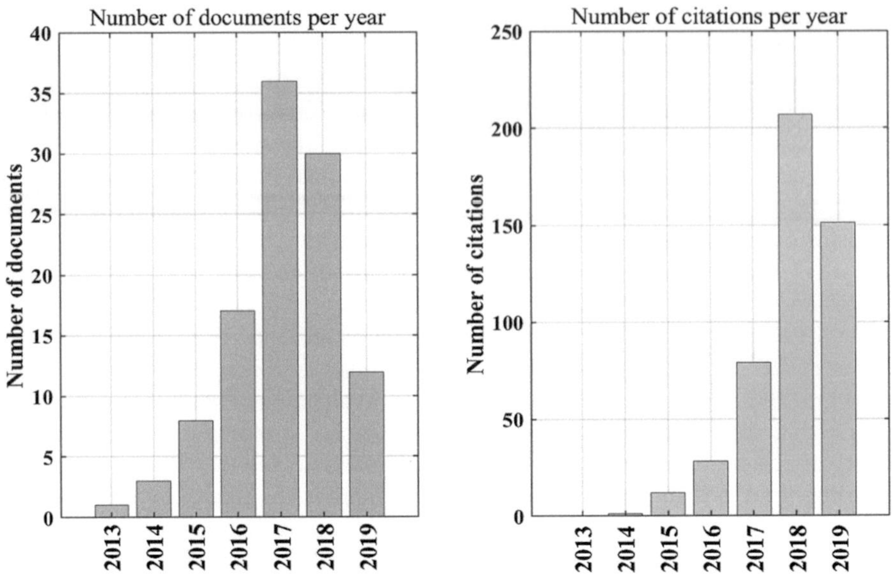

Fig. 1.5 Citation report of subject retrieval on "TS= (smart cities) AND TS= (big data forecasting OR big data prediction)"

The relationship map of paper network of subject retrieval is clustering with keywords as the target and each color represents a different research direction, in which the sky blue cluster represents the "SMART CITY CONSTRUCTION" direction, the navy blue cluster represents the " TRANSPORTATION SAFETY" direction, the green cluster represents the "TRANSPORTATION AND ENVIRONMENT" direction, the red cluster represents the "SOCIOLOGY" direction, the orange cluster represents the "DEEP LEARNING" direction, the yellow cluster represents the "ALGORITHM SCIENCE" direction, the purple cluster represents the "SMART GRID AND BUILDING" direction, and the brown cluster represents the "BIG DATA APPLICATION" direction. The size of the ball indicates the number of citations of the literature, and the distance between the two balls roughly indicates the relationship between the two documents in the aspect of co-citation links. Generally speaking, the closer the position of the two documents is, the stronger the correlation between them is.

Figure 1.7 shows the specific research field applied in the smart cities, it can be seen that there are some similarities between the network relationship map and the research field map.

In the overlay map of document, the colors from blue to green to yellow correspond to the publication year of the document from far to near, as shown in the lower right corner of Fig. 1.8. Combined with network relationship map in Fig. 1.6, it can be concluded that algorithm science and technology such as machine learning and deep learning have been applied to the humanities, society, and economy of smart cities in recent years. This is also the main foothold for the development of

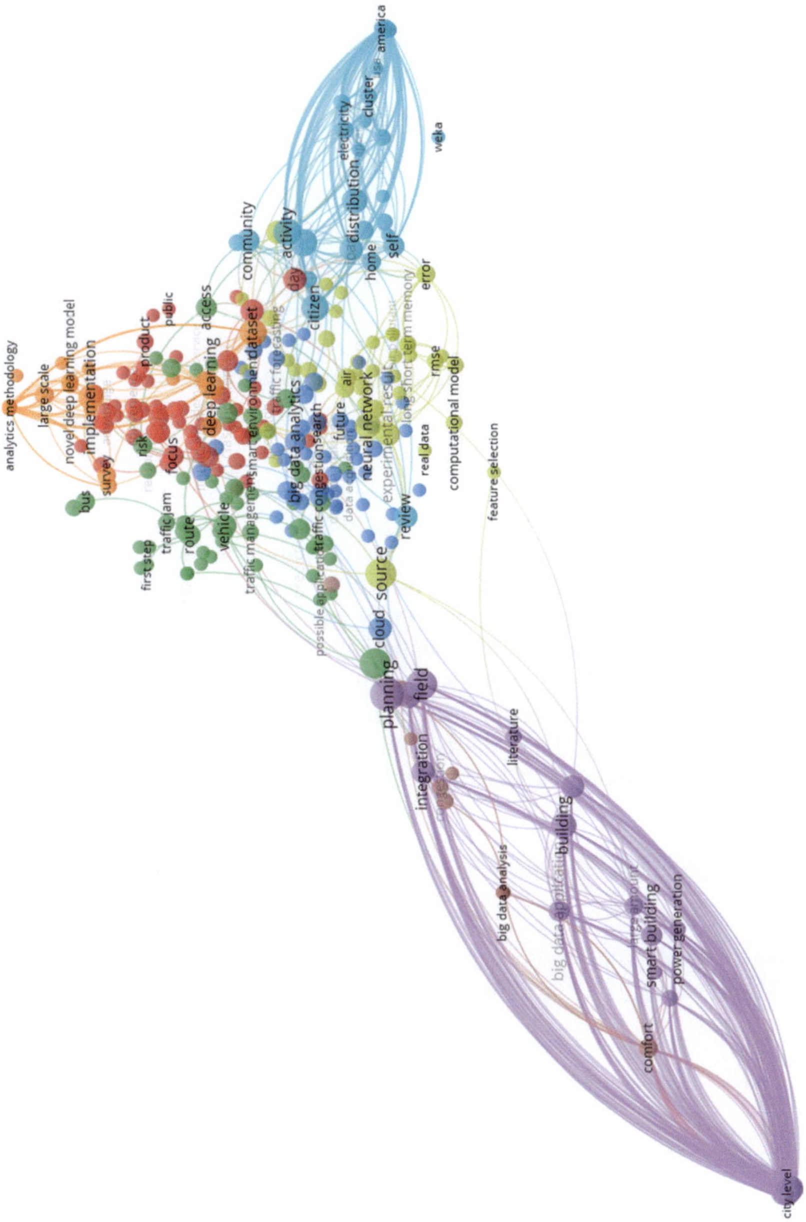

Fig. 1.6 The network diagram of documents based on the subject retrieval of "TS= (smart cities) AND TS= (big data forecasting OR big data prediction)"

big data processing technology and artificial intelligence technology (Figs. 1.9 and 1.10).

Each point in the item density map of document represents a research subject. The more documents near the point are, the higher the weight of the adjacent literature is, and the closer the color of the point is to yellow. On the other hand, the smaller the number of documents near a point is, the lower the weight of adjacent items is and the closer the color of the point is to blue. In the cluster density map, the weight given to the color of a cluster is determined by the number of items belonging to that cluster in the neighborhood of the point.

From the above literature analysis, it can be concluded that the current smart city-related research is mainly focused on energy, transportation, environment, and other fields, which is also one of the key research objects of the book.

1.5.3 Review on Big Data Forecasting in Smart Gird and Buildings

On the *Web of Science* platform, taking "TS= (smart grid OR smart buildings) AND TS= (big data forecasting OR big data prediction)" as the subject retrieval, 158 articles can be searched by July 2019, including 80 papers in *ARTICLE*, 72 papers in *MEETING*, 6 papers in *REVIEW*. On this basis, 2 articles can be obtained by using high-level papers in the field as the refining basis. However, when the title was searched by "TI= (smart grid OR smart buildings) AND TI= (big data forecasting OR big data prediction)", only 3 articles could be searched. The annual publication volume of the main types of literature is shown in Fig. 1.11.

As can be seen from Fig. 1.11, the research on smart power grid or smart buildings began to start gradually after 2010, although it has shown a steady upward

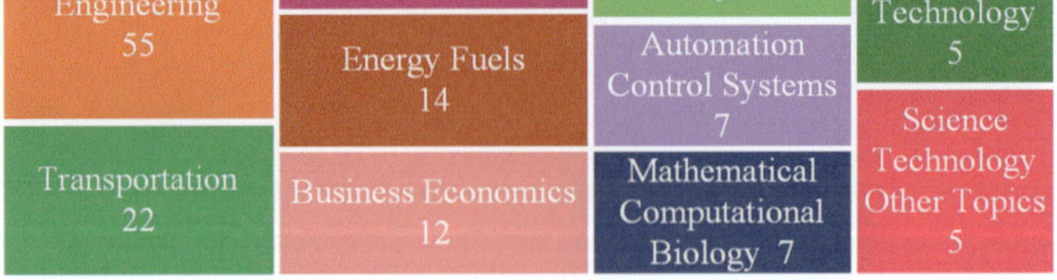

Fig. 1.7 Research direction map of document based on the subject retrieval of "TS= (smart cities) AND TS= (big data forecasting OR big data prediction)"

Fig. 1.8 The overlay map of document based on the subject retrieval of "TS= (smart cities) AND TS= (big data forecasting OR big data prediction)"

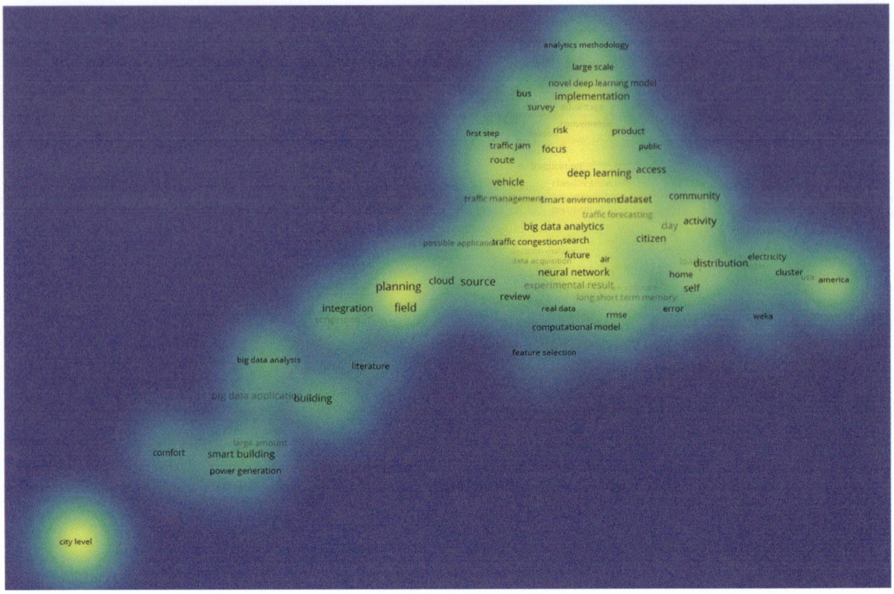

Fig. 1.9 The item density map of document based on the subject retrieval of "TS= (smart cities) AND TS= (big data forecasting OR big data prediction)"

Fig. 1.10 The cluster density map of document based on the subject retrieval of "TS= (smart cities) AND TS= (big data forecasting OR big data prediction)"

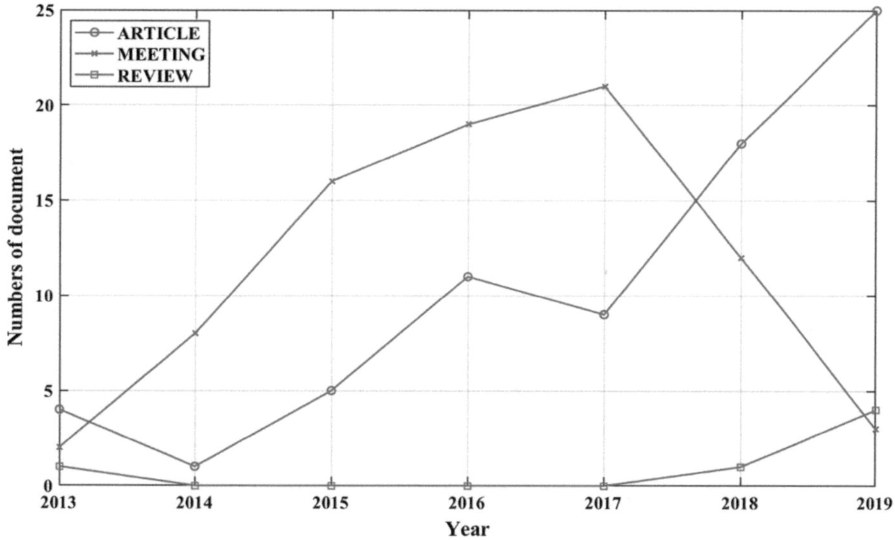

Fig. 1.11 Subject retrieval of annual publication volume of various types of literature on "TS= (smart grid OR smart buildings) AND TS= (big data forecasting OR big data prediction)"

trend in recent years. However, there are still some problems, such as the lack of the number of achievements, especially the high display degree, unsystematic, etc. The number of periodical articles is increasing, but the number of conference papers has a downward trend in the past 2 years, which is a sign of uneven results.

1.5.4 Review on Big Data Forecasting in Smart Traffic Systems

On the *Web of Science* platform, taking "TS= (smart traffic OR smart transportation) AND TS= (big data forecasting OR big data prediction)" as the subject retrieval, 73 articles can be searched by July 2019, including 35 papers in *ARTICLE*, 35 papers in *MEETING*, 2 papers in *REVIEW*, 1 paper in *EDITORIAL*. On this basis, no article can be obtained by using high-level papers in the field as the refining basis. However, when the title was searched by "TI= (smart traffic OR smart transportation) AND TI= (big data forecasting OR big data prediction)", only 2 articles could be searched. The annual publication volume of the main types of literature is shown in Fig. 1.12.

As can be seen from the Fig. 1.12, the research on smart transportation big data prediction technology is similar to that of smart power grid and buildings, and it began to study after 2010, so there are not many high-level research results at present. However, the related sub-field research has been in full swing.

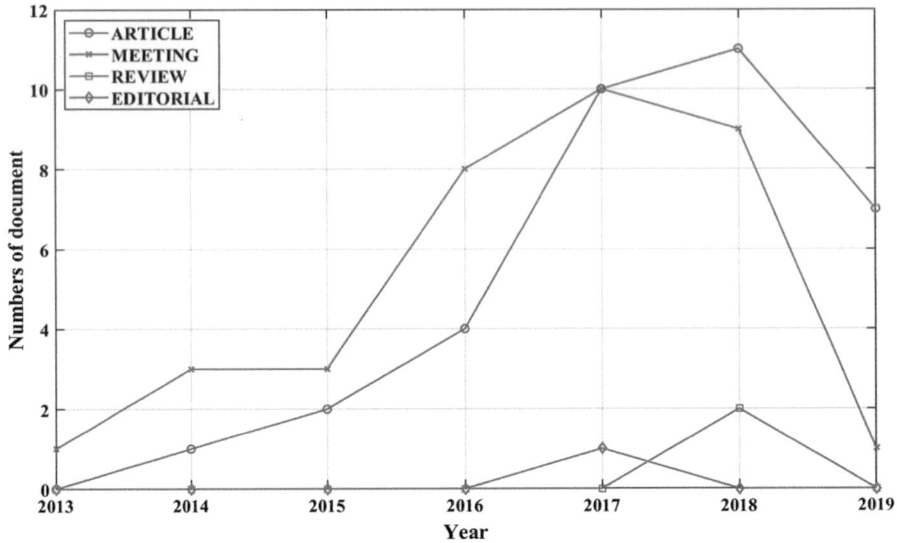

Fig. 1.12 Subject retrieval of annual publication volume of various types of literature on "TS= (smart traffic OR smart transportation) AND TS= (big data forecasting OR big data prediction)"

1.5.5 Review on Big Data Forecasting in Smart Environment

On the *Web of Science* platform, taking "TS= (smart environment) AND TS= (big data forecasting OR big data prediction)" as the subject retrieval, 103 articles can be searched by July 2019, including 54 papers in *ARTICLE*, 42 papers in *MEETING*, 4 papers in *REVIEW*, 1 paper in *EDITORIAL*, 1 paper in *EARLY ACCESS*, 1 paper in *OTHER PAPER*. On this basis, 2 articles can be obtained by using high-level papers in the field as the refining basis. However, when the title was searched by "TI= (smart environment) AND TI= (big data forecasting OR big data prediction)," only 1 article could be searched. The annual publication volume of the main types of literature is shown in Fig. 1.13.

As can be seen from Fig. 1.13, the research on the environment has always been a hot topic, and the start time of the research is earlier than that of smart grid and smart transportation, and the types of results produced are more abundant. Because the environmental problem is a livelihood problem, which is closely related to human life, especially in recent years, the environmental protection problem is becoming more and more serious. With the introduction of the concept of smart city, the environmental problem has become more and more concerned. On the whole, the research results of smart environment are on the rise.

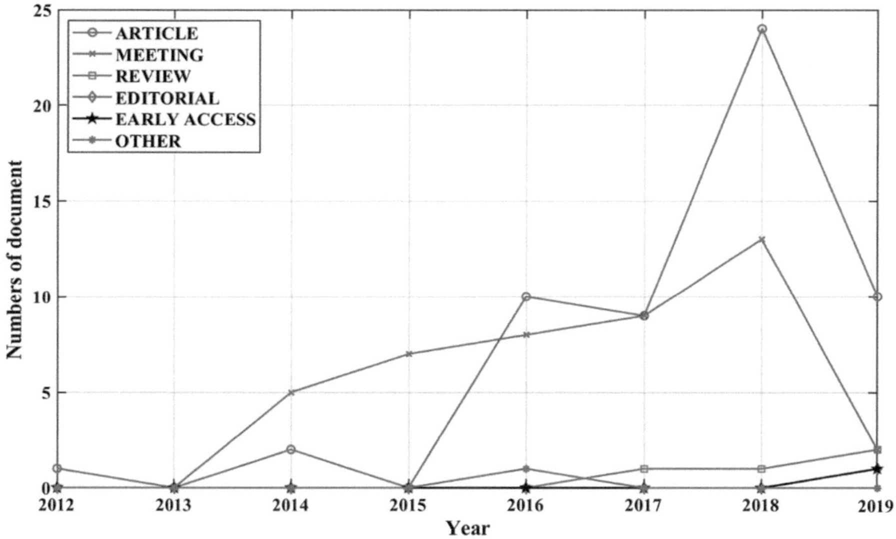

Fig. 1.13 Subject retrieval of annual publication volume of various types of literature on "TS= (smart environment) AND TS= (big data forecasting OR big data prediction)"

References

Adankon MM, Cheriet M (2002) Support Vector Machine. Computer Science 1(4):1–28

Akaike H (1969) Fitting autoregressive models for prediction. Annals of the Institute of Statistical Mathematics 21(1):243–247

Akaike H (1987) Factor analysis and AIC. Psychometrika 52(3):317–332

Akhtar S, Rozi S (2009) An autoregressive integrated moving average model for short-term prediction of hepatitis C virus seropositivity among male volunteer blood donors in Karachi, Pakistan. World Journal of Gastroenterology 15(13):1607–1612

Aldegheishem A, Yasmeen H, Maryam H, Shah MA, Mehmood A, Alrajeh N et al (2018) Smart road traffic accidents reduction strategy based on intelligent transportation systems (TARS). Sensors 18(7):1983

Alvear O, Calafate C, Cano JC, Manzoni P (2018) Crowdsensing in smart cities: Overview, platforms, and environment sensing issues. Sensors 18(2):460

Anthopoulos L, Fitsilis P (2013) Using classification and roadmapping techniques for smart city viability's realization. Electronic Journal of e-Government 11(2):326–336

Arafah Y, Winarso H (2017) Redefining smart city concept with resilience approach. IOP Conference Series: Earth Environmental Science 70:012065

Cocchia, A. (2014). Smart and digital city: A systematic literature review.

Eren L, Devaney MJ (2004) Bearing damage detection via wavelet packet decomposition of the stator current. IEEE Transactions on Instrumentation Measurement 53(2):431–436

Gao XZ, Gao XM, Ovaska SJ (1996) A modified Elman neural network model with application to dynamical systems identification. In: IEEE International Conference on Systems

Graves, A. (2012). Long Short-Term Memory.

Gupta P, Chauhan S, Jaiswal MP (2019) Classification of smart city research - A descriptive literature review and future research agenda. Information Systems Frontiers 21(1)

Huang GB, Zhu QY, Siew CK (2006) Extreme learning machine: Theory and applications. Neurocomputing 70(1):489–501

Huang NE, Zheng S, Long SR, Wu MC, Shih HH, Zheng Q et al (1998) The empirical mode decomposition and the Hilbert spectrum for nonlinear and non-stationary time series analysis. Proceedings Mathematical Physical Engineering Sciences 454(1971):903–995

Huang R, Lifeng XI, Xinglin LI, Liu RC, Qiu H et al (2007) Residual life predictions for ball bearings based on self-organizing map and back propagation neural network methods. Mechanical Systems Signal Processing 21(1):193–207

Kennedy, J., & Eberhart, R. (2002). Particle swarm optimization. In *Icnn95-international Conference on Neural Networks.*

Khajenasiri I, Estebsari A, Verhelst M, Gielen G (2017) A review on internet of things solutions for intelligent energy control in buildings for smart city applications. Energy Procedia 111:770–779

Krizhevsky, A., Sutskever, I., & Hinton, G. E. (2012). ImageNet classification with deep convolutional neural networks. In *International Conference on Neural Information Processing Systems.*

Krogh A, Larsson B, Von HG, Sonnhammer EL (2001) Predicting transmembrane protein topology with a hidden Markov model: Application to complete genomes. Journal of Molecular Biology 305(3):567–580

Kyriazopoulou, C. (2015). Smart city technologies and architectures: A literature review. In *International Conference on Smart Cities & Green ICT Systems.*

Liu Y, Zhang S, Liu L, Wang X, Huang H (2015) Research on urban flood simulation: A review from the smart city perspective. Progress in Geography 34(4):494–504

Maulik U, Bandyopadhyay S (2000) Genetic algorithm-based clustering technique. Pattern Recognition 33(9):1455–1465

Neusser, K. (2016). Autoregressive moving-average models.

Öberg C, Graham G, Hennelly P (2017) Smart cities. IMP Journal 11(3):468–484

Pablo C, Alfonso G-B, Sara R, Corchado JM (2018) Tendencies of technologies and platforms in smart cities: A state-of-the-art review. Wireless Communications Mobile Computing 2018:1–17

Pati, Y. C., Rezaiifar, R., & Krishnaprasad, P. S. (2002). Orthogonal matching pursuit: Recursive function approximation with applications to wavelet decomposition. In *Conference on Signals, Systems & Computers.*

Shakarami MR, Davoudkhani IF (2016) Wide-area power system stabilizer design based on Grey Wolf Optimization algorithm considering the time delay. Electric Power Systems Research 133:149–159

Silvestre BS, Dyck B (2017) Enhancing socio-ecological value creation through Sustainable Innovation 2.0: Moving away from maximizing financial value capture. Journal of Cleaner Production 171:1593–1604

Souza JTD, Francisco ACD (2019) Data mining and machine learning to promote smart cities: A systematic review from 2000 to 2018. Sustainability 11

Stephan KE, Wddaunizeau P (2014) Bayesian model selection for group studies. Neuroimage 84(4):971–985

Svetnik V, Liaw A, Tong C, Culberson JC, Sheridan RP, Feuston BP (2003) Random forest: A classification and regression tool for compound classification and QSAR modeling. Journal of Chemical Information Computer Sciences 43(6):1947

Van Bastelaer, B. (1998). Digital cities and transferability of results. In *4th EDC Conference on digital cities, Salzburg* (pp. 61–70).

Winden WV, Buuse DVD (2017) Smart city pilot projects: exploring the dimensions and conditions of scaling up. Journal of Urban Technology 24(4):1–22

Wold S, Esbensen K, Geladi P (1987) Principal component analysis. Chemometrics Intelligent Laboratory Systems 2(1):37–52

Yan, M., Li, X., Lai, L. L., & Xu, F. (2017). Energy internet in smart city review. In *International Conference on Wavelet Analysis & Pattern Recognition.*

Part II
Smart Grid and Buildings

With the continuous development of big data, artificial intelligence, and the continuous advancement of modern science and technology, the application of big data in the power industry is more and more extensive and in-depth. The concept of a smart grid comes into being. The concept of smart grid does not currently form a unified definition. It mainly refers to the use of rich and valuable information to guide and promote the operation of the power sector, improve the utilization of electric energy, and achieve scientific resource allocation and consumption. The smart grid can be based on massive, heterogeneous, and polymorphic power big data, applying modern artificial intelligence and big data processing solutions to realize the integration and analysis of massive data, and useful information extraction to assist decision-making.

The core function of smart grid big data is the analysis and interpretation of data. Intelligence means that it has a relatively complete computing system and thinking operation mode. The data analysis of big data is mainly divided into two aspects: data analysis and data interpretation. Big data analytics is the centralized integration of multiple types of data, designed to study the patterns behind the data, explore the mutual information between the data, and re-detect and analyze the available value of the data. Data analysis is only a preliminary work. Data interpretation can transform data so that the information hidden in the data can be further clarified and applied reasonably. Therefore, the data interpretation of the smart grid is essentially a process of deep analysis and stereoscopic display of the data.

For one thing, this part explores the inherent laws of regional correlation and time-domain changes in power load data. For another thing, the relationship between electricity load data and environmental factors in the context of big data is explored. At the same time, this part also establishes energy consumption data prediction models based on the corresponding influencing factors and time-series analysis methods. Through the above analysis and verification using actual data, it provides a series of solutions for the deep interpretation of smart grid big data.

Chapter 2
Electrical Characteristics and Correlation Analysis in Smart Grid

Abstract The power grid system contains a large number of data sources, the amount of data and data types are very large, and the correlation between different kinds of data is also very complex. How to extract effective data sources from massive data, simplify the identification process, and effectively improve the accuracy of data processing is a necessary way to realize power grid intelligence. In this chapter, taking the air-conditioning circuit system of a complete building as an example, 11 features and 3 dependent variables are extracted, the correlation among features is analyzed by using three correlation analysis methods, and one dependent variable is selected. Based on the rough extraction of all the features, three feature selection methods are used to search for the optimal features, and finally, the optimal feature subset based on the selected dependent variables is obtained by comprehensive comparative analysis.

2.1 Introduction

Smart grid is a digital energy self-healing system that transmits electricity from generating sources (including distributed renewable energy) to the consumer side. Because of its observable, controllable, automatic, and integrated capabilities, the smart grid can optimize the power supply, promote the two-way communication of the power grid, realize the energy management of terminal users, minimize the interruption of power supply, and transmit electricity on demand. As a result, the costs borne by power plants and customers are reduced, the reliability of power supply is improved, and carbon emissions are greatly reduced. Smart building is an optimized combination structure that takes the building as the platform and integrates information, service, and security. The power grid system of smart buildings refers to the automatic power system which maintains the smooth operation and demand–supply of the whole building platform.

H. Liu, *Smart Cities: Big Data Prediction Methods and Applications*,
https://doi.org/10.1007/978-981-15-2837-8_2

While smart power grid and buildings are moving towards informatization and intelligence, there are also a series of problems that need to be solved. From the point of view of the book, smart grid and buildings face the problems of massive data storage and processing, mainly showing the characteristics of large amounts of data, low data utilization, lack of connection between data, etc. Taking the air-conditioning power system under the smart buildings as an example, the chapter analyzes the correlation between the features and characteristics of the massive electrical data of the air-conditioning power system under the background of big data, as well as the feature selection based on the target variables. It can provide dimension reduction for the process of power grid data identification and processing under smart buildings, and provide guidance for the development of smart city big data prediction technology.

The commonly used correlation analysis methods include mutual information (Su et al. 2018), Pearson correlation coefficient (Ly et al. 2018), Kendall correlation coefficient (Sembiring et al. 2017), Spearman correlation coefficient (Ashok and Abirami 2018), etc. The three correlation coefficients reflect the changing trend and connection degree between two variables in statistics, and they reflect and complement each other. The feature selection methods are more extensive. According to the generation process of feature subset, it can be divided into three categories: full search, heuristic search, and random search. Full search includes Breadth First Search (Rahim et al. 2018), Branch and Bound Search (Atashpaz-Gargari et al. 2017), Beam Search (Murrieta-Mendoza et al. 2017), and Best First Search (Lei et al. 2002). Heuristic search mainly includes Sequence Forward Selection (San-Chuan and Li-Li 2018), Sequence Backward Selection (Zhu and Jie 2013), Bidirectional Search (Zhang et al. 2015), Plus L Minus R Selection (Sabeti et al. 2007), Sequence Floating Selection (Rodriguez-Galiano et al. 2018), Decision Tree Method (Sun et al. 2019), RelieF (Abbasi and Rahmani 2019), etc. Random search includes Las Vegas Filter (Nandi 2011), Las Vegas Wrapper (Xiong et al. 2019), Simulated Annealing (Mafarja and Mirjalili 2017), Genetic Algorithm (Aliane et al. 2017), and other optimization algorithms.

In the chapter, taking the air-conditioning circuit system of a complete building as an example, 11 features and 3 dependent variables are extracted, the correlation between each two features is analyzed by using three correlation analysis methods, and one dependent variable is selected. Based on the rough extraction of all the features, three feature selection methods are used to search for the optimal features, and finally, the optimal feature subset based on the selected dependent variables is obtained by comprehensive comparative analysis.

2.2 Extraction of Building Electrical Features

In the chapter, taking an overall building that maintains normal interaction with the outside world as an example, the common electrical characteristics of its construction power grid and the main factors affecting it are analyzed. As we all know, the air-conditioning and ventilation system in a complete building is one of the objects that can best reflect the characteristics of the electrical system, which is also the focus of this section. It includes a comprehensive analysis of many influencing factors, such as electricity consumption, indoor heat disturbance, system load, meteorological factors, etc.

2.2.1 *Analysis of Meteorological Elements*

In the section, three meteorological factors affecting the energy consumption of building air-conditioning electrical systems such as dry bulb, moisture, and total solar radiation are selected. The sampling interval of original meteorological factors series is 1 h and the sampling span is 1 year, from 1st January to 31st December, including 8760 sample points. Figure 2.1 shows the original time series of the three meteorological factors, denoted as $\{X_{DB}\}$, $\{X_M\}$, and $\{X_{TSR}\}$, respectively. It can be seen from the diagram that the three kinds of meteorological parameters are in accordance with the normal meteorological variation process in the actual situation, that is, the northern hemisphere has low temperature, dry, and little rain in winter, not

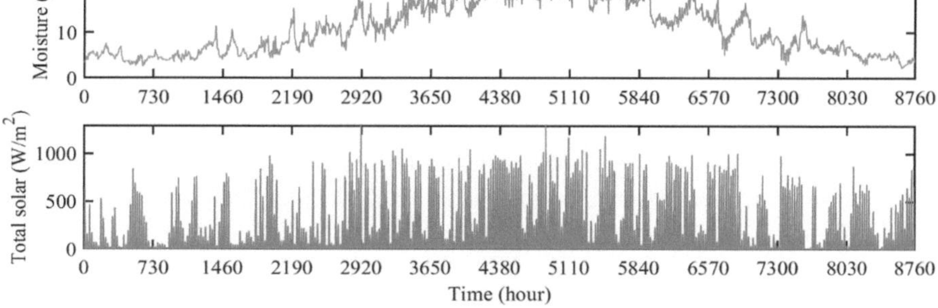

Fig. 2.1 Original meteorological factors series

Table 2.1 Statistical characteristics of the original meteorological factors series

Original series	Minimum	Maximum	Mean value	Standard deviation	Skewness	Kurtosis
$\{X_{DB}\}$ (°C)	−3.2000	38.2000	17.0603	8.9597	−0.0248	1.9795
$\{X_M\}$ (g/kg)	2.2500	27.1900	11.2034	5.7999	0.3516	1.7874
$\{X_{TSR}\}$ (W/m^2)	0	1293.94	123.8463	214.1274	2.1180	7.0230

directly exposed to the sun, and the total solar radiation is low, while high temperature and rainy, by the direct exposure of sun, the total solar radiation is high in summer.

In order to comprehensively analyze the characteristics of meteorological factors series, and further calculate the common statistical characteristics of the three groups of time series, such as extremum, mean value, standard deviation, skewness, and kurtosis, Table 2.1 is the statistical characteristics of the three groups of time series adopted in the chapter. Where, skewness is the third-order ratio of central moment and standard deviation of time series, which reflects the deviation of time series data distribution relative to symmetric distribution. When skewness value is negative, its distribution is skewed to the left, when skewness value is positive, its distribution is skewed to the right, and when skewness value is 0, its distribution is symmetrical. Kurtosis is the ratio of the fourth-order central moment of time series to the fourth power of standard deviation, which reflects the outliers of data. The kurtosis value adopted is the kurtosis value relative to the standard normal distribution. When the kurtosis value is positive, it indicates that the dispersion degree of the distribution is greater than the standard normal distribution.

The skewness values of the original dry bulb series are negative, indicating that its distribution is skewed to the left relative to the symmetrical distribution, while the distributions of original moisture and total solar time series are all skewed to the right relative to the symmetrical distribution, and their statistical distribution characteristics are closer to the symmetrical distribution. The kurtosis values of the three groups of time series are all positive, indicating that the distribution of sample points relative to the mean is relatively scattered, and the probability of occurrence of extreme values in sample points is higher than that in normal distribution.

2.2.2 *Analysis of System Load*

In the section, three system loads affecting the energy consumption of building air-conditioning electrical system such as heat load, cooling load, and humidity are selected. The sampling interval of original system load factors series is 1 h and the sampling span is 1 year, from 1st January to 31st December, including 8760 sample points. Figure 2.2 shows the original time series of the three system load factors, denoted as $\{X_{HL}\}$, $\{X_{CL}\}$, and $\{X_H\}$, respectively. It can be seen from the diagram that the variation law of cooling and heating load accords with the process of cold and heat exchange in the actual process, that is, the demand for heating load is high in winter and the demand for cooling load is high in summer.

Similarly, in order to comprehensively analyze the characteristics of the system load factors series, and further calculate the common statistical characteristics of the three groups of time series, Table 2.2 is the statistical characteristics of the three groups of time series adopted in the chapter.

The skewness values of the original cooling load series are negative, indicating that its distribution is skewed to the left relative to the symmetrical distribution,

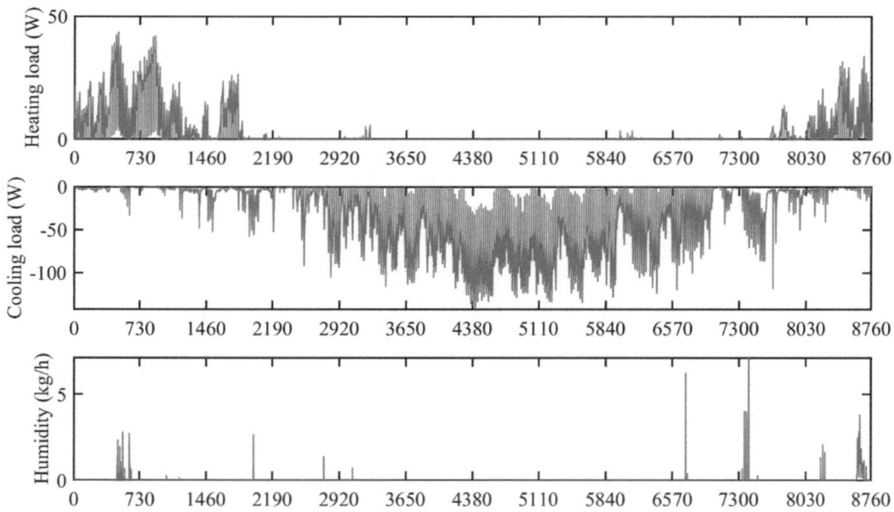

Fig. 2.2 Original system load factors series

Table 2.2 Statistical characteristics of the original system load factors series

Original series	Minimum	Maximum	Mean value	Standard deviation	Skewness	Kurtosis
$\{X_{HL}\}$ (kW)	0	43.5400	3.1101	7.0868	2.6785	9.8576
$\{X_{CL}\}$ (kW)	−143.2900	0	−32.4353	34.7669	−0.8659	2.5614
$\{X_{H}\}$ (kg/h)	0	7.1300	0.0248	0.2267	14.6230	290.0951

while the distributions of original heating load and humidity time series are all skewed to the right relative to the symmetrical distribution, and their statistical distribution characteristics are closer to the symmetrical distribution. The kurtosis values of the three groups of time series are all positive, indicating that the distribution of sample points relative to the mean is relatively scattered, and the probability of occurrence of extreme values in sample points is higher than that in normal distribution. Particularly, the kurtosis of humidity is so large that can also be intuitively observed in Fig. 2.2.

2.2.3 Analysis of Thermal Perturbation

In the section, three thermal perturbations affecting the energy consumption of building air-conditioning electrical system such as personnel calorific value, personnel humidity, personnel fresh air volume, light calorific value, and device calorific value are selected. The sampling interval of original thermal perturbation series is 1 h and the sampling span is 1 year, from 1st January to 31st December, including 8760 sample points. Figure 2.3 shows the first 800 points of original time

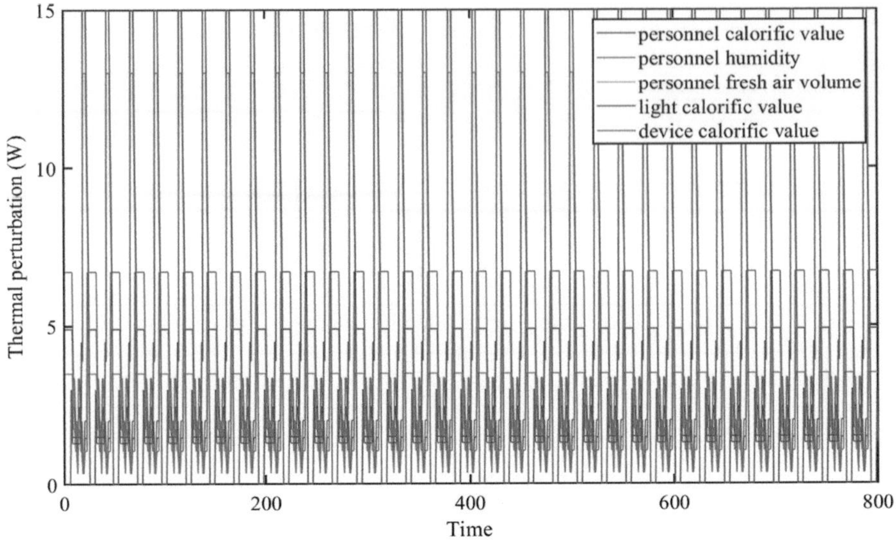

Fig. 2.3 Original thermal perturbation indoors series

Table 2.3 Statistical characteristics of the original thermal perturbation indoors series

Original series	Minimum	Maximum	Mean value	Standard deviation	Skewness	Kurtosis
$\{X_{\text{pCA}}\}$ (W)	0.4900	4.9000	2.9400	1.7383	0.0672	1.3173
$\{X_{\text{pH}}\}$ (W)	0.6700	6.7200	4.0329	2.3833	0.0665	1.3186
$\{X_{\text{pFAV}}\}$ (W)	0.3500	3.5000	2.1000	1.2416	0.0672	1.3173
$\{X_{\text{lCA}}\}$ (W)	0	15	4.3125	5.3618	1.2081	2.8438
$\{X_{\text{dCA}}\}$ (W)	0	13	3.7375	4.6469	1.2081	2.8438

series of the five thermal perturbation factors, denoted as $\{X_{\text{pCA}}\}$, $\{X_{\text{pH}}\}$, $\{X_{\text{pFAV}}\}$, $\{X_{\text{lCA}}\}$, and $\{X_{\text{dCA}}\}$, respectively. It can be seen from the diagram that the main thermal perturbation indoors is the light thermal perturbation and the device thermal perturbation, and the influence of the personnel thermal perturbation is relative low.

Similarly, in order to comprehensively analyze the characteristics of thermal perturbation factors series, and further calculate the common statistical characteristics of the five groups of time series, Table 2.3 is the statistical characteristics of the five groups of time series adopted in the chapter.

The skewness values of the five groups of original thermal perturbation series are all positive, indicating that their distributions are all skewed to the right relative to the symmetrical distribution, and their statistical distribution characteristics are closer to the symmetrical distribution. The kurtosis values of the five groups of time series are all positive, indicating that the distribution of sample points relative to the mean is relatively scattered, and the probability of occurrence of extreme values in sample points is higher than that in normal distribution. Combined with the analysis of the charts, although the amplitudes of the light thermal perturbation and the device thermal perturbation are large, but at the same time, the fluctuation is also very large, and the amplitude and frequency of the personnel thermal perturbation are relatively concentrated, so the comprehensive effects of the three kinds of thermal perturbation are basically the same.

2.3 Cross-Correlation Analysis of Electrical Characteristics

2.3.1 Cross-Correlation Analysis Based on MI

2.3.1.1 The Theoretical Basis of MI

The mutual information (MI) is a measurement index of the mutual relationship among variables, also known as transinformation. Different from the correlation coefficient, the MI determines the similarity of the joint distribution $q(A, B)$ and the product of the edge distribution $q(A)q(B)$. Mutual information is usually defined as the following form (Barman and Kwon 2017):

$$\mathrm{MI}(A,B)=\sum_{\beta\in B}\sum_{\alpha\in A}q(\alpha,\beta)\log\left(\frac{q(\alpha,\beta)}{q(\alpha)q(\beta)}\right) \tag{2.1}$$

where A and B represent the two discrete random variables, $q(\alpha, \beta)$ is the joint distribution functions of A and B, while $q(\alpha)$ and $q(\beta)$ are the edge distribution of A and B, respectively.

When the variables are continuous random variables, the calculation of the MI is in the form of integral (Koizumi et al. 2017).

$$\mathrm{MI}(A,B)=\int_B\int_A p(\alpha,\beta)\log\left(\frac{p(\alpha,\beta)}{p(\alpha)p(\beta)}\right)d\alpha d\beta \tag{2.2}$$

where A, B, $p(\alpha, \beta)$, $p(\alpha)$, and $p(\beta)$ have the same meaning as above.

Only if A and B are two random variables with independence, the MI between them will be zero, that is, $\mathrm{MI}(A, B) = 0$. From a view of statistical point, it is that the joint distribution of the two variables equals to the product of the edge distribution, that is, $p(\alpha, \beta) = p(\alpha)p(\beta)$. In addition, MI also has the characteristics of non-negativity and symmetry.

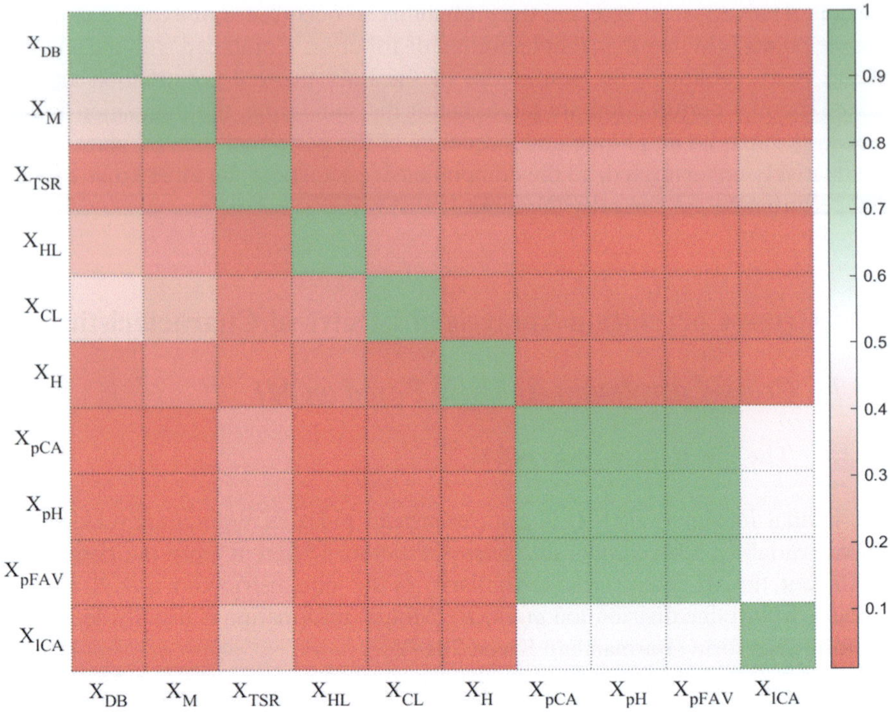

Fig. 2.4 Heat map of cross-correlation result based on MI

Table 2.4 Cross-correlation coefficient based on MI

	X_{DB}	X_{M}	X_{TSR}	X_{HL}	X_{CL}	X_{H}	X_{pCA}	X_{pH}	X_{pFAV}	X_{ICA}	X_{dCA}
X_{DB}	1	0.35	0.08	0.27	0.37	0.07	0.02	0.02	0.02	0.02	0.02
X_{M}	0.35	1	0.06	0.18	0.26	0.11	0.01	0.01	0.01	0.01	0.01
X_{TSR}	0.08	0.06	1	0.05	0.10	0.06	0.19	0.19	0.19	0.26	0.26
X_{HL}	0.27	0.18	0.05	1	0.15	0.08	0.01	0.01	0.01	0.02	0.02
X_{CL}	0.37	0.26	0.10	0.15	1	0.04	0.03	0.03	0.03	0.03	0.03
X_{H}	0.07	0.11	0.06	0.08	0.04	1	0.02	0.02	0.02	0.02	0.02
X_{pCA}	0.02	0.01	0.19	0.01	0.03	0.02	1	1	1	0.47	0.47
X_{pH}	0.02	0.01	0.19	0.01	0.03	0.02	1	1	1	0.47	0.47
X_{pFAV}	0.02	0.01	0.19	0.01	0.03	0.02	1	1	1	0.47	0.47
X_{ICA}	0.02	0.01	0.26	0.02	0.03	0.02	0.47	0.47	0.47	1	0.47
X_{dCA}	0.02	0.01	0.26	0.02	0.03	0.02	0.47	0.47	0.47	0.47	1

2.3.1.2 Cross-Correlation Result of Electrical Characteristics

The results obtained by using the mutual information for variable correlation analysis are shown below. Figure 2.4 is a heat map drawn based on mutual information values, and Table 2.4 is the mutual information values among all variables.

It can be analyzed in combination with Fig. 2.4 and Table 2.4 that there is a strong correlation among the five influencing factors of thermal perturbation indoors, and the independence of variables is weak, showing a significant correlation. In particular, the change trend of personnel calorific value $\{X_{pCA}\}$, personnel humidity $\{X_{pH}\}$, and personnel fresh air volume $\{X_{pFAV}\}$ contained in personnel thermal perturbation are completely consistent, and the variables are completely positively correlated. The correlations between other variables are less significant.

2.3.2 *Cross-Correlation Analysis Based on Pearson Coefficient*

2.3.2.1 The Theoretical Basis of Pearson Coefficient

The Pearson correlation coefficient is utilized to measure the linear correlation between two variables, and the standard deviation is not zero. Its calculation method is as follows (Mu et al. 2018).

$$P_{x,y} = \frac{\sum x \cdot y - \frac{\sum x \cdot \sum y}{M}}{\sqrt{\left(\sum x^2 - \frac{(\sum x)^2}{M}\right) \cdot \left(\sum y^2 - \frac{(\sum y)^2}{M}\right)}} \tag{2.3}$$

where x and y are the variables and M is the number of variables.

The Pearson correlation coefficient value varies from −1 to 1, which is usually divided into three grades, $|P_{x,y}| \le 0.4$ represents the weak linear correlation, $0.4 < |P_{x,y}| \le 0.7$ is strong correlation, and $0.7 < |P_{x,y}| \le 1$ is highly linear correlation.

2.3.2.2 Cross-Correlation Result of Electrical Characteristics

The results obtained by using the Pearson coefficient for variable correlation analysis are shown below. Figure 2.5 is a heat map drawn based on Pearson coefficient, and Table 2.5 is the Pearson coefficient among all variables.

It can be analyzed in combination with Fig. 2.5 and Table 2.5 that the five variables contained in thermal perturbation indoors are still the strongest correlation, which can be used to perform feature extraction in advance for the subsequent feature selection process. In addition, the correlation coefficient distribution between variables is more uniform by using Pearson correlation coefficient to do correlation

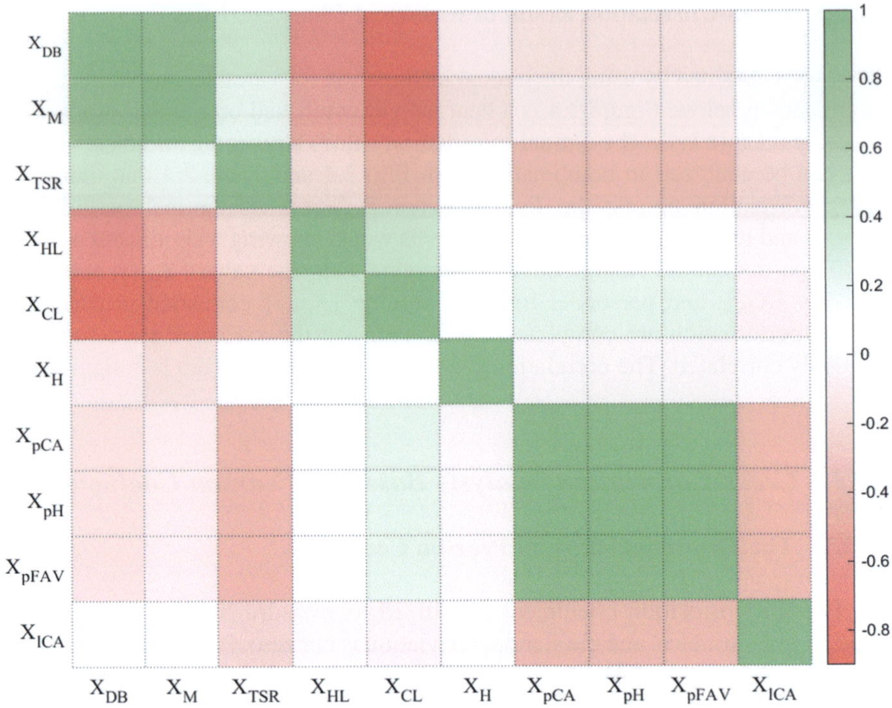

Fig. 2.5 Heat map of cross-correlation result based on Pearson coefficient

Table 2.5 Cross-correlation coefficient based on Pearson coefficient

	X_{DB}	X_{M}	X_{TSR}	X_{HL}	X_{CL}	X_{H}	X_{pCA}	X_{pH}	X_{pFAV}	X_{ICA}	X_{dCA}
X_{DB}	1	0.92	0.37	−0.64	−0.90	−0.10	−0.16	−0.16	−0.16	0.06	0.06
X_{M}	0.92	1	0.18	−0.50	−0.85	−0.15	−0.08	−0.08	−0.08	0.02	0.02
X_{TSR}	0.37	0.18	1	−0.19	−0.44	0.06	−0.41	−0.41	−0.41	−0.24	−0.24
X_{HL}	−0.64	−0.50	−0.19	1	0.39	0.08	0.08	0.08	0.08	−0.04	−0.04
X_{CL}	−0.90	−0.85	−0.44	0.39	1	0.05	0.22	0.22	0.22	−0.06	−0.06
X_{H}	−0.10	−0.15	0.06	0.08	0.05	1	−0.04	−0.04	−0.04	−0.01	−0.01
X_{pCA}	−0.16	−0.08	−0.41	0.08	0.22	−0.04	1	1	1	−0.40	−0.40
X_{pH}	−0.16	−0.08	−0.41	0.08	0.22	−0.04	1	1	1	−0.40	−0.40
X_{pFAV}	−0.16	−0.08	−0.41	0.08	0.22	−0.04	1	1	1	−0.40	−0.40
X_{ICA}	0.06	0.02	−0.24	−0.04	−0.06	−0.01	−0.40	−0.40	−0.40	1	−0.40
X_{dCA}	0.06	0.02	−0.24	−0.04	−0.06	−0.01	−0.40	−0.40	−0.40	−0.40	1

analysis, which can be seen more intuitively in the correlation coefficient heat map. Different from the non-negative characteristics of the mutual information method, the Pearson correlation coefficient can be negative, so the positive and negative correlation between variables can be obtained obviously, which provides a certain reference for the later feature selection process.

2.3.3 Cross-Correlation Analysis Based on Kendall Coefficient

2.3.3.1 The Theoretical Basis of Kendall Coefficient

The Kendall correlation coefficient is utilized to test the statistical correlation between two random variables (Doorn et al. 2018). The calculation method is as follows:

$$K_{x,y} = \frac{A - B}{\sqrt{(N_3 - N_1)\cdot(N_3 - N_2)}} \tag{2.4}$$

$$\begin{cases} N_3 = N\cdot(N-1)/2 \\ N_1 = \sum_{i=1}^{s} P_i\cdot(P_i - 1)/2 \\ N_2 = \sum_{j=1}^{t} Q_j\cdot(Q_j - 1)/2 \end{cases} \tag{2.5}$$

where A represents the number of consistent logarithm pairs of elements in x or y (two elements are a pair); B means the number of inconsistent logarithm pairs of elements in x or y; N is the number of variables. The same elements in x are combined into small sets, s represents the number of small sets owned in variable x, P_i represents the number of elements contained in the ith smallest set; t and Q_i are calculated based on variable y in the same way.

2.3.3.2 Cross-Correlation Result of Electrical Characteristics

The results obtained by using the Kendall coefficient for variable correlation analysis are shown below. Figure 2.6 is a heat map drawn based on Kendall coefficient, and Table 2.6 is the Kendall coefficient among all variables.

It can be analyzed in combination with Fig. 2.6 and Table 2.6 that the results of Kendall correlation analysis are basically consistent with those of Pearson correlation analysis, but there are some differences in the value of correlation coefficient. The variables with significant correlation and negative correlation are the same as Pearson correlation analysis.

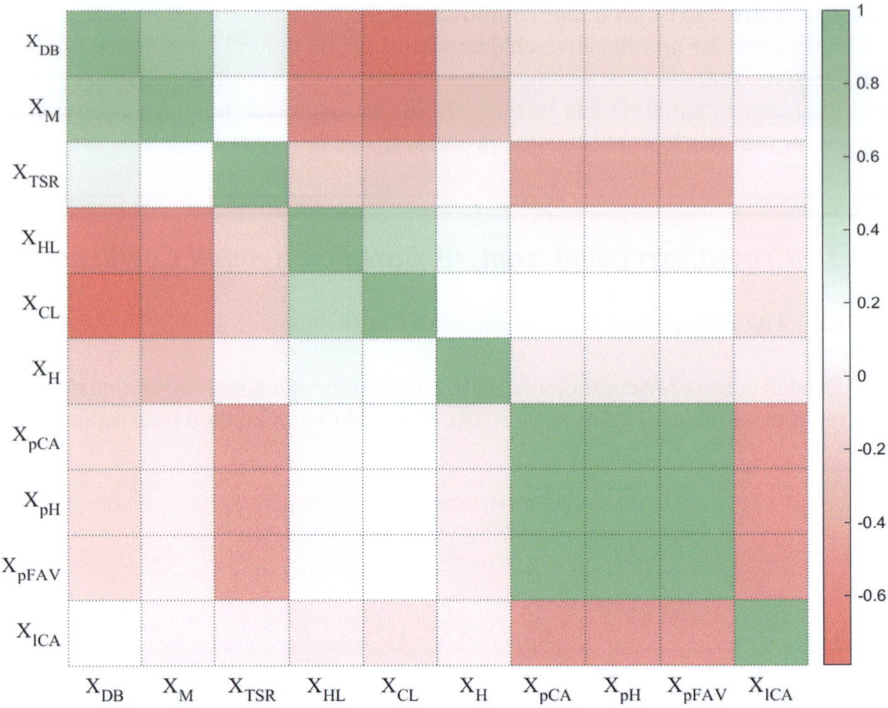

Fig. 2.6 Heat map of cross-correlation result based on Kendall coefficient

Table 2.6 Cross-correlation coefficient based on Kendall coefficient

	X_{DB}	X_M	X_{TSR}	X_{HL}	X_{CL}	X_H	X_{pCA}	X_{pH}	X_{pFAV}	X_{ICA}	X_{dCA}
X_{DB}	1	0.78	0.22	−0.65	−0.79	−0.18	−0.11	−0.11	−0.11	0.07	0.07
X_M	0.78	1	0.10	−0.58	−0.67	−0.25	−0.01	−0.01	−0.01	0.01	0.01
X_{TSR}	0.22	0.10	1	−0.17	−0.20	0.04	−0.44	−0.44	−0.44	0.01	0.01
X_{HL}	−0.65	−0.58	−0.17	1	0.52	0.20	0.05	0.05	0.05	−0.03	−0.03
X_{CL}	−0.79	−0.67	−0.20	0.52	1	0.16	0.14	0.14	0.14	−0.07	−0.07
X_H	−0.18	−0.25	0.04	0.20	0.16	1	−0.02	−0.02	−0.02	−0.01	−0.01
X_{pCA}	−0.11	−0.01	−0.44	0.05	0.14	−0.02	1	1	1	−0.48	−0.48
X_{pH}	−0.11	−0.01	−0.44	0.05	0.14	−0.02	1	1	1	−0.48	−0.48
X_{pFAV}	−0.11	−0.01	−0.44	0.05	0.14	−0.02	1	1	1	−0.48	−0.48
X_{ICA}	0.07	0.01	0.01	−0.03	−0.07	−0.01	−0.48	−0.48	−0.48	1	−0.48
X_{dCA}	0.07	0.01	0.01	−0.03	−0.07	−0.01	−0.48	−0.48	−0.48	−0.48	1

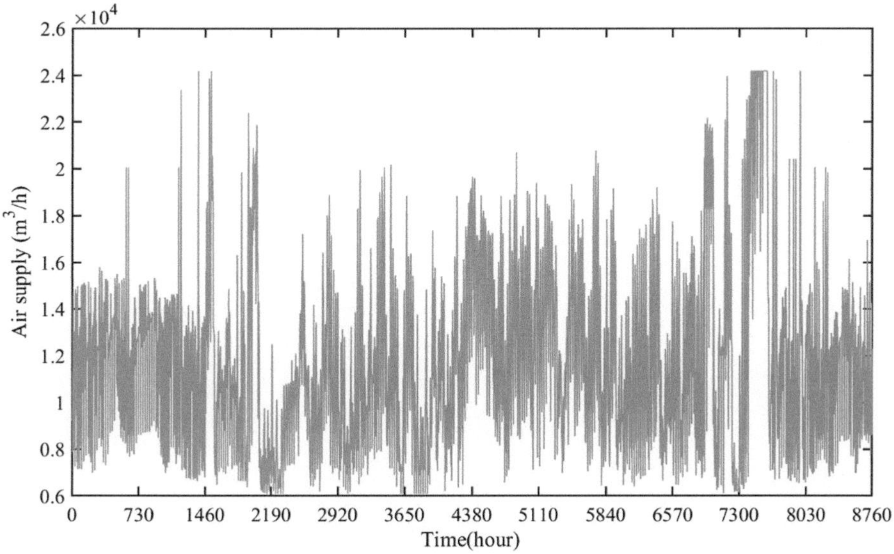

Fig. 2.7 Original supply air volume series

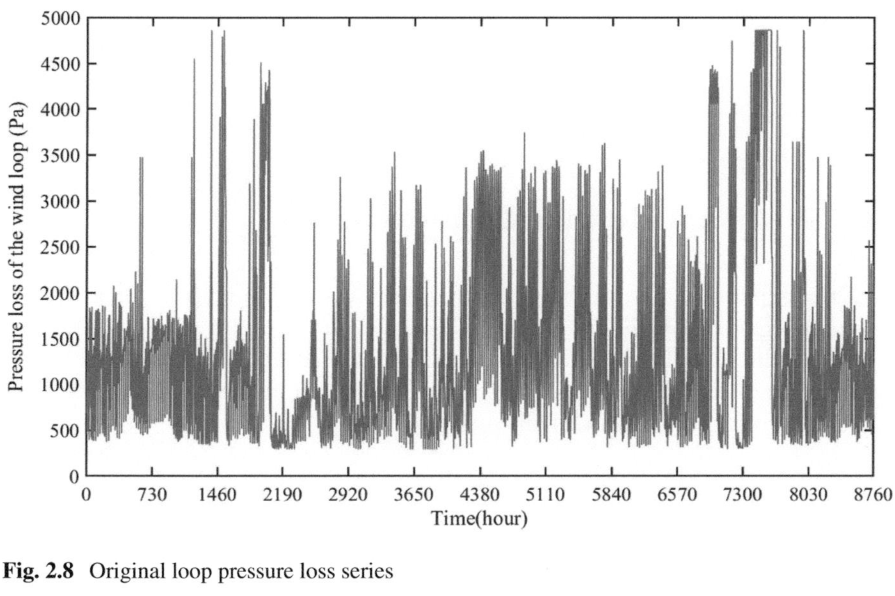

Fig. 2.8 Original loop pressure loss series

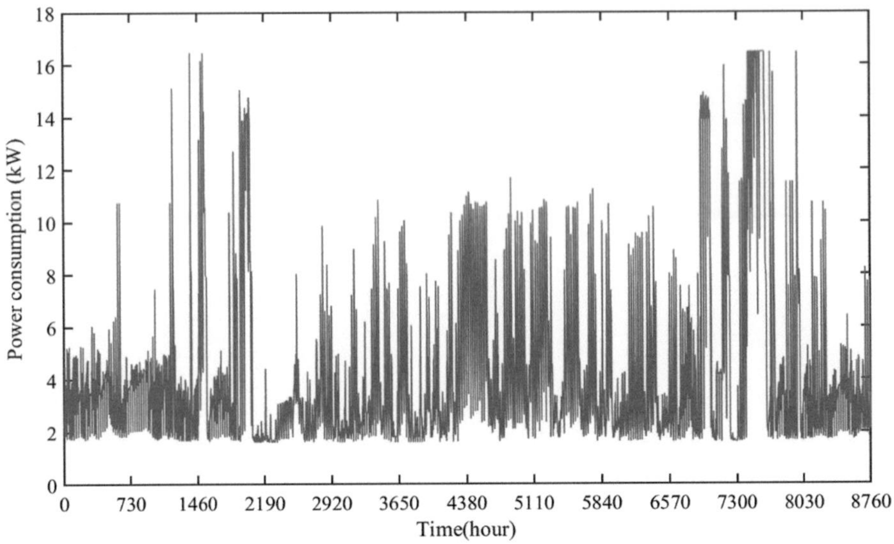

Fig. 2.9 Original power consumption series

Table 2.7 Statistical characteristics of the original electrical characteristics series

Original series	Minimum	Maximum	Mean value	Standard deviation	Skewness	Kurtosis
$\{Y_{SAV}\}$ (m^3/h)	6088.6101	24180.8208	11967.3869	3663.7208	0.9801	4.1395
$\{Y_{LPL}\}$ (Pa)	288.7048	4860.6251	1379.8335	982.1385	1.6204	5.5003
$\{Y_{PC}\}$ (kW)	1.5912	16.4673	4.2550	3.1271	2.1317	7.4345

2.4 Selection of Electrical Characteristics

The three common electrical characteristics of the air-conditioning electrical system of a complete open building selected in the section are analyzed as dependent variables, including supply air volume, loop pressure loss, and power consumption. And based on the three dependent variables, the feature selection is carried out according to the 11 influencing factors described above.

2.4.1 *Electrical Characteristics of Construction Power Grid*

In the section, the sampling interval of three groups of original series is 1 h and the sampling span is 1 year, from 1st January to 31st December, including 8760 sample points. Figures 2.7, 2.8, and 2.9 show the original time series of supply air volume, loop pressure loss, and power consumption, respectively, denoted as $\{Y_{SAV}\}$, $\{Y_{LPL}\}$, and $\{Y_{PC}\}$, respectively. From the waveform of the original series, it can be seen that the three groups of series all have strong randomness and non-stationarity. And there is a certain correlation between the three groups of sequences.

Similarly, in order to comprehensively analyze the characteristics of the three groups of time series, and further calculate the common statistical characteristics of them, Table 2.7 is the statistical characteristics of the three groups of time series adopted in the section.

The skewness values of the three groups of original electrical characteristics series are all positive, indicating that their distributions are all skewed to the right relative to the symmetrical distribution, and their statistical distribution characteristics are closer to the symmetrical distribution. The kurtosis values of the three groups of time series are all positive, indicating that the distribution of sample points relative to the mean is relatively scattered, and the probability of occurrence of extreme values in sample points is higher than that in normal distribution.

2.4.2 *Feature Selection Based on Spearman Correlation Coefficient*

Through the analysis of Sect. 2.3, it can be seen that there is a complete positive correlation among the three variables of personnel calorific value, personnel humidity, personnel fresh air volume contained in the thermal perturbation indoors. Therefore, in the feature selection of this section, these two variables are eliminated. That is, the feature selection of 9 influencing factors is carried out.

2.4.2.1 The Theoretical Basis of Spearman Coefficient

The Spearman correlation coefficient, also called the Spelman rank correlation coefficient, represents the correlation between the two variables in the form of monotone function (Dikbaş 2018). The calculation process is as follows.

Suppose X and Y are two random variables and the number of them is M, x_i and y_i denote the ith value in X and Y, respectively. The values in X and Y are sorted in the same way, and two new variables X' and Y' are obtained, where the values x_i' and y_i' are sorted by x_i in X and y_i in Y, respectively. Subtract the values in X' and Y' to get a sorted difference set d, where $d = x_i' - y_i'$. The Spearman correlation coefficient is calculated as follows:

$$S = 1 - \frac{6\sum_{i=1}^{M} d_i^2}{M \cdot \left(M^2 - 1\right)}, \quad i = 1, 2, \ldots, M \tag{2.6}$$

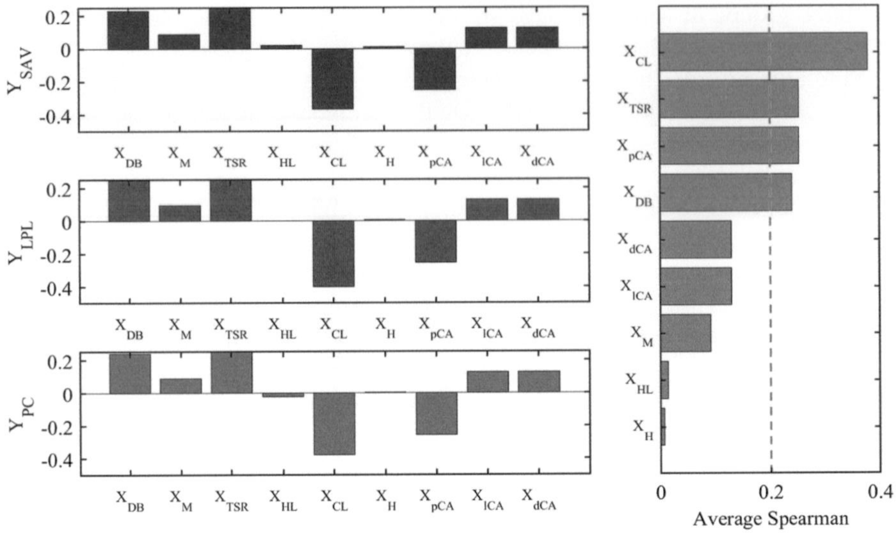

Fig. 2.10 Feature selection result based on Spearman correlation analysis

Table 2.8 Spearman coefficient of feature selection

	X_{DB}	X_M	X_{TSR}	X_{HL}	X_{CL}	X_H	X_{pCA}	X_{ICA}	X_{dCA}
Y_{SAV}	0.2291	0.0875	0.2507	0.0213	−0.3648	0.0093	−0.2504	0.1254	0.1254
Y_{LPL}	0.2496	0.0977	0.2541	0.0019	−0.3976	0.0097	−0.2550	0.1313	0.1313
Y_{PC}	0.2405	0.0893	0.2553	−0.0199	−0.3748	0.0037	−0.2541	0.1294	0.1294

2.4.2.2 Result of Feature Selection

The results obtained by using the Spearman coefficient for feature selection are shown below. Figure 2.10 shows the Spearman correlation coefficients between 9 groups of factors and three dependent variables, and the figure on the right shows the results of 9 groups of factors sorted by correlation coefficient. Table 2.8 shows the correlation values between all influencing factors and dependent variables.

Combined with Fig. 2.10 and Table 2.8, it can be concluded that when the average correlation analysis of all influencing features to the three dependent variables is carried out by using the Spearman coefficient, and 0.2 is selected as the correlation coefficient threshold for feature selection in the section, $\{X_{CL}\}$, $\{X_{TSR}\}$, $\{X_{pCA}\}$, and $\{X_{DB}\}$ are the four groups of features with the strongest correlation, while the correlation of other features is weak. In addition, there is a complete positive correlation between $\{X_{ICA}\}$ and $\{X_{dCA}\}$, and one of the features can be removed for further rough feature selection in the follow-up research.

2.4.3 *Feature Selection Based on CFS*

2.4.3.1 The Theoretical Basis of CFS

Correlation-based feature selection (CFS) first calculates the correlation coefficient matrix among features and the correlation coefficient matrix between feature and dependent variable from the sample set and then uses forward selection, backward elimination method, or best first search method to search for best feature subset. In the section, the forward selection method is utilized to increase the features greedily in the order of the correlation coefficient and the average Spearman correlation coefficient of Sect. 2.4.2, respectively. The process of CFS is shown in Fig. 2.11.

The formula for its search calculation is as follows:

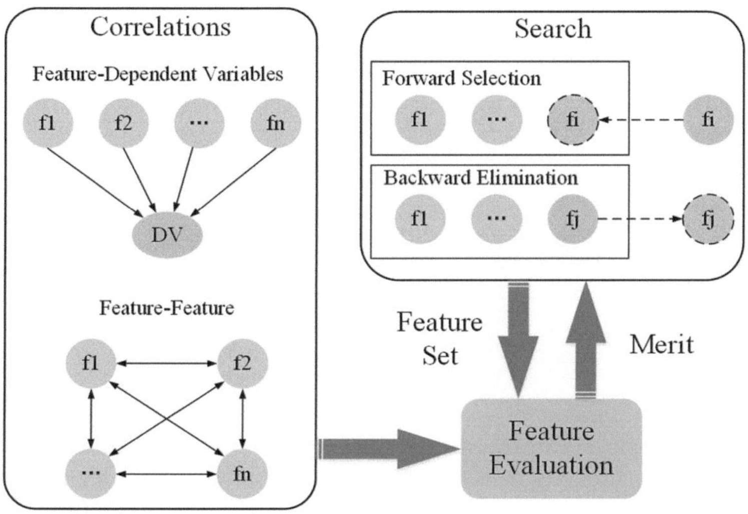

Fig. 2.11 Flowchart of CFS

Table 2.9 Correlation coefficient of 8 groups of variables

	X_{DB}	X_M	X_{TSR}	X_{HL}	X_{CL}	X_H	X_{pCA}	X_{dCA}	Y_{SAV}
X_{DB}	1	0.6011	0.2592	−0.7786	−0.6115	−0.0372	−0.3057	0.1497	0.0475
X_M		1	−0.2097	−0.3441	−0.3764	−0.4016	−0.0010	0.0397	0.1078
X_{TSR}			1	−0.3323	−0.1607	0.2318	−0.3444	−0.2335	−0.0844
X_{HL}				1	0.3784	0.0050	0.1827	−0.0766	0.3077
X_{CL}					1	−0.2811	0.1315	−0.0277	−0.3246
X_H						1	−0.0810	−0.0091	0.0585
X_{pCA}							1	−0.0405	−0.1829
X_{dCA}								1	0.1314

Table 2.10 Calculation result based on forward selection search

Feature set	k	$\overline{\lvert r_{cf} \rvert}$	$\overline{\lvert r_{ff} \rvert}$	Merit
[]	0	N/A	N/A	0
[X_{CL}]	1	0.3246	1.0000	0.3246
[X_{HL}]	1	0.3077	1.0000	0.3077
[X_{CL} X_{HL}]	2	0.3162	0.3784	0.3808
[X_{CL} X_{HL} X_{pCA}]	3	0.2717	0.2309	0.3893
[X_{CL} X_{HL} X_{pCA} X_{dCA}]	**4**	**0.2366**	**0.1396**	**0.3974**
[X_{CL} X_{HL} X_{pCA} X_{dCA} X_{M}]	5	0.2109	0.1599	0.3683
[X_{CL} X_{HL} X_{pCA} X_{dCA} X_{M} X_{TSR}]	6	0.1898	0.1919	0.3321

Table 2.11 Calculation result based on backward elimination search

Feature set	k	$\overline{\lvert r_{cf} \rvert}$	$\overline{\lvert r_{ff} \rvert}$	Merit
[]	0	N/A	N/A	0
[X_{CL}]	1	0.3246	1.0000	0.3246
[X_{TSR}]	1	0.0844	1.0000	0.0844
[X_{CL} X_{TSR}]	2	0.2045	0.1607	0.2684
[X_{CL} X_{TSR} X_{pCA}]	**3**	**0.1973**	**0.2122**	**0.2863**
[X_{CL} X_{TSR} X_{pCA} X_{DB}]	4	0.1599	0.3022	0.2315
[X_{CL} X_{TSR} X_{pCA} X_{DB} X_{dCA}]	5	0.1542	0.2264	0.2497
[X_{CL} X_{TSR} X_{pCA} X_{DB} X_{dCA} X_{M}]	6	0.1464	0.2328	0.2438

$$\text{Merit}_Q = \frac{k \times \overline{\lvert r_{cf} \rvert}}{\sqrt{k + k \times (k-1) \times \overline{\lvert r_{ff} \rvert}}} \tag{2.7}$$

where Merit_Q represents the evaluation value parameter of the features subset Q and the length is k, $\overline{\lvert r_{cf} \rvert}$ denotes the average correlation coefficient between features and dependent variables, and $\overline{\lvert r_{ff} \rvert}$ represents the average correlation coefficient between features. All variables need to be standardized before calculation.

2.4.3.2 Result of Feature Selection

According to the analysis results of Sect. 2.4.2, the $\{X_{ICA}\}$ factor is eliminated in this section, and $\{Y_{SAV}\}$ is selected as the dependent variable to select the features of the remaining 8 influencing factors. First, the correlation coefficient between the eight features and the correlation coefficient with $\{Y_{SAV}\}$ are calculated, and the results are shown in Table 2.9.

The forward selection method in the order of the correlation coefficient and the average Spearman correlation coefficient of Sect. 2.4.2 are used to search the optimal feature subset, respectively, and the results are shown in Tables 2.10 and 2.11.

According to the forward selection search method based on the descending order of correlation coefficients, when the feature subset includes the four features of $\{X_{CL}\}$, $\{X_{HL}\}$, $\{X_{pCA}\}$, and $\{X_{dCA}\}$, the evaluation value parameters of the feature subset are optimal.

According to the forward selection search method based on the descending order of the average Spearman correlation coefficient, when the feature subset includes the three features of $\{X_{CL}\}$, $\{X_{TSR}\}$, and $\{X_{pCA}\}$, the evaluation value parameters of the feature subset are optimal.

Combining with Fig. 2.10, Table 2.10, and Table 2.11, it can be concluded that $\{X_{CL}\}$ and $\{X_{pCA}\}$ are the common elements of optimal feature subset, but the optimal feature subsets obtained by the three feature selection methods are still different. The contingency of feature selection by calculating the Spearman correlation coefficient between features and dependent variables is too large, the accuracy of the results cannot be guaranteed, and the forward selection method is easy to fall into local optimization, thus the global optimal feature subset cannot be found.

2.4.4 Feature Selection Based on Global Search-ELM

In order to solve the defect that the two feature selection methods proposed above are easy to fall into local optimization, the section carries out a global search for all feature subsets based on ELM predictors and takes the RMSE index of the predicted results as the evaluation function. Thus, the global optimal feature subset is searched.

2.4.4.1 The Theoretical Basis of Global Search-ELM

The basic principle of the proposed global search feature selection method is that by using the method of random sampling, n features are randomly selected from all N features ($n \leq N$) to form a feature subset, and the corresponding input sample set is formed according to the labels of the n feature subset. The dependent variable is used as the output sample set. All the samples are divided into training set and testing set, which are imported into the ELM predictor, and the RMSE value predicted by the model is taken as the evaluation function of the selected feature subset. Finally, the feature subset with the smallest RMSE value is found. Thus the global optimal feature subset under the proposed method is obtained. The process of the Global Search-ELM method is shown in Fig. 2.12.

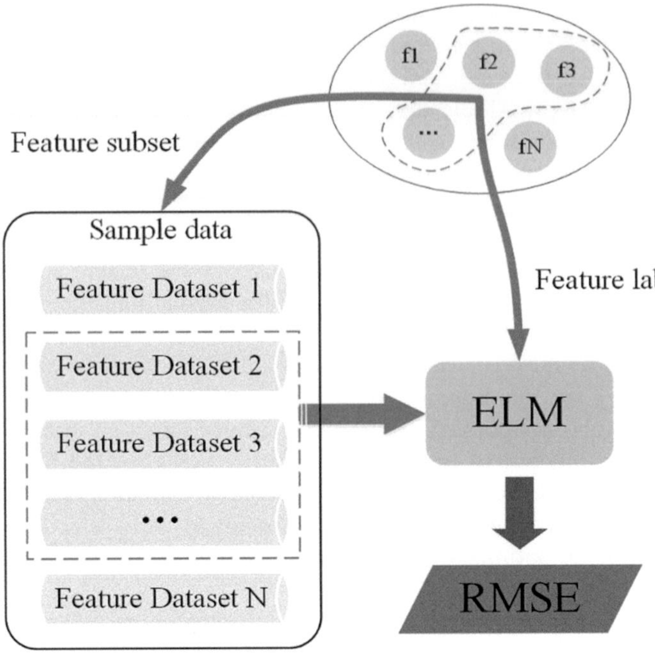

Fig. 2.12 Flowchart of Global Search-ELM

2.4.4.2 Result of Feature Selection

The results obtained by using the Global Search-ELM for feature selection are shown in Fig. 2.13. The above figure shows the change of the evaluation function value for all the feature subsets composed with different numbers of features, and the following figure shows the evaluation function value of the optimal feature subset whose number of features is 1–8, respectively.

According to the results of Global Search-ELM, when the feature subset contains the three features of $\{X_{TSR}\}$, $\{X_{HL}\}$, and $\{X_{pCA}\}$, the evaluation function of the feature subset is optimal, which is the optimal feature subset. This is different from the optimal feature subset obtained by the previous methods. It is mainly due to the accuracy index of the prediction results of the predictors used in the selection of evaluation functions, and there will be differences in each testing process. It can be improved by performing multiple global searches to find the mean.

2.5 Conclusion

A comprehensive analysis of the four feature selection methods used in the chapter is carried out, which are feature selection based on Spearman correlation, feature selection based on CFS (including forward selection based on correlation coeffi-

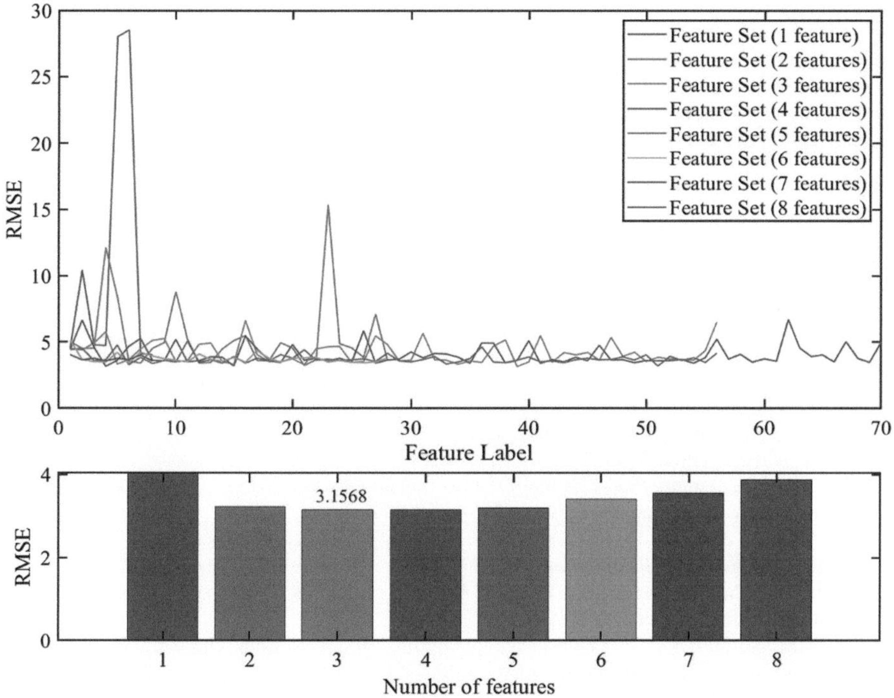

Fig. 2.13 Feature selection result based on Global Search-ELM

cient order and forward selection based on Spearman correlation order), and feature selection based on Global Search-ELM, the following conclusions can be obtained.

(a) $\{X_{\text{pCA}}\}$ is the common element of the optimal feature subset obtained by the four feature selection methods, which is most closely related to the dependent variable $\{Y_{\text{SAV}}\}$.
(b) $\{X_{\text{TSR}}\}$ and $\{X_{\text{CL}}\}$ can be found as the optimal feature subset by the three feature selection methods, and their relationship with the dependent variable $\{Y_{\text{SAV}}\}$ is relatively high.
(c) $\{X_{\text{HL}}\}$, $\{X_{\text{DB}}\}$, and $\{X_{\text{dCA}}\}$ can only be found by two or one feature selection methods to find the elements of the optimal feature subset, and the relationship between them and dependent variables is weak.
(d) $\{X_{\text{M}}\}$ and $\{X_{\text{H}}\}$ cannot be searched by either feature selection method, and these features cannot be taken into account in the process of identification and prediction of dependent variable $\{Y_{\text{SAV}}\}$.
(e) Synthesizing the above conclusions, it can be considered that for the dependent variable $\{Y_{\text{SAV}}\}$ in the chapter, the optimal feature subset is $[\{X_{\text{pCA}}\}\ \{X_{\text{TSR}}\}\ \{X_{\text{CL}}\}]$.
(f) The feature selection is for dependent variable $\{Y_{\text{SAV}}\}$ in the chapter, and follow-up research work can be focused on $\{Y_{\text{LPL}}\}$ and $\{Y_{\text{PC}}\}$ or other dependent variables for feature selection.

References

Abbasi Z, Rahmani M (2019) An instance selection algorithm based on reliefF. International Journal of Artificial Intelligence Tools 28(01):1950001

Aliane, A. A., Aliane, H., Ziane, M., & Bensaou, N. (2017). A genetic algorithm feature selection based approach for Arabic Sentiment Classification. In *Computer Systems & Applications*.

Ashok KJ, Abirami S (2018) Aspect-based opinion ranking framework for product reviews using a Spearman's rank correlation coefficient method. Information Sciences 460:23–41

Atashpaz-Gargari E, Reis MS, Braga-Neto UM, Barrera J, Dougherty ER (2017) A fast branch-and-bound algorithm for U-curve feature selection. Pattern Recognition 73:172–188

Barman S, Kwon YK (2017) A novel mutual information-based Boolean network inference method from time-series gene expression data. PLoS One 12(2):1–19

Dikbaş F (2018) A new two-dimensional rank correlation coefficient. Water Resources Management 32(5):1–15

Doorn JV, Ly A, Marsman M, Wagenmakers EJ (2018) Bayesian inference for Kendall's rank correlation coefficient. American Statistician 72(4):303–308

Koizumi Y, Niwa K, Hioka Y, Koabayashi K, Ohmuro H (2017) Informative acoustic feature selection to maximize mutual information for collecting target sources. IEEE/ACM Transactions on Audio Speech & Language Processing 25(4):768–779

Lei, X., Yan, P., & Tong, C. (2002). Best first strategy for feature selection. In *International Conference on Pattern Recognition*.

Ly A, Marsman M, Wagenmakers EJ (2018) Analytic posteriors for Pearson's correlation coefficient. Statistica Neerlandica 72(1):4–13

Mafarja MM, Mirjalili S (2017) Hybrid Whale Optimization Algorithm with simulated annealing for feature selection. Neurocomputing 260:302–312

Mu Y, Liu X, Wang L (2018) A Pearson's correlation coefficient based decision tree and its parallel implementation. Information Sciences 435:40–58

Murrieta-Mendoza A, Beuze B, Ternisien L, Botez RM (2017) New reference trajectory optimization algorithm for a flight management system inspired in beam search. Chinese Journal of Aeronautics 30(4):1459–1472

Nandi, G. (2011). An enhanced approach to Las Vegas Filter (LVF) feature selection algorithm. In *Emerging Trends & Applications in Computer Science*.

Rahim R, Abdullah D, Nurarif S, Ramadhan M, Anwar B, Dahria M et al (2018) Breadth first search approach for shortest path solution in Cartesian area. Journal of Physics Conference Series 1019(1):012038

Rodriguez-Galiano VF, Luque-Espinar JA, Chica-Olmo M, Mendes MP (2018) Feature selection approaches for predictive modelling of groundwater nitrate pollution: An evaluation of filters, embedded and wrapper methods. Science of the Total Environment 624:661–672

Sabeti M, Boostani R, Katebi SD, Price GW (2007) Selection of relevant features for EEG signal classification of schizophrenic patients. Biomedical Signal Processing & Control 2(2):122–134

San-Chuan LI, Li-Li WU (2018) Forward sequence feature selection algorithm based on correlation search. Communications Technology 51(12):2920–2924

Sembiring, P., Sinulingga, U., Situmorang, M., & Sembiring, S. (2017) Representative model the graph theory in calculations Kendall correlation coefficient. In *International Conference on Information and Communication Technology* 930.

Su X, Li L, Shi F, Qian H (2018) Research on the fusion of dependent evidence based on mutual information. IEEE Access 6:71839–71845

Sun L, Fu S, Wang F (2019) Decision tree SVM model with Fisher feature selection for speech emotion recognition. EURASIP Journal on Audio, Speech, and Music Processing 2019(1):2

Xiong CZ, Su M, Jiang Z, Jiang W (2019) Prediction of hemodialysis timing based on LVW feature selection and ensemble learning. Journal of Medical Systems 43(1):18

Zhang K, Dong Y, Andrew B (2015) Feature selection by merging sequential bidirectional search into relevance vector machine in condition monitoring. Chinese Journal of Mechanical Engineering 28(6):1248–1253

Zhu M, Jie S (2013) An embedded backward feature selection method for MCLP classification algorithm ☆. Procedia Computer Science 17:1047–1054

Chapter 3
Prediction Model of City Electricity Consumption

Abstract Power load forecasting is a series of forecasting work for power load. From the point of view of forecasting objects, power load forecasting includes the prediction of future power demand and the prediction of load curve, which provides a reliable decision-making basis for power system planning and operation. This chapter focuses on the analysis and prediction of short-term electricity consumption and long-term electricity consumption data in urban areas. The short-term prediction of electricity consumption is carried out by using the time-series model, and the periodic characteristics of the long-term power consumption series are further mined by the seasonal time-series model. On this basis, the uncertainty interval prediction of electricity consumption data is realized by using heteroscedasticity model, and the distributed computing strategies of these methods under the framework of big data are provided.

3.1 Introduction

Electricity big data include power grid construction data, electric equipment data, power marketing data, and power customer service data. The power grid construction data is the construction site information describing the electricity grid construction project, such as infrastructure pictures. The electric equipment data is the quality information describing various equipment, recorded in real-time and used to accurately identify the quality of equipment. The power marketing data is described as intelligent and interactive. Service data exist in power information collection and measurement, power verification and distribution, etc. Power customer service data mainly refers to service data collected from three channels: intelligent interactive website, business support system, and basic support platform. It includes structured data such as files, transactions, and GIS, as well as unstructured data such as voice, text, and images (Jaffel et al. 2017).

H. Liu, *Smart Cities: Big Data Prediction Methods and Applications*,
https://doi.org/10.1007/978-981-15-2837-8_3

Electricity load forecasting is a series of forecasting work performed on the power load. From the perspective of forecasting objects, electricity load forecasting includes prediction of future power demand and prediction of load curve. It provides a reliable decision basis for power system planning and operation. The electricity market forecast is mainly to comprehensively analyze and evaluate the current situation of the power market and the past development of the electricity market, so as to effectively predict the future development direction of the power market and the market trend. Electricity disaster prediction refers to the power grid failure warning caused by natural disasters such as ice storms, earthquakes, and other power accidents.

The traditional electricity load forecasting method is easier to use, but the final prediction result is not reliable. Therefore, the traditional method is suitable for situations where the accuracy of the prediction result is relatively low. The traditional prediction method is as follows (Chhay et al. 2018):

(a) Trend extrapolation

The trend extrapolation method is to speculate on the future according to the entire development trend of things that have already occurred. The load data itself has a certain regularity. Using trend extrapolation to analyze the overall change in historical load data can predict future loads. The trend extrapolation method is often used to have a polynomial method, a horizontal method, and a speculation method based on seasonal changes (Zhou et al. 2012). All of the above methods only consider the trend of the load itself and do not consider some factors related to the load and affect the load. Although these methods are simple in structure and easy to apply, their prediction accuracy is very low, so they are rarely used in load forecasting.

(b) Time-series method

This method converts historical load values into a series of time series. When acquiring the load prediction sample, it is generally measured in a certain time period, and usually, the sampling interval is 1 h (Brockwell and Davis 2009). A set of load data obtained by this method can be regarded as a set of series, and the series is variable in time. The sequence is used as a research object when a mathematical model adapted to the object is matched, and the load can be predicted using the model. The problem with this method is that it still not consider the relevant factors affecting the load.

(c) Regression analysis

Regression analysis takes the impact of raw data and other factors into account. This method is used to analyze the relationship between the load and the influencing factors, so as to find a mathematical model corresponding to it and achieve the final prediction. However, in load prediction, since the load change is a nonlinear change, the nonlinear analysis method is used to perform load prediction. The disadvantage is that the calculation is more complicated facing a large amount of data (Bianchi et al. 2017).

With the development of big data, machine learning, and artificial intelligence, the predictive models and methods of electricity load are gradually transforming into intelligent machine learning. Some excellent machine learning algorithms such as fuzzy logic, genetic algorithms, and neural networks are used in power load. The predictions show excellent results (Persio and Honchar 2017). As one of the most widely used artificial intelligence technologies in big data technology (Shi et al. 2017), neural networks have parallel processing information capabilities, nonlinear mapping capabilities, self-learning capabilities, fault tolerance, and robustness. Among the many models of neural networks, the most commonly used is the BP network. In recent years, the load forecast based on Elman network has been increasing, which is due to its dynamic characteristics, which has significantly improved the prediction effect. At the same time, the combination of two or more basic models into new models is also increasingly appearing in short-term load forecasting, such as neural networks (Sadaei et al. 2017).

This chapter focuses on the analysis and prediction of short-term electricity consumption and long-term electricity consumption data in urban areas. At present, many experts and scholars around the world have already carried out research work in the field of big data load forecasting, and have also achieved some results. The earliest application of neural networks to load forecasting is *Professor DC Park*. Since then, load forecasting has changed from manual intervention to machine-autonomous learning (Saeed Madani 1991). In his research, weather factors were added to the input variables of the model to obtain nerves. The conclusion shows that neural network prediction is more accurate than traditional methods. *Sadaei* proposed a short-term load forecasting method with fuzzy time series and seasonality (Sadaei et al. 2017). Since the seasonal changes in different regions are different, time variables are particularly important in the load forecasting model. In this paper, the fuzzy time series was combined with the seasonal differential autoregressive moving average model to establish a prediction model and consider the necessary parameters. Some researchers used a priority vector method to build a load forecasting model that selected the day type and season as the influencing factors (Rahman and Shrestha 1991). According to the adaptability of the priority vector method, the relationship between variables is automatically generated inside the system. Kadir Kavaklioglu established the model to predict the electricity consumption of Turkey with SVR (Kavaklioglu 2011). This article used the e-SVR model for prediction. To facilitate predicting power consumption, it used a separate SVR model with the current and past values of power consumption as input variables. Experiments showed that the scheme had higher accuracy for the prediction of Turkey's electricity consumption. Wu et al. used gray convex correlation analysis to describe the relationship between power consumption and its related factors. A multivariate grey forecasting model considering the total population was established to realize the electricity consumption forecast in Shandong Province, China. The actual data test results showed that the model has a better significant prediction performance than the traditional gray prediction model (Wu et al. 2018). Oprea et al. established a mixed neural network model based on NARX network and FITNETs to forecast the residential electricity consumption (Oprea et al. 2018).

3.2 Original Electricity Consumption Series

3.2.1 Regional Correlation Analysis of Electricity Consumption Series

The original sequences used in this chapter are preliminary integrated hourly loads calculated from raw telemetry data. The data are derived from historical electricity consumption data collected by eight regions served by PJM in the USA. These eight regions include AEP, AP, ATSI, DAY, DEOK, DOM, DUQ, and EKPC. Figure 3.1 shows the long-term electricity consumption series of eight different regions, whose sampling period is 1 day. It can be seen from the figure that the long-term electricity consumption series in the eight regions have a high degree of cyclic correlation although the amplitudes are different. The overall fluctuation trend of each series is the same, with similar peaks and troughs. This shows that users in each region have similar electricity usage habits as a whole.

In order to further accurately analyze the correlation between the power consumption data of each region, this chapter uses Pearson, Kendall, and Spearman correlation coefficients to calculate the cross-correlation coefficient of different regions (Mu et al. 2018). The calculation results are shown in Tables 3.1, 3.2, and 3.3.

It can be concluded from the tables that the three types of correlation coefficients have approximate calculation results. The three types of correlation coefficients between any two regions are greater than 0.6 and are generally greater than 0.75. The above phenomenon further confirms that the long-term electricity consumption series in different regions is highly correlated. The period of the overall fluctuation trend of long-term electricity consumption series, peak and valley characteristics meet the general law (Hilbert 2017).

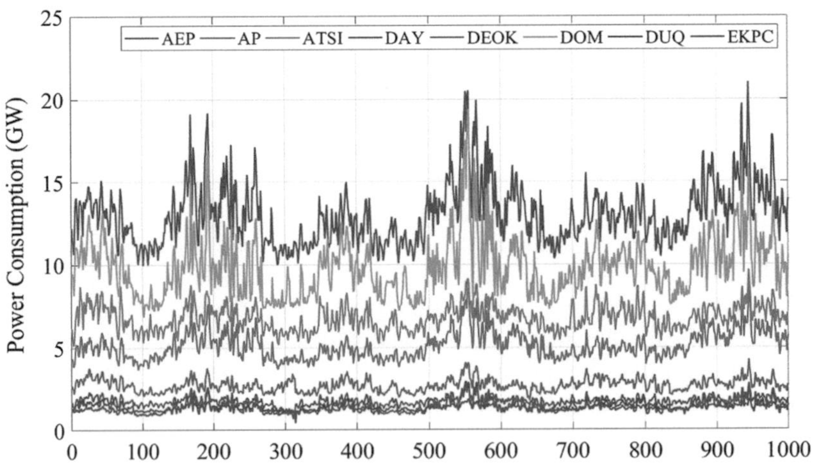

Fig. 3.1 Long-term electricity consumption series of different regions

Table 3.1 Cross-correlation coefficient of different regions based on Pearson

Region	AEP	AP	ATSI	DAY	DEOK	DOM	DUQ	EKPC
AEP	1.0000	0.9617	0.8826	0.8973	0.7942	0.8675	0.7249	0.9364
AP	0.9617	1.0000	0.8224	0.8244	0.6908	0.8547	0.6465	0.9041
ATSI	0.8826	0.8224	1.0000	0.9089	0.8358	0.7580	0.8604	0.7446
DAY	0.8973	0.8244	0.9089	1.0000	0.7757	0.7365	0.7680	0.7902
DEOK	0.7942	0.6908	0.8358	0.7757	1.0000	0.7524	0.8241	0.7349
DOM	0.8675	0.8547	0.7580	0.7365	0.7524	1.0000	0.7561	0.7997
DUQ	0.7249	0.6465	0.8604	0.7680	0.8241	0.7561	1.0000	0.5986
EKPC	0.9364	0.9041	0.7446	0.7902	0.7349	0.7997	0.5986	1.0000

Table 3.2 Cross-correlation coefficient of different regions based on Kendall

Region	AEP	AP	ATSI	DAY	DEOK	DOM	DUQ	EKPC
AEP	1.0000	0.9617	0.8826	0.8973	0.7942	0.8675	0.7249	0.9364
AP	0.9617	1.0000	0.8224	0.8244	0.6908	0.8547	0.6465	0.9041
ATSI	0.8826	0.8224	1.0000	0.9089	0.8358	0.7580	0.8604	0.7446
DAY	0.8973	0.8244	0.9089	1.0000	0.7757	0.7365	0.7680	0.7902
DEOK	0.7942	0.6908	0.8358	0.7757	1.0000	0.7524	0.8241	0.7349
DOM	0.8675	0.8547	0.7580	0.7365	0.7524	1.0000	0.7561	0.7997
DUQ	0.7249	0.6465	0.8604	0.7680	0.8241	0.7561	1.0000	0.5986
EKPC	0.9364	0.9041	0.7446	0.7902	0.7349	0.7997	0.5986	1.0000

Table 3.3 Cross-correlation coefficient of different regions based on Spearman

Region	AEP	AP	ATSI	DAY	DEOK	DOM	DUQ	EKPC
AEP	1.0000	0.9617	0.8826	0.8973	0.7942	0.8675	0.7249	0.9364
AP	0.9617	1.0000	0.8224	0.8244	0.6908	0.8547	0.6465	0.9041
ATSI	0.8826	0.8224	1.0000	0.9089	0.8358	0.7580	0.8604	0.7446
DAY	0.8973	0.8244	0.9089	1.0000	0.7757	0.7365	0.7680	0.7902
DEOK	0.7942	0.6908	0.8358	0.7757	1.0000	0.7524	0.8241	0.7349
DOM	0.8675	0.8547	0.7580	0.7365	0.7524	1.0000	0.7561	0.7997
DUQ	0.7249	0.6465	0.8604	0.7680	0.8241	0.7561	1.0000	0.5986
EKPC	0.9364	0.9041	0.7446	0.7902	0.7349	0.7997	0.5986	1.0000

3.2.2 Original Sequences for Modeling

The original sequences for modeling include two sets of time series, short-term electricity consumption series and long-term electricity consumption series. The data represent the total amount of electricity consumed per unit of time in the AEP area.

(a) *Short-term electricity consumption series*

The short-term electricity consumption series is shown in Fig. 3.2. It is marked as an experimental sample $\{X_t\}$. The sample size of the time series is 1200, the

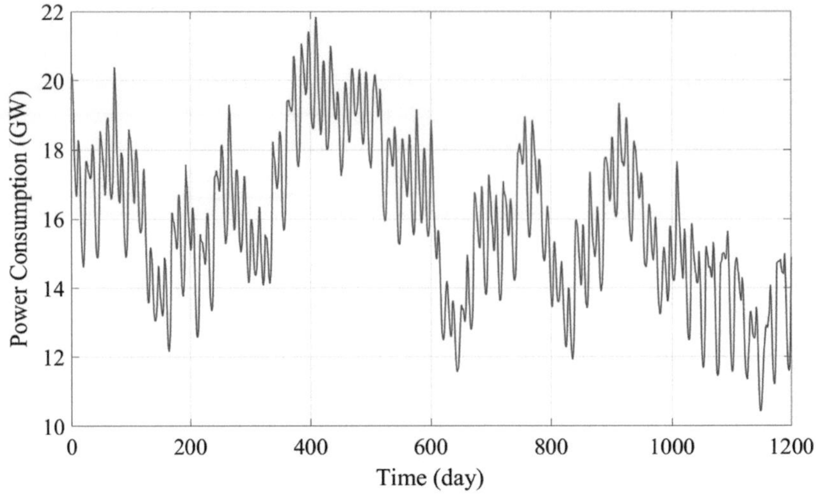

Fig. 3.2 Short-term electricity consumption series

sampling period is 1 h, and the unit of electricity consumption is GW. As can be seen from Fig. 3.1, it is obvious that the short-term electricity consumption series, on the one hand, shows strong non-stationary and randomness. Whether viewed on a smaller time scale or on a longer time scale, the waveforms of short-term series have strong volatility, and the amplitude of the fluctuation is large. On the other hand, judging from the figure, the short-term series does not have strong periodicity.

(b) *Long-term electricity consumption series*

The long-term electricity consumption series is shown in Fig. 3.3. It is marked as an experimental sample $\{Y_t\}$. The sample size of the time series is 1200, the sampling period is 1 day, and the unit of electricity consumption is GW. Similar to the short-term series, long-term electricity consumption series, on the one hand, also shows strong non-stationary and randomness. The waveforms of long-term series have strong volatility with large amplitude of the fluctuation no matter on a smaller time scale or on a longer time scale. However, on the other hand, judging from the figure, the long-term series shows significant periodicity.

3.2.3 Separation of Sample

In this chapter, the ARIMA model and the SARIMA model will be used to perform regression analysis on the electricity consumption series and construct a predictive model. At the same time, this chapter will also build an electricity consumption series non-deterministic interval prediction model based on ARIMA model prediction results and ARCH algorithm. In order to better realize the training and testing

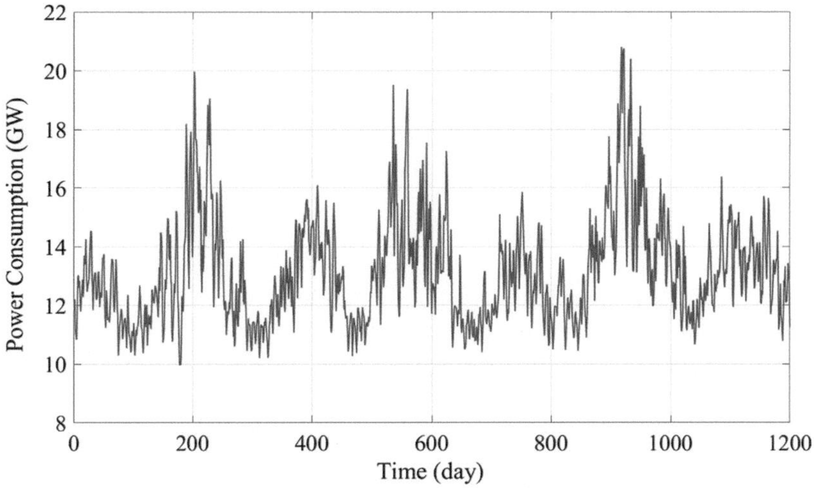

Fig. 3.3 Long-term electricity consumption series

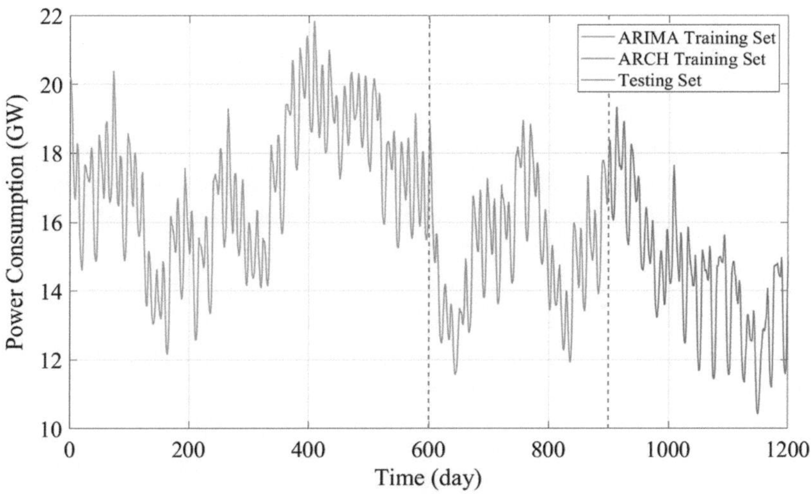

Fig. 3.4 Separation of short-term electricity consumption series

of the above various models, it is necessary to rationally separate the original time series. The separations of short-term electricity consumption series and long-term electricity consumption series are shown in Figs. 3.4 and 3.5.

In this chapter, two types of electricity consumption series are separated into three parts: D_1, D_2, and D_3. Among them, D_1 includes the first 600 sample points for training deterministic prediction models based on the ARIMA model and SARIMA model. The D_2 and D_3 parts are used as test sets for deterministic prediction models to evaluate model prediction performance. D_2 includes the 601st~900th sample points, and the prediction residuals are obtained after the prediction of the ARIMA model and SARIMA model, as a training set of non-deterministic prediction models. D_3 includes the 901st to 1200th sample points as a test set for the final non-deterministic prediction model.

The purpose of using D_2 as a non-deterministic predictive model training set is to improve the generalization performance of the model. If D_1 and D_2 are combined as a common training set of the deterministic prediction model and the non-deterministic prediction model, the prediction accuracy of the deterministic prediction model for the training set is relatively high, while the residual fluctuation is too small. The non-deterministic prediction model may underestimate the residual fluctuations in the real situation so that the final prediction interval is too small to effectively cover the actual data, thus making the model show low generalization.

3.3 Short-Term Deterministic Prediction of Electricity Consumption Based on ARIMA Model

3.3.1 Model Framework of ARIMA

ARIMA analysis mainly consists of time-series pre-processing, model identification, model order, parameter estimation, and lead prediction. Among them, the recognition and ordering of the model is the most important part of time-series modeling. The ARIMA predictive model modeling process is shown in Fig. 3.6.

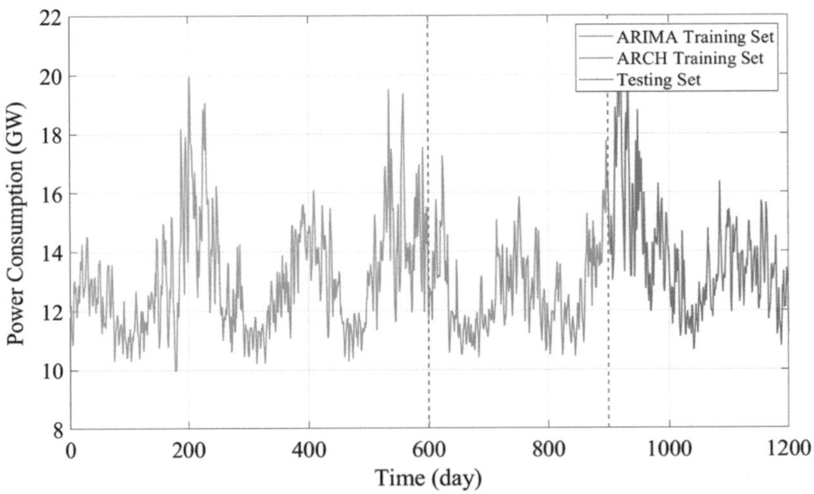

Fig. 3.5 Separation of long-term electricity consumption series

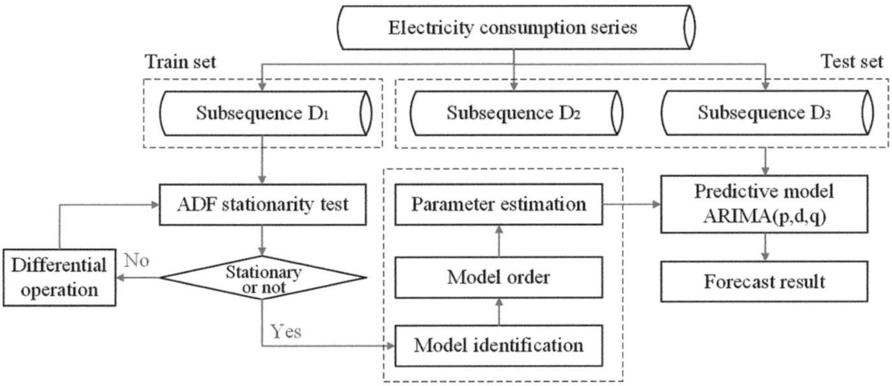

Fig. 3.6 Modeling process of ARIMA predictive model

3.3.2 *Theoretical Basis of ARIMA*

Common time-series models can be divided into autoregressive model, moving average model, autoregressive moving average model, and autoregressive integrated moving average model.

(a) *Autoregressive model*

The AR model is a linear regression model that uses a linear combination of observations during the previous times to describe the variables at a later time. The AR(p) model can be expressed as follows:

$$x_t = \alpha_0 + \alpha_1 x_{t-1} + \alpha_2 x_{t-2} + \ldots + \alpha_p x_{t-p} + \varepsilon_t \tag{3.1}$$

where x_t is the observed value at the previous moment, α_0 is a constant, $\alpha_1, \alpha_2, \ldots, \alpha_p$ means the autoregressive coefficient to be estimated, and the random error term $\varepsilon_t \sim N(0, \sigma^2)$ means a white noise sequence that is independent of each other. The random error term is not related to the lag variable sequence. The AR model is mainly used to model the past performance of the time series. Its expression is similar to the expression of the time-series reversibility condition. The difference is that the AR model has the largest hysteresis range p, while the linear conditional expression has the hysteresis range to infinity.

(b) *Moving Average model*

A linear combination of current and previous random error terms is taken by the MA model to express the current predicted value. The MA(q) model can be defined as follows:

$$x_t = \overline{x} + \varepsilon_t + \lambda_1 \varepsilon_{t-1} + \lambda_2 \varepsilon_{t-2} + \ldots + \lambda_q \varepsilon_{t-q} \tag{3.2}$$

where $\varepsilon_1, \varepsilon_2, \ldots, \varepsilon_p$ is the random error term at the previous moment, $\overline{x}$ is the mean value of the sequence, and $\lambda_1, \lambda_2, \ldots, \lambda_p$ is the moving average coefficient to be estimated. The MA model is often used to characterize the impact effect, and its expression is similar to the linear conditional expression of the time series. The difference is that the MA model has a maximum hysteresis range of q, while the linear conditional expression has a hysteresis range of infinity.

(c) *Autoregressive Moving Average model*

The ARMA model is a combination of an autoregressive model and a moving average model. The ARMA model uses observations at several previous times and random error terms to predict the current value. The ARMA (p,q) model can be defined as follows:

$$x_t = \alpha_0 + \alpha_1 x_{t-1} + \alpha_2 x_{t-2} + \ldots + \alpha_p x_{t-p} + \varepsilon_t - \lambda_1 \varepsilon_{t-1} - \lambda_2 \varepsilon_{t-2} - \ldots - \lambda_q \varepsilon_{t-q} \quad (3.3)$$

where $x_{t-1}, x_{t-2}, \ldots, x_{t-p}$ is the observation at the previous moment, $\varepsilon_t, \varepsilon_{t-1}, \ldots, \varepsilon_{t-p}$ is the random error term at the previous moment, $\alpha_1, \alpha_2, \ldots, \alpha_p$ is the autoregressive coefficient to be estimated, and $\lambda_1, \lambda_2, \ldots, \lambda_p$ is the moving average coefficient to be estimated. The ARMA model has a wider range of applications than the AR and MA models.

(d) *Autoregressive Integrated Moving Average model*

The above model can only handle smooth time series, but the actual time series is often unstable. For the non-stationary time series, the above model cannot be used directly, but the time series needs to be smoothed by difference and then the above model is used. The ARIMA(p,d,q) model first performs a d-order difference on the time series and then applies an autoregressive moving average model to analyze the time series.

The ARIMA model belongs to the ARMA model but only one more difference operation, so only the AR model, the MA model, and the ARMA model are analyzed here. When the AR model and the MA model lags are infinite, the three models are completely equivalent to a stable and reversible time series. The parameters of another model can be derived from the parameters of one model.

3.3.3 Modeling Steps of ARIMA Predictive Model

(a) *Electricity consumption series pre-processing and non-stationary testing*

After the original sequence is divided, the ADF test method is needed to judge the stability of the original training sequence. If it is not stable, the difference processing is performed until the sequence is stable. After testing, both the short-term electricity consumption series and the long-term electricity consumption series are not stable, and a first-order difference is required. The first-order differential post-sequence is shown in Figs. 3.7 and 3.8. The differential power consumption

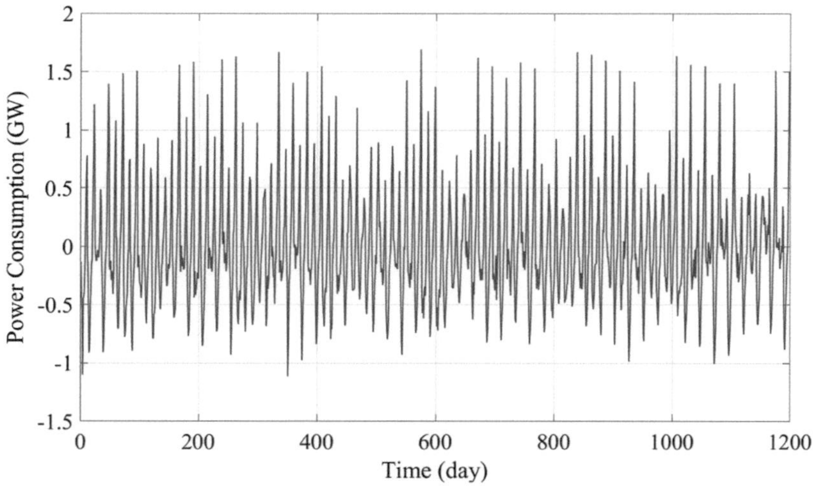

Fig. 3.7 Short-term electricity consumption first-order difference series

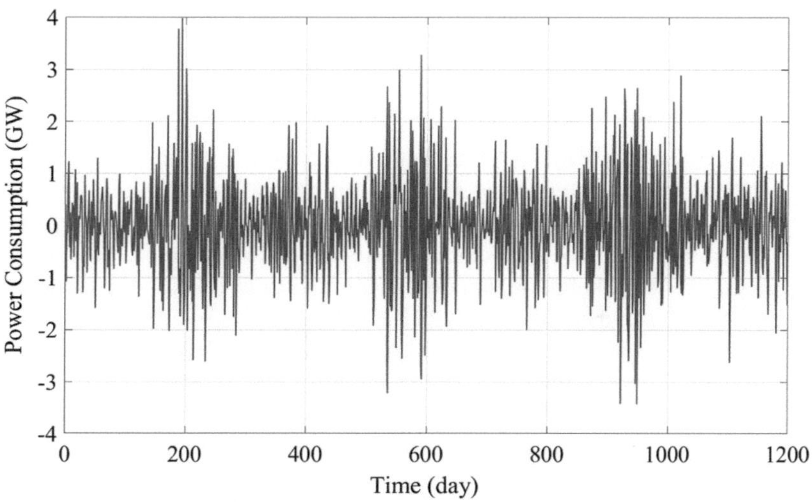

Fig. 3.8 Long-term electricity consumption first-order difference series

sequences are denoted as $\left\{\hat{X}_t\right\}$ and $\left\{\hat{Y}_t\right\}$, respectively. Further verifying the stationarity of the sequences $\left\{\hat{X}_t\right\}$ and $\left\{\hat{Y}_t\right\}$, the sequences are found to be stationary.

(b)*Identification of the ARIMA model*

Generally speaking, in the time-series model identification, the autocorrelation coefficient and the partial autocorrelation coefficient are used to determine the

Table 3.4 Correlation coefficient characteristics of ARIMA model

First-order difference sequence	Models	Autocorrelation coefficient	Partial autocorrelation coefficient
$\{\hat{X}_t\}$	AR(p)	Trailing	p-order censored
	MA(q)	q-order censored	Trailing
	ARMA(p,q)	Trailing	Trailing
$\{\hat{Y}_t\}$	AR(p)	Trailing	p-order censored
	MA(q)	q-order censored	Trailing
	ARMA(p,q)	Trailing	Trailing

coefficients and order of the model. The specific identification and discrimination process is shown in Table 3.4.

Figures 3.9 and 3.10 show the calculation results of the autocorrelation coefficients and the partial autocorrelation coefficients of the first 20 samples after the first-order difference between short-term electricity consumption series and long-term electricity consumption series. By analyzing Figs. 3.9 and 3.10, both types of difference sequences exhibit the following characteristics: the autocorrelation coefficients values exhibit a trailing state, and the partial autocorrelation coefficients exhibit a truncated state. It can be preliminarily judged that both types of difference sequences are in accordance with the ARMA model.

(c) *Order of the ARIMA model*

The chapter uses the Bayesian Information Criterion (BIC) to achieve the order of the ARIMA model. The guiding ideology for choosing the optimal model is to examine two aspects: one is the maximization of the likelihood function, and the other is the minimization of the number of unknown parameters in the model. The larger the value of the likelihood function is, the better the effect of the model fitting will be, but the pros and cons of the model with the accuracy of the fitting cannot be measured simply. This leads to more and more unknown parameters in the model, and the model becomes more and more complicated. It will cause overfitting. Therefore, a good model should be a comprehensive optimization of the fitting accuracy and the number of unknown parameters (Tamura et al. 2010).

When training the ARIMA model, increasing the number of parameters, that is, increasing the complexity of the model, on the one hand, increases the likelihood function, but on the other hand, it leads to overfitting. In response to this problem, both AIC and BIC introduce penalty terms related to the number of model parameters. The penalty of BIC is larger than that of AIC. Considering the number of samples, if the number of samples is too large, the model complexity can be effectively prevented from being too high due to excessive model accuracy. Therefore, when the sample size is large, the model selected using the AIC criterion does not converge with the real model, which usually has more unknown parameters than the real model. BIC makes up for the shortcomings of AIC.

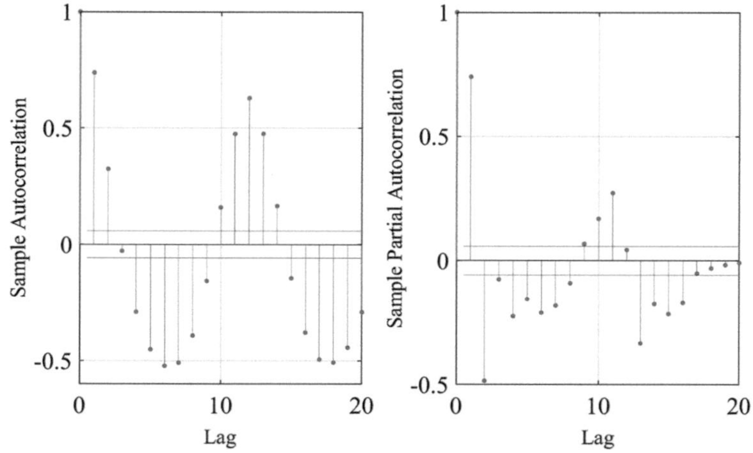

Fig. 3.9 Short-term series calculation results of autocorrelation coefficient and the partial autocorrelation coefficient

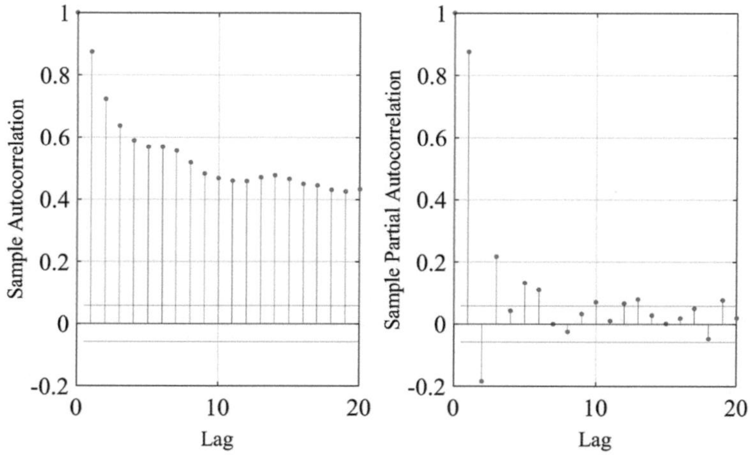

Fig. 3.10 Long-term series calculation results of autocorrelation coefficient and the partial autocorrelation coefficient

The BIC function is defined as follows:

$$\mathrm{BIC}(n) = \ln(n)t - 2\ln\left(\hat{C}\right) \tag{3.4}$$

where $\hat{C}$ is the maximum value of the model likelihood function, n is the number of observation points in the sequence, and t is the number of parameters estimated in the model.

It can be derived that the BIC criterion consists of two items. The first item reflects the number of model parameters and increases with the increase of the order. The second item reflects the quality of the model fitting and becomes smaller as the order increases. The BIC criterion means finding a model that has a good fit and is not too complicated to cause overfitting. Therefore, when n exceeds the optimal model order n_b of the world sequence, continuing to increase n will reduce accuracy. The principle of the BIC criterion is to find the parameter that minimizes the BIC as the optimal order of the model, which is expressed as follows:

$$n_b = \underset{0 \le n \le N}{\arg\min}\, \mathrm{BIC}(n) \tag{3.5}$$

where n_b is the upper bound of the model order. In order to balance the calculation amount and the criterion discrimination accuracy, this chapter takes $n_b = 50$ for calculation. Judging from the calculation results, the best fit model for the sequence $\{X_t\}$ is ARIMA(3,1,3), and the best fit model for the sequence $\{Y_t\}$ is ARIMA(3,1,5).

3.3.4 Forecasting Results

The ARIMA advanced prediction model is trained by using the D_1 portion of the short-term electricity consumption series and long-term electricity consumption series shown above as a training set. The model is verified by using the $D_2 + D_3$ part as the test set. The predicted results of the advanced prediction accuracy evaluation indicators are shown in Tables 3.5 and 3.6.

Figures 3.11, 3.12, 3.13, 3.14, 3.15, and 3.16 show partial previews of the short-term electricity consumption series and long-term electricity consumption series. By comparing the results of the graphs and tables, the following conclusions can be drawn:

(a) Through longitudinal comparison analysis of the calculation results of different advanced prediction steps under the same evaluation index, it can be seen that

Table 3.5 The forecasting performance indices of the ARIMA model for series 1

Model	Advance step	MAPE (%)	MAE(GW)	RMSE(GW)	SDE(GW)	P
ARIMA	1-step	1.3930	0.2087	0.2859	0.2859	0.9883
	2-step	3.3342	0.4996	0.6537	0.6536	0.9387
	3-step	5.2681	0.7836	1.0047	1.0046	0.8524

Table 3.6 The forecasting performance indices of the ARIMA model for series 2

Model	Advance step	MAPE (%)	MAE(GW)	RMSE(GW)	SDE(GW)	P
ARIMA	1-step	4.8321	0.6562	0.8831	0.8831	0.8782
	2-step	7.0083	0.9465	1.2529	1.2529	0.7449
	3-step	7.8883	1.0598	1.3942	1.3942	0.6781

with the increase of the number of advance prediction steps, the evaluation indices of the two types of electricity consumption series, the mean absolute error (MAE), the mean absolute percentage error (MAPE), root mean square error (RMSE), and standard deviation of error (SDE) are significantly improved, while the Pearson's correlation coefficient P is significantly reduced. Taking the short-term electricity consumption series as an example, the MAPE, MAE, RMSE, and SDE of the one-step prediction are 1.3930%, 0.2087GW, 0.2859GW, and 0.2859GW, respectively. The value is extremely low, which proves that the

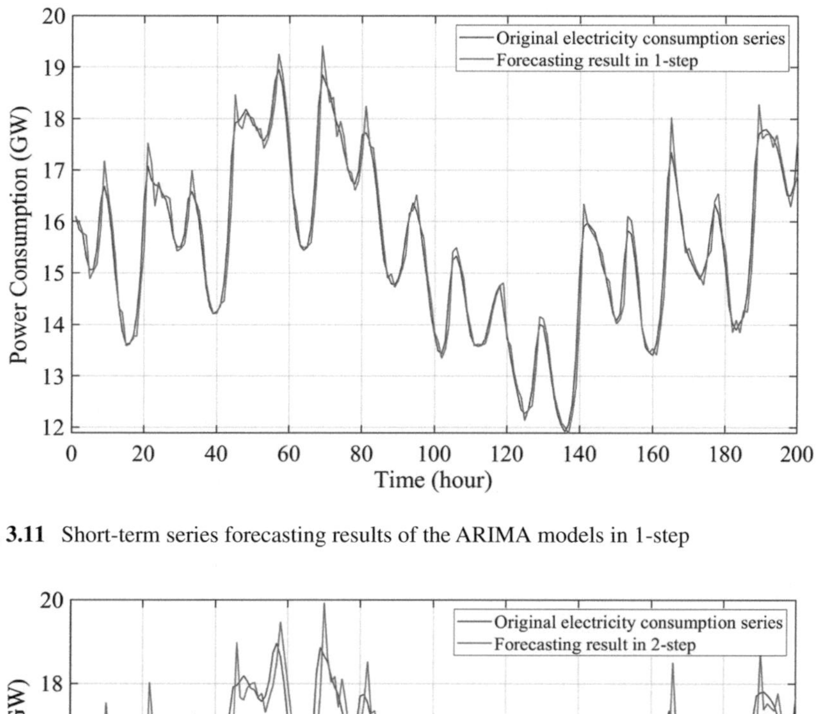

Fig. 3.11 Short-term series forecasting results of the ARIMA models in 1-step

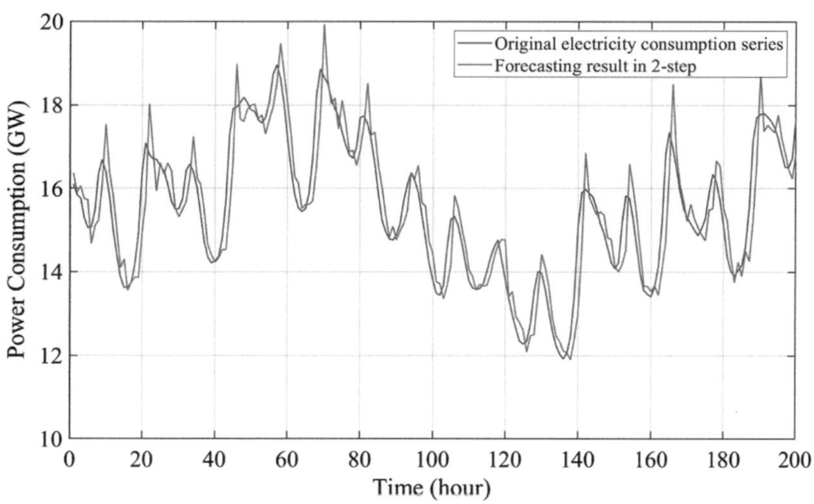

Fig. 3.12 Short-term series forecasting results of the ARIMA models in 2-step

model has measured data and predicted data when the number of advanced prediction steps is low. There is extremely low relative and absolute prediction error. The Pearson correlation coefficient is 0.9883, which indicates that there is a very high linear correlation between the measured data and the predicted data. When the number of advance steps reaches three steps, the first four evaluation indicators are significantly increased to 5.2681%, 0.7836, 1.0047, and 1.0046, respectively, and the Pearson correlation coefficient falls to 0.8524, indicating a linear correlation between the measured data and the predicted data. The model performance has a certain degree of decline, and the prediction accuracy of the model shows a trend of overall deterioration with the increase of the number of prediction steps.

(b) By comparing the results of the various evaluation indicators under the same number of advanced prediction steps in the two tables, it can be seen that the ARIMA model has a higher prediction accuracy for the short-term electricity consumption series than the long-term electricity consumption series. In the prediction of the previous step, the MAPE, MAE, RMSE, SDE, and P indices of the $\{X_t\}$ sequence are 1.3930%, 0.2087GW, 0.2859GW, 0.2859GW, and 0.9883, respectively, while the MAPE, MAE, RMSE, SDE, and P indicators of the $\{Y_t\}$ sequence are 4.8311%, 0.6562GW, 0.8831GW, 0.8831GW, and 0.8782, respectively; in the first three steps of prediction, the MAPE, MAE, RMSE, SDE, and P indices of the $\{X_t\}$ sequence are 5.2681%, 0.7836GW, 1.0047GW, 1.046GW, and 0.8524, respectively, while the MAPE, MAE, RMSE, SDE, and P of the $\{Y_t\}$ sequence are 7.8833%, 1.0598GW, 1.3942GW, 1.3942GW, and 0.6781, respectively. The evaluation indices of the long-term electricity consumption series are significantly inferior to the short-term electricity consumption series. By analyzing Figs. 3.5 and 3.6, it can be concluded that the above

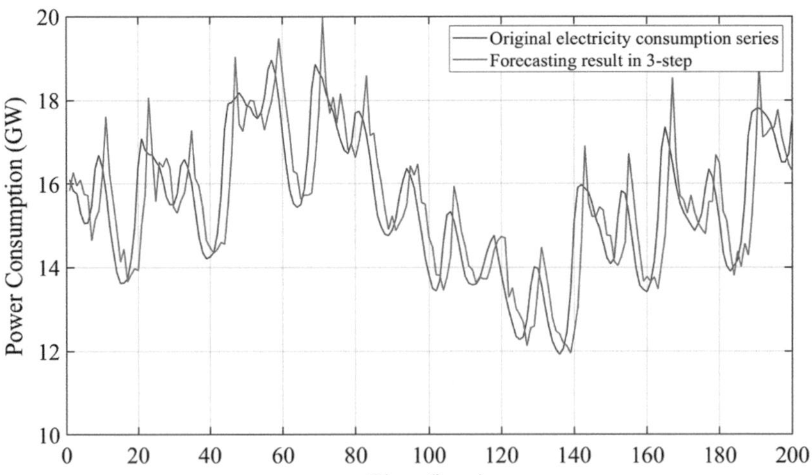

Fig. 3.13 Short-term series forecasting results of the ARIMA models in 3-step

phenomenon is mainly caused by the fact that, long-term electricity consumption first-order difference series shows a higher fluctuation frequency than the short-term electricity consumption first-order difference series. With the amplitude of the fluctuation, the overall instability is stronger, and the fitting effect of the ARIMA model is relatively worse.

(c) By visually analyzing Figs. 3.11, 3.12, 3.13, 3.14, 3.15, and 3.16, the following conclusions can be drawn. The short-term electricity consumption series predicts the original sequence well when the number of advanced prediction steps

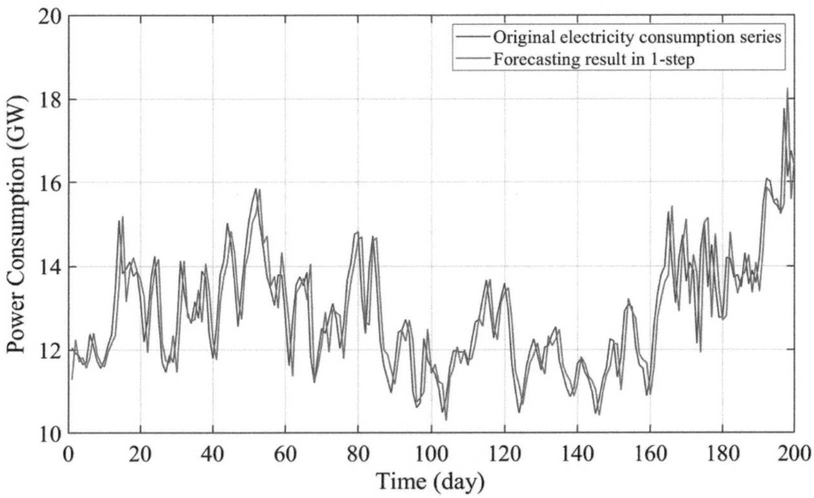

Fig. 3.14 Long-term series forecasting results of the ARIMA models in 1-step

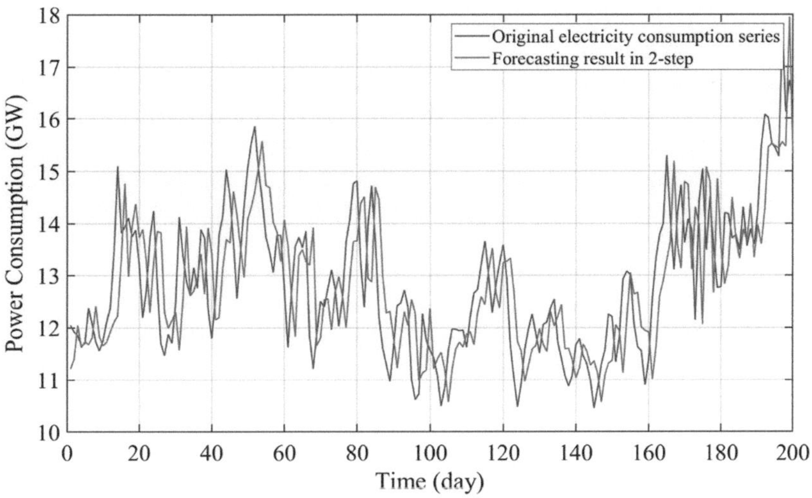

Fig. 3.15 Long-term series forecasting results of the ARIMA models in 2-step

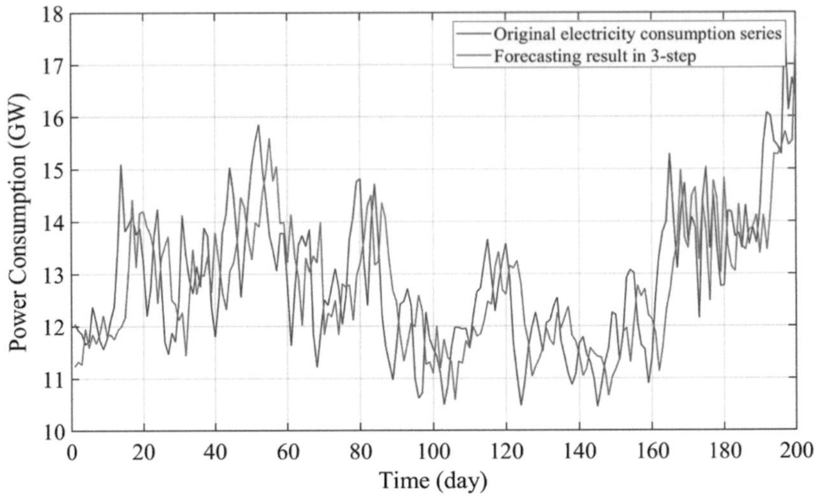

Fig. 3.16 Long-term series forecasting results of the ARIMA models in 3-step

is low. With the increase of the number of advance prediction steps, a certain phase shift and obvious mutation point fitting error appear between the prediction result and the original sequence, but the overall prediction trend can still be consistent with the original sequence. Relatively speaking, the long-term electricity consumption series can better fit the original sequence when the number of advanced prediction steps is low. However, when the number of advanced prediction steps is high, the prediction result deteriorates severely. The phase shift is very significant, and the error of the break point is high. The predicted results cannot meet the actual needs.

According to the comprehensive analysis, the ARIMA model shows a high difference for the prediction results of the two types of electricity consumption series. The short-term electricity consumption series uses the ARIMA model for predictive effects with high prediction accuracy. However, the performance of long-term electricity consumption series using the ARIMA model is not satisfactory. When the number of advanced prediction steps is 1, it can meet the requirements, and there is a high linear correlation between measured data and predicted data. When the number of advanced prediction steps is high, the linear correlation is greatly reduced, the errors are significantly increasing, and the misalignment and fluctuation are obvious. At the same time, the prediction accuracy of the ARIMA model of the two types of electricity consumption series is significantly reduced with the increase of the number of advance prediction steps.

3.4 Power Consumption Interval Prediction Based on ARIMA-ARCH Model

3.4.1 Model Framework of ARCH

This section uses the ARIMA model prediction results as the predicted mean of the interval prediction. The residual sequence of the prediction result and the real value is calculated, and the interval estimation model based on the ARCH model is fitted by the residual sequence. The final ARIMA-ARCH mixed interval prediction model is obtained. The prediction ability of the mixed interval prediction model is evaluated by using PIP, PINAW, and CWC indicators. The flowchart of the interval prediction model is shown in Fig. 3.17:

3.4.2 The Theoretical Basis of the ARCH

When using ARIMA and other regression models to fit non-stationary sequences, there is generally an important assumption for the residuals of predicted and real values. The residual sequence is a white noise sequence with a mean of zero, which is the assumption of homogeneity of variance. However, in the real use environment, the homogeneity of variance is often not established, and the variance of the residual sequence is a non-constant variable with time. Therefore, this section intro-

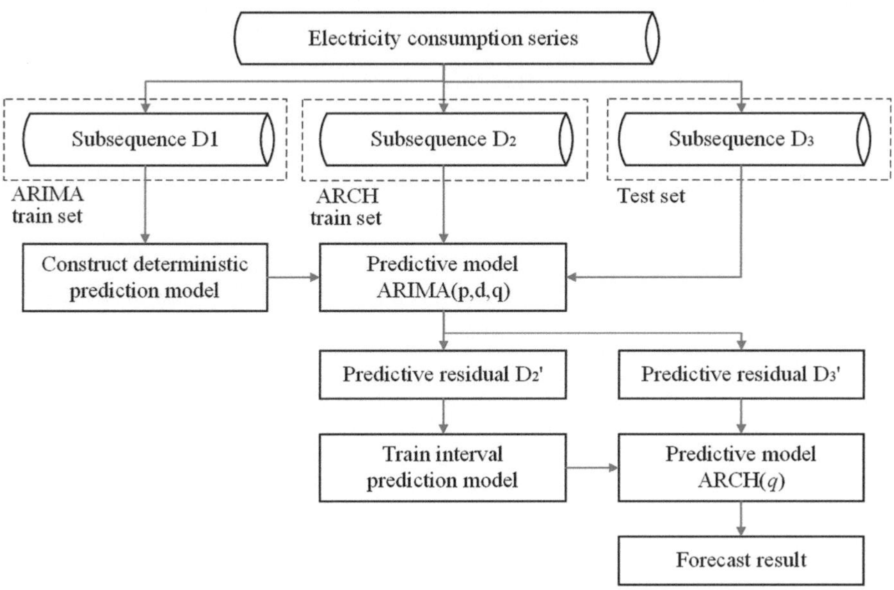

Fig. 3.17 Modeling process of the ARCH predictive model

duces the conditional heteroscedasticity model ARCH to realize the variance fitting of the ARIMA deterministic prediction model prediction residual and constructs the interval prediction model based on the conditional heteroscedasticity.

The autoregressive conditional heteroscedasticity (ARCH) model was originally proposed by Professor Engel R.F. in 1982. It is often used for time-series data regression analysis and prediction with clustering and variance volatility. Practice has proved that the ARCH model has achieved good results in related applications. The basic idea of using the ARCH model for non-deterministic interval prediction is to use the corresponding autoregressive model to fit the variance variability of time-series residuals based on historical volatility information of time series. Compared with the homoscedasticity model, the ARCH model can better reflect the fluctuation characteristics of the sequence over time.

The fitting of the ARCH model to the conditional variance follows the moving average law, and the conditional variance is different according to the information of the historical moment. When performing ARCH(q) test and model fitting on time series, the formula is expressed as:

$$\begin{cases} x_t = f(x_t) + \lambda_t \\ \lambda_t = \sigma_t e_t \\ \sigma_t^2 = \xi + \sum_{j=1}^{q} \omega_j \lambda_{t-j}^2 \end{cases} \tag{3.6}$$

where $f(x_t)$ is the predicted value obtained by the ARIMA model at time t, and is used as the mean of the interval prediction model. σ_t denotes the variance, λ_t denotes the residual, ω_t is the obtained autoregressive coefficient, and ξ is the constant obtained. e_t is an independent and identically distributed white noise sequence.

3.4.3 Modeling Steps of ARIMA-ARCH Interval Predictive Model

The D_2 portion of the two types of electricity consumption series is predicted using the trained ARIMA model, and the predicted residuals are shown in Figs. 3.18 and 3.19.

It can be seen from the figures that the prediction residual sequences of the two types of electricity consumption series both have strong heteroscedasticity. The heteroscedasticity of the sequences are further verified by the ARCH-LM test. The test results are in accordance with the above judgment. The ARCH model is then used to fit the residuals and the BIC criteria are used to select the model structure.

After obtaining the above heteroscedasticity model, the variance of the D_3 partial data can be inferred. Assuming that the volatility follows a normal distribution, the point estimate volatility range can be derived from the variance and the different saliency levels. Further, combined with the deterministic prediction results obtained by the ARIMA model described above, the final interval prediction result is obtained.

3.4.4 Forecasting Results

The ARIMA-ARCH interval prediction model is trained and tested using the $D_2 + D_3$ part of short-term electricity consumption series and long-term electricity consumption series, respectively, and based on the influence of three different significance levels of $\alpha = 0.1$, $\alpha = 0.05$, and $\alpha = 0.01$, 1-step advanced, 2-step advanced, 3-step advanced interval prediction. Three types of interval prediction evaluation indicators are used to evaluate the reliability and accuracy of model interval prediction: PI coverage probability (PICP), PI normalized average width (PINAW), coverage width-based criterion (CWC). The prediction accuracy and prediction results are shown in Tables 3.7, 3.8, and 3.9 and Figs. 3.20, 3.21, 3.22, 3.23, 3.24, and 3.25. A comprehensive analysis of the results of the comparison chart can initially lead to the following conclusions:

(a) By comparing the interval prediction results of two types of electricity consumption series, it can be intuitively seen that whether it is a deterministic prediction or an uncertain prediction, the forecast performance of short-term electricity consumption series is better than the long-term electricity consumption series. The main reason is that the differential sequence of short-term electricity consumption series has lower high-frequency volatility. At the same

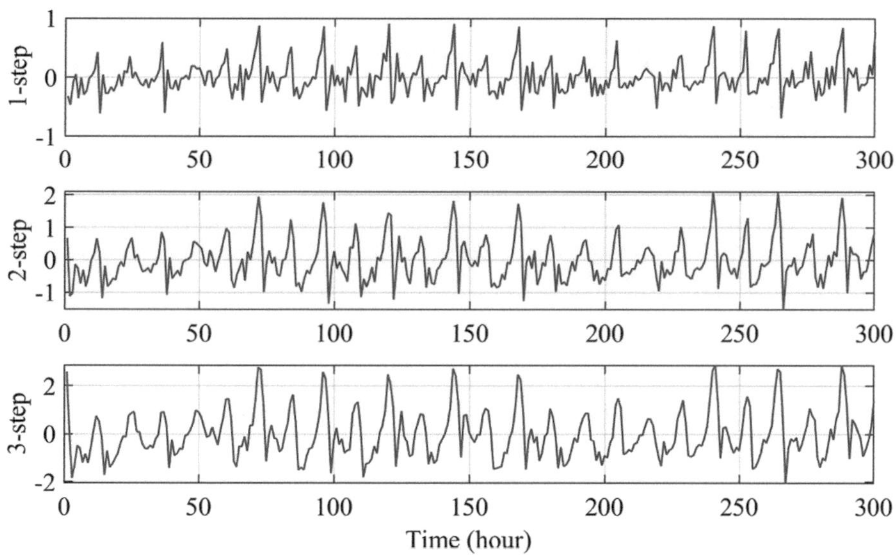

Fig. 3.18 Short-term series predicted residuals series

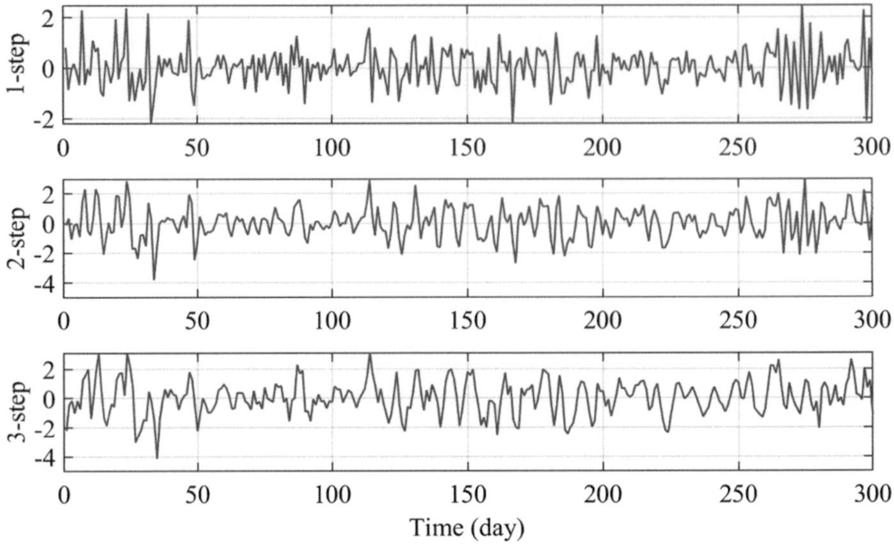

Fig. 3.19 Long-term series predicted residuals series

Table 3.7 The forecasting performance indices of the ARIMA-ARCH model in 1-step

Original series	Significance level	PICP	PINAW	CWC
$\{X_t\}$	$\alpha = 0.1$	0.8767	0.1019	0.4290
	$\alpha = 0.05$	0.9100	0.1214	1.0184
	$\alpha = 0.01$	0.9800	0.1595	0.4226
$\{Y_t\}$	$\alpha = 0.1$	0.8267	0.2579	10.3484
	$\alpha = 0.05$	0.8967	0.3073	4.7305
	$\alpha = 0.01$	0.9567	0.4039	2.5424

time, the deterministic prediction result of the predicted mean can better fit the true value of the test set. The ratio of the predicted residual amplitude to the true value is low so that the interval prediction accuracy is high. The mean of long-term electricity consumption series prediction values is larger than the real value residual. Especially when advancing multi-step prediction, there is a significant offset between the predicted value and the real value. This results in its interval prediction results being weaker than the short-term electricity consumption series interval prediction results, whether it is the confidence interval width or the true value coverage rate. For example, when it is 1-step advanced predicted by the significance level of $\alpha = 0.1$, the PPIC, PINAW, and CWC indicators of the short-term electricity consumption series are 0.8767, 0.1019, and 0.4290, respectively. The indicators of the long-term electricity consumption series were 0.8267, 0.2579, and 10.3484, respectively, and the accuracy indicators of the latter are significantly inferior to the former.

(b) The interval coverage probability is generally lower than the pre-set confidence. PI coverage probability (PICP) is used to evaluate the coverage of real-time data in prediction intervals, which is important for evaluating the robustness of the interval prediction model. The larger the PICP value is, the better the robustness of the model is. By analyzing the table data, it can be concluded that for two types of electricity consumption series, the PICP index decreases with the increase of the significance level, and the PICP value is generally lower than the pre-set reliability, which indicates the accuracy and robustness of the ARIMA-ARCH model. It is difficult to fully meet the pre-set requirements. Especially for long-term electricity consumption series, the gap between PICP indicators and pre-set reliability is more obvious.

(c) The three types of indicators have different trends with the increase of the number of advance prediction steps and the level of significance. Take the short-term electricity consumption series interval prediction result as an example. When the significance level is $\alpha = 0.1$ and the ARIMA-ARCH model is predicted 1-step advanced, the PICP, PINAW, and CWC indicators are 0.8767, 0.1019, and 0.4290, respectively. When the two-step prediction is advanced, the PICP, PINAW, and CWC indicators are 0.8800, 0.2305, and 0.8572, respectively. The coverage rate tends to be optimized, the interval width tends to deteriorate, and the comprehensive index tends to deteriorate, indicating that the comprehensive precision is reduced due to the increase of the predicted steps. In the three-step forecast, the PPIC, PINAW, and CWC indicators are 0.8967, 0.3654, and 0.7970, respectively. Although the interval width tends to deteriorate, the coverage and overall accuracy tend to be optimized. The comprehensive comparison shows that the short-term electricity consumption series has the best comprehensive performance in the 3-step interval prediction, while the comprehensive

Table 3.8 The forecasting performance indices of the ARIMA-ARCH model in 2-step

Original series	Significance level	PICP	PINAW	CWC
$\{X_t\}$	$\alpha = 0.1$	0.8800	0.2305	0.8572
	$\alpha = 0.05$	0.9367	0.2747	0.8098
	$\alpha = 0.01$	0.9767	0.3610	1.0642
$\{Y_t\}$	$\alpha = 0.1$	0.8267	0.3836	15.3901
	$\alpha = 0.05$	0.8933	0.4571	8.2283
	$\alpha = 0.01$	0.9700	0.6007	2.2336

Table 3.9 The forecasting performance indices of the ARIMA-ARCH model in 3-step

Original series	Significance level	PICP	PINAW	CWC
$\{X_t\}$	$\alpha = 0.1$	0.8967	0.3654	0.7970
	$\alpha = 0.05$	0.9433	0.4354	1.0430
	$\alpha = 0.01$	0.9933	0.5722	0.5722
$\{Y_t\}$	$\alpha = 0.1$	0.8333	0.4156	12.0663
	$\alpha = 0.05$	0.8900	0.4952	10.4426
	$\alpha = 0.01$	0.9700	0.6509	2.4201

performance in the 1-step and 2-step advanced interval prediction is relatively the worst. Correspondingly, the PICP indicator deteriorates as the level of significance increases, and the PINAW indicator is optimized as the level of significance increases. The two have opposite trends, so the relationship between the CWC index and the significance level is more complicated. Overall, it performs better at $\alpha = 0.1$ and $\alpha = 0.01$, while at $\alpha = 0.05$ it performs poorly. However, for the long-term electricity consumption series, when the significance level is at $\alpha = 0.1$, the comprehensive index CWC is 2.5424, 2.2336, and

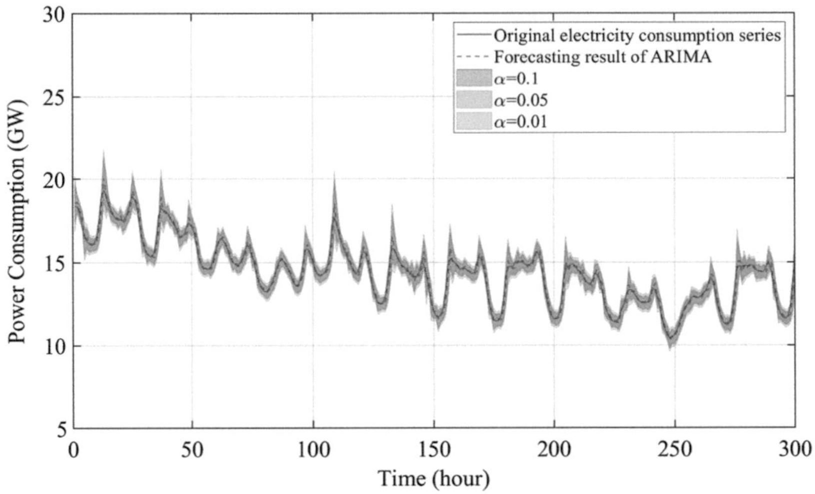

Fig. 3.20 Short-term series interval forecasting results of the ARIMA models in 1-step

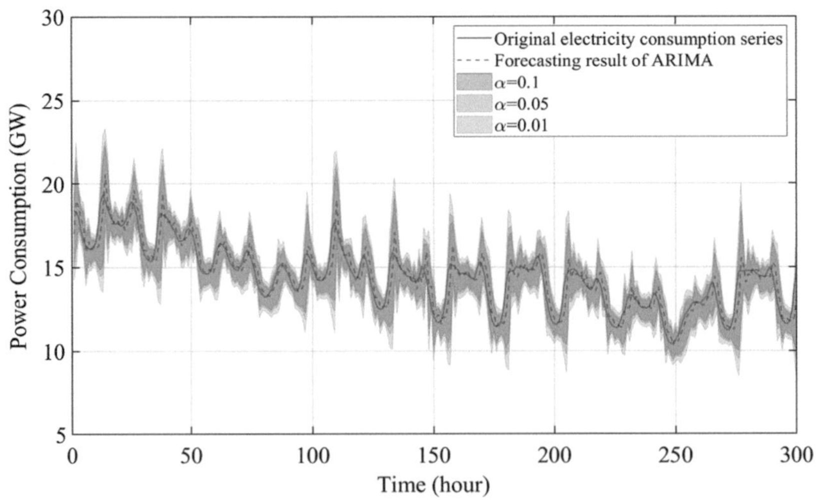

Fig. 3.21 Short-term series interval forecasting results of the ARIMA models in 2-step

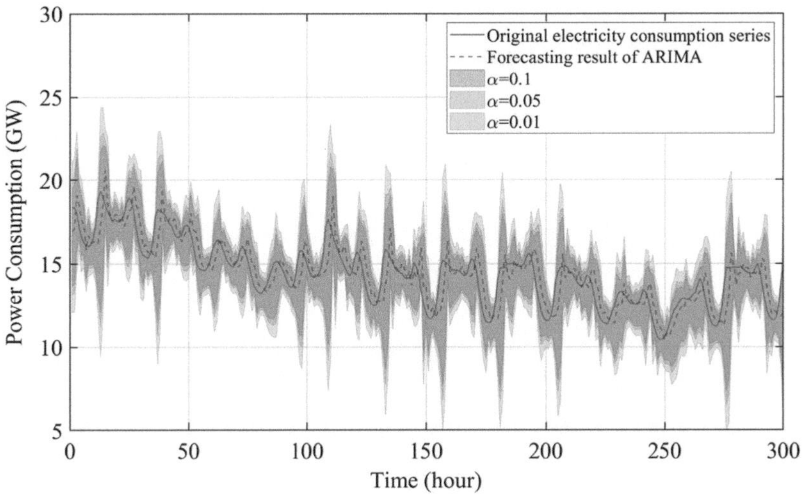

Fig. 3.22 Short-term series interval forecasting results of the ARIMA models in 3-step

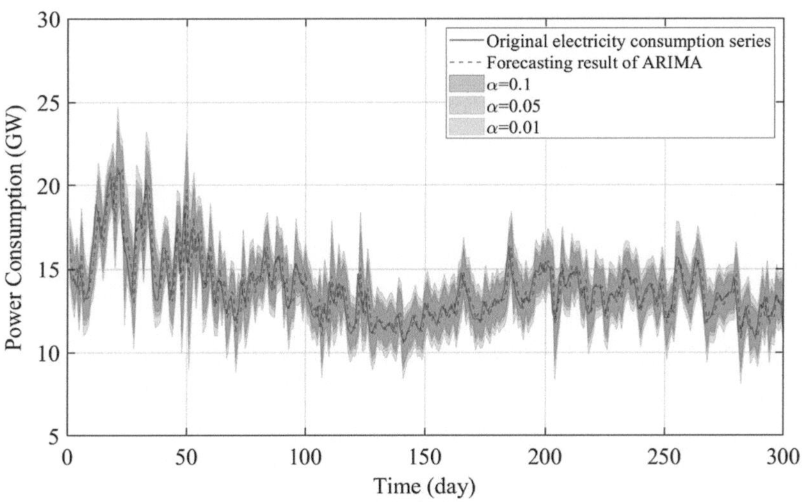

Fig. 3.23 Long-term series interval forecasting results of the ARIMA models in 1-step

2.4201, respectively, in the prediction of 1-step advanced, 2-step advanced, and 3-step advanced. They are optimal results at different levels of significance. This indicates that for a long-term electricity consumption series with a large relative deviation between the predicted mean and the true value, the wide prediction interval brought by the low significance level can procedurally improve the prediction accuracy of the interval model. However, it can be seen from the analysis of the prediction results graph that the practicality of the interval prediction model is insufficient.

3.5 Long-Term Electricity Consumption Prediction Based on the SARIMA Model

3.5.1 Model Framework of the SARIMA

The SARIMA model is basically similar to the ARIMA model. The difference is that the SARIMA model incorporates seasonal term parameters in parameter estimation and model order. The SARIMA predictive model modeling process is shown in Fig. 3.26.

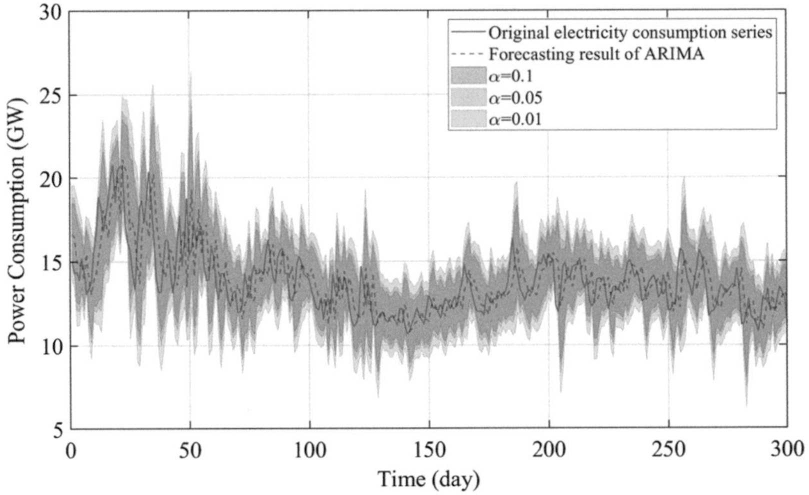

Fig. 3.24 Long-term series interval forecasting results of the ARIMA models in 2-step

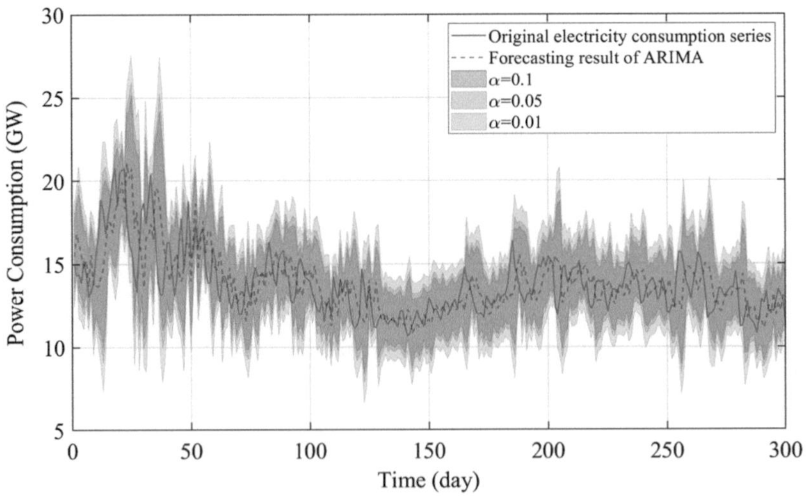

Fig. 3.25 Long-term series interval forecasting results of the ARIMA models in 3-step

3.5.2 The Theoretical Basis of the SARIMA

In real life, many time series itself have strong seasonality. The non-seasonal ARIMA model contains only non-seasonal feature parameters, while the seasonal SARIMA model further incorporates parameters for processing seasonal characteristics. The seasonal time-series model is also called the product season model because the final form of the model is expressed in terms of factor multiplication.

The SARIMA model is a variant of the ARIMA model. It is a model obtained by seasonal differentiation of the ARIMA model. The period step of the long-term sequence is denoted by *T*. First, the seasonal variation is used to eliminate periodic changes. The seasonal difference operator can be expressed as:

$$\Delta_T = 1 - H^T \tag{3.7}$$

where *H* represents the hysteresis operator. The first-order seasonal difference is expressed as:

$$\Delta_T x_t = \left(1 - H^S\right) x_t = x_t - x_{t-T} \tag{3.8}$$

where if the original long-term sequence contains *N* observations, the sequence will lose $T + 1$ observations after one seasonal difference and one non-seasonal difference.

The seasonal SARIMA model can be expressed as follows:

$$\begin{aligned} &\left(1-\alpha_1 H - \ldots - \alpha_p H^p\right)\left(1-\gamma_1 H^T - \ldots - \gamma_p H^{PT}\right)\left(\Delta^d \Delta_T^D x_t\right) \\ &= \left(1+\lambda_1 H + \ldots + \lambda_q H^q\right)\left(1+\eta_1 H^T + \ldots + \eta_Q H^{QT}\right)\varepsilon_t \end{aligned} \tag{3.9}$$

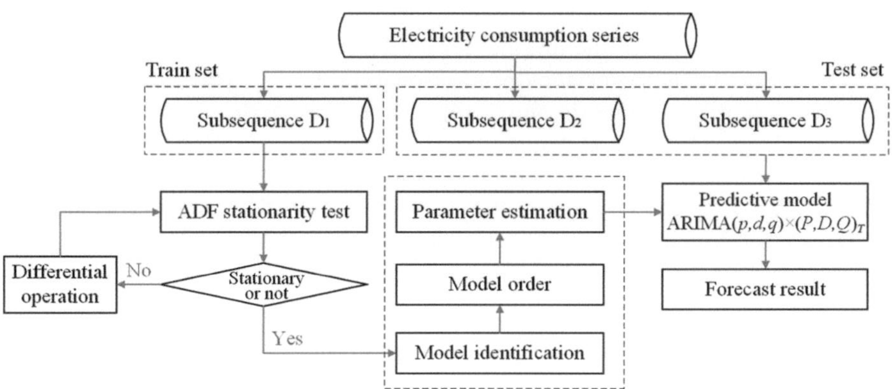

Fig. 3.26 Modeling process of SARIMA predictive model

where Δ represents non-seasonal difference and Δ_T means seasonal difference operators. d and D represent the non-seasonal difference order and the seasonal difference order, respectively. ε_t is a random error term. $1 - \alpha_1 H - \ldots - \alpha_p H^p$ and $1 - \gamma_1 H^T - \ldots - \gamma_p H^{PT}$ are non-seasonal and seasonal AR feature polynomials, respectively. $1 + \lambda_1 H + \ldots + \lambda_q H^q$ and $1 + \eta_1 H^T + \ldots + \eta_Q H^{QT}$ are non-seasonal and seasonal MA feature polynomials, respectively. The subscripts p, P, q, and Q are the maximum lag order of the autoregressive and moving average feature polynomials of the non-seasonal and seasonal terms, respectively. The above seasonal time-series model can be expressed as ARIMA$(p,d,q) \times (P,D,Q)_T$.

3.5.3 Modeling Steps of the SARIMA Predictive Model

As can be seen from the foregoing, the original short-term electricity consumption series is a non-stationary sequence, and the sequence after the first-order difference is a stationary sequence. However, from the principle of the SARIMA model, the SARIMA method is divided into two steps for the differential process of the original short-term electricity consumption series: the trend d-order difference and the seasonal D-order difference. Therefore, when the SARIMA model is employed for the original short-term electricity consumption series, how the differential processing is allocated is temporarily undetermined. The specific difference method needs to obtain the optimal model parameter assignment result by performing BIC detection on models with different differential coefficients. It can be concluded from Fig. 3.2 that the original short-term electricity consumption series has no obvious seasonal effect and its period step S cannot be directly determined. Therefore, the period step of the original short-term electricity consumption series also needs to be determined by BIC detection to obtain the optimal model.

The original long-term electricity consumption series is a non-stationary sequence. The trend difference term, seasonal difference term, and residual parameters of the SARIMA model also need to be determined by BIC detection. According to the observation and analysis in Fig. 3.3 and the actual life experience, the period step S of the original long-term electricity consumption series is 365. However, due to the existence of seasonal differences, excessive period values require extremely large training samples to match, so the period value should not be too large. This section defines the period of short-term electricity consumption series and long-term electricity consumption series to be 50~100. This section then calculates the minimum value of the BIC value that can be achieved with different SARIMA parameter changes when using a certain period. Figures 3.27 and 3.28 show the relationship between minimum BIC detection and period for short-term electricity consumption series and long-term electricity consumption series.

After experimental comparison and analysis, this chapter selects a period of 84 for the short-term electricity consumption series, and the selected SARIMA model is $\text{ARIMA}(1,0,1) \times (1,1,1)_{84}$. The period of long-term electricity consumption series is 56, and the selected SARIMA model is $\text{ARIMA}(1,0,2) \times (1,1,2)_{56}$.

3.5.4 Forecasting Results

The SARIMA prediction model is trained and tested in the same manner as the ARIMA model using short-term electricity consumption series and long-term electricity consumption series, respectively. The predicted results of the advanced prediction accuracy evaluation indicators obtained are shown in Tables 3.10 and 3.11.

Figures 3.29, 3.30, 3.31, 3.32, 3.33, and 3.34 show some of the predictions of the short-term electricity consumption series and the long-term electricity consumption series. By comparing the results of the graphs and tables, the following conclusions can be drawn:

(a) As the number of advance prediction steps increases, as with the ARIMA model, the various evaluation indicators of the SARIMA model prediction results are significantly degraded. Taking short-term electricity consumption series as an example, MAPE, MAE, RMSE, SDE, and P are 0.8308%, 0.1227GW, 0.1698GW, 0.1698GW, and 0.9958, respectively. All evaluation indicators have excellent performance. It shows that the prediction results can fit the real sequence well. The prediction error is extremely small, and the prediction sequence has a strong correlation with the real sequence. When the number of advanced step reaches 3-step, the evaluation indicators are somewhat degraded to 2.2801%, 0.3374GW, 0.4594GW, 0.4593GW, and 0.9697, respectively. Although the indicators also perform better, compared with the 1-step

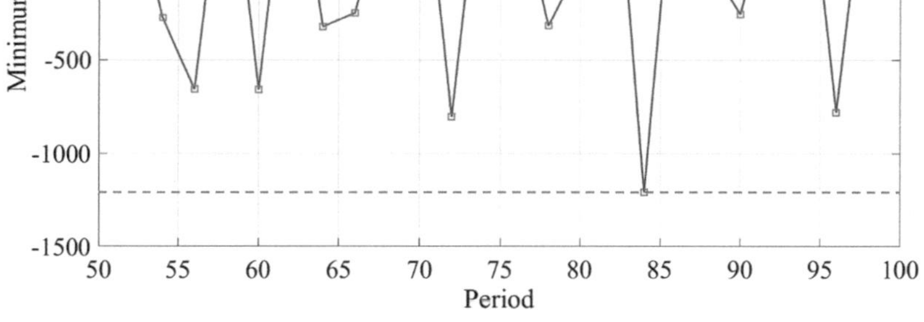

Fig. 3.27 Short-term series relationship between minimum BIC detection and period

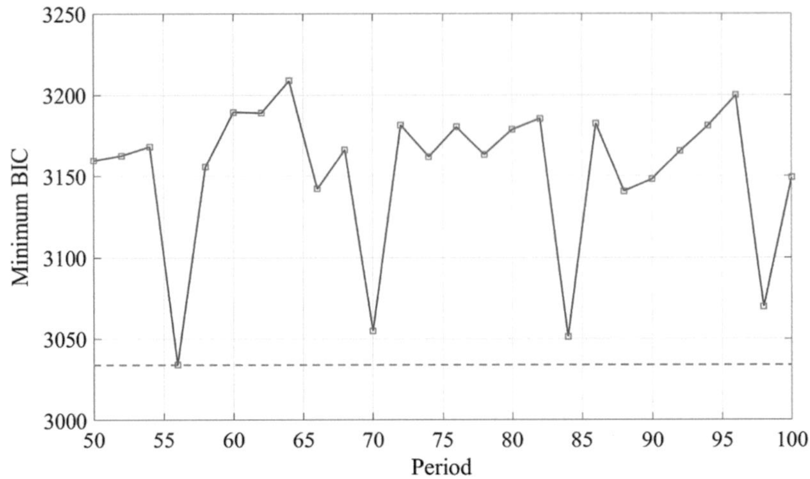

Fig. 3.28 Long-term series relationship between minimum BIC detection and period

Table 3.10 The forecasting performance indices of the SARIMA model for series 1

Model	Advance step	MAPE (%)	MAE (GW)	RMSE (GW)	SDE (GW)	P
SARIMA	1-step	0.8308	0.1227	0.1698	0.1698	0.9958
	2-step	1.5446	0.2288	0.3147	0.3147	0.9857
	3-step	2.2801	0.3374	0.4594	0.4593	0.9697

prediction, the linear correlation between the measured data and the predicted data shows an overall downward trend, and the error is increasing. The prediction accuracy of the model shows a trend of overall deterioration as the number of prediction steps increase. The above phenomenon is mainly due to the fact that when the multi-step prediction is advanced, the prediction residual of the advanced step will accumulate to the 2-step, thereby affecting the accuracy of the guess.

(b) Like the ARIMA model, the SARIMA model has higher prediction accuracy for the short-term electricity consumption series than the long-term electricity consumption series. The MAPE, MAE, RMSE, SDE, and P of the short-term electricity consumption series are 0.8308%, 0.1227GW, 0.1698GW, 0.1698GW, and 0.9958, respectively, while the long-term electricity consumption series for MAPE, MAE, RMSE, SDE, and the P indicators are 4.5487%, 0.6174GW, 0.8282GW, 0.8282GW, and 0.8865, respectively. In the 3-step, the MAPE, MAE, RMSE, SDE, and P of the short-term electricity consumption series are 2.2801%, 0.3374GW, 0.4594GW, 0.4593GW, and 0.9697, respectively, while the MAPE, MAE, RMSE, SDE, and P of the long-term electricity consumption series are 7.0424%, 0.9489GW, 1.2473GW, 1.2473GW, and 0.7196, respectively. At the same time, a comprehensive comparative analysis of Figs. 3.26, 3.27, 3.28, 3.29, 3.30, and 3.31 can lead to the following conclusions. For the short-term electricity

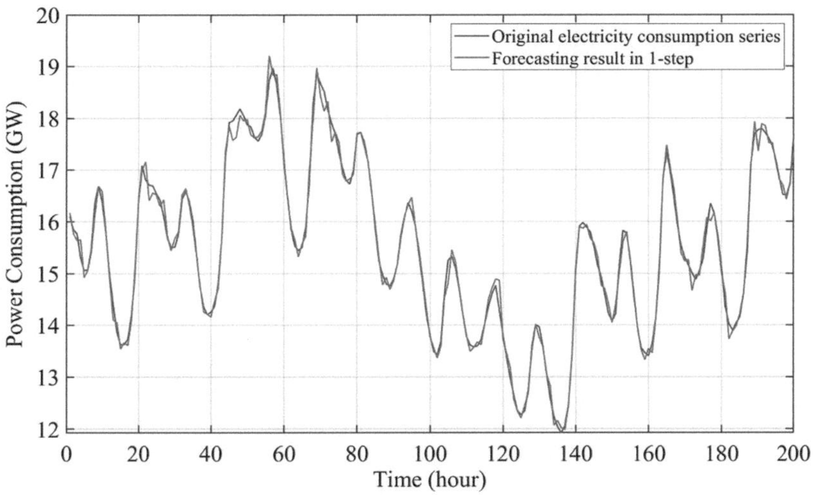

Fig. 3.29 Short-term series forecasting results of the SARIMA models in 1-step

Table 3.11 The forecasting performance indices of the SARIMA model for series 2

Model	Advance step	MAPE (%)	MAE(GW)	RMSE(GW)	SDE(GW)	P
SARIMA	1-step	4.5487	0.6174	0.8282	0.8282	0.8865
	2-step	6.3245	0.8549	1.1364	1.1364	0.7737
	3-step	7.0424	0.9489	1.2473	1.2473	0.7196

consumption series test set, the predicted values of the SARIMA model fit well with the true values and perform well in trend follow-up and extreme point deviation. Moreover, even in the prediction of the 3-step, the prediction results still have higher prediction accuracy. Conversely, for the short-term electricity consumption series test set, the SARIMA model predicts in advance that the prediction results can better fit the true value. However, when the multi-step prediction is advanced, the prediction accuracy is extremely degraded. The predicted value cannot effectively fit the true value, and there is obvious prediction delay and extreme point deviation. The prediction accuracy deteriorates seriously when the multi-step prediction is advanced.

3.6 Big Data Prediction Architecture of Household Electric Power

The short-term electricity consumption series and long-term electricity consumption series prediction methods described in the chapter are all for predictive analysis of data from a single location. In the actual electricity consumption forecasting application, developers often need to analyze and predict the massive power data of different regions. A sheer volume of data has extremely high requirements on the performance of the processor and the prediction accuracy of a single model. Therefore, the usage of big data framework to deal with massive electricity consumption data in multiple regions has become an efficient and reasonable solution.

The chapter proposes a scheme to realize the multi-region electricity consumption series prediction using Hadoop. The Hadoop architecture optimizes computing resources by building different subsystems such as real-time distributed databases, distributed offline computing frameworks, and distributed resource management

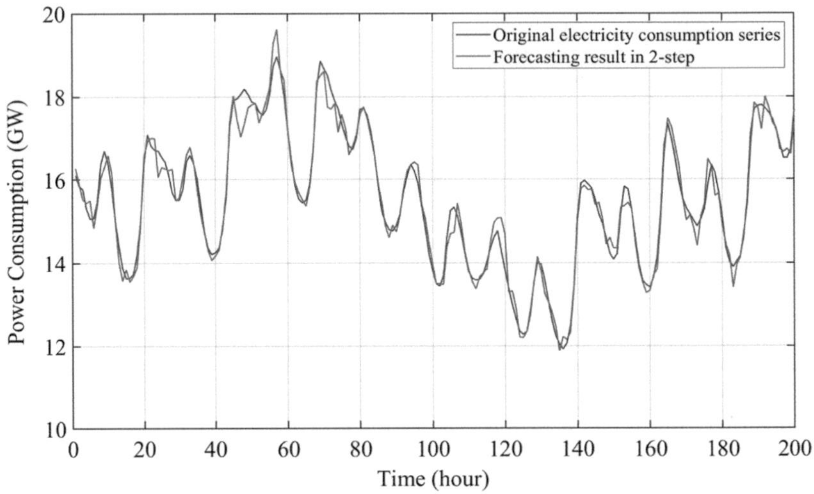

Fig. 3.30 Short-term series forecasting results of the SARIMA models in 2-step

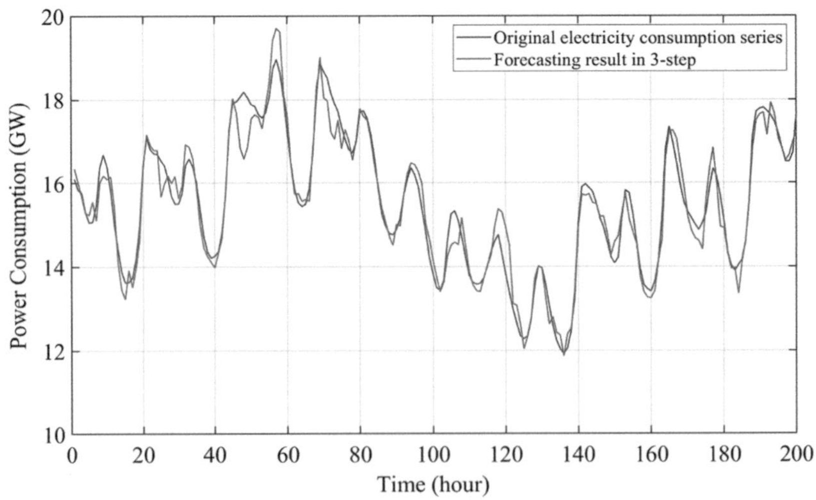

Fig. 3.31 Short-term series forecasting results of the SARIMA models in 3-step

systems. By simplifying the design process of distributed programs, Hadoop transforms multiple parallel computing processes into common modules and implements them in the system. This development model improves the development efficiency of distributed computing. Users only need to consider the logical implementation of their own applications.

The prediction of electricity consumption using the Hadoop big data platform is mainly carried out in MapReduce mode. The overall process is shown in Fig. 3.35. First, the data of N regions are divided into M_1 data segments according to the amount of data. A MapTask is assigned to each data segment for parallel prediction processing to obtain prediction results. The prediction results are reclassified and

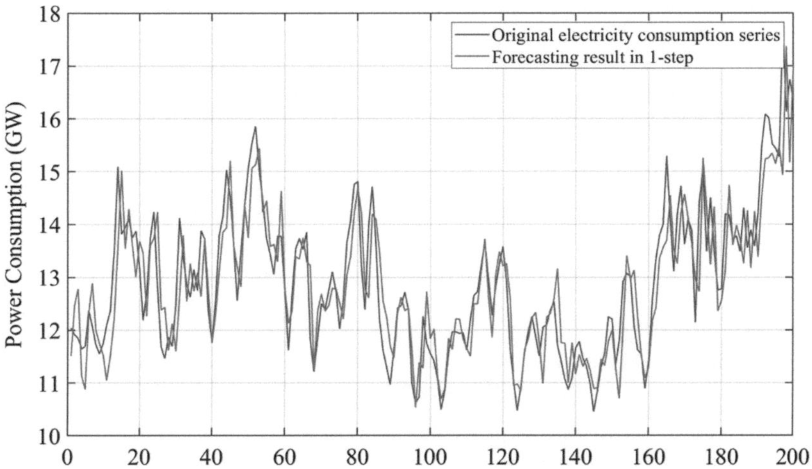

Fig. 3.32 Long-term series forecasting results of the SARIMA models in 1-step

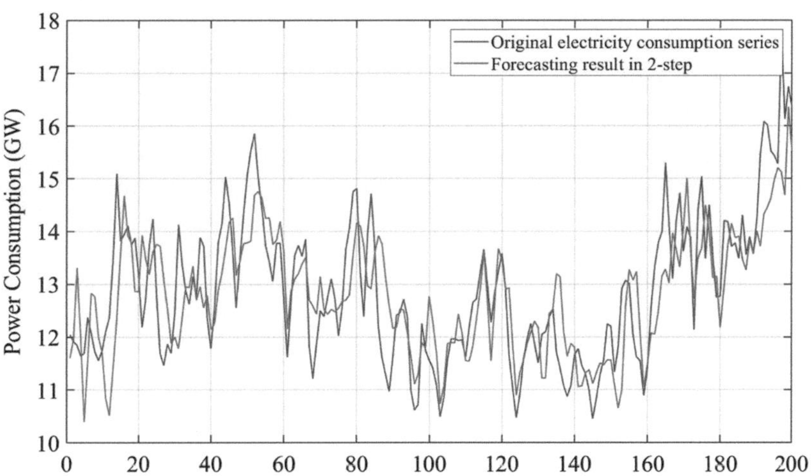

Fig. 3.33 Long-term series forecasting results of the SARIMA models in 2-step

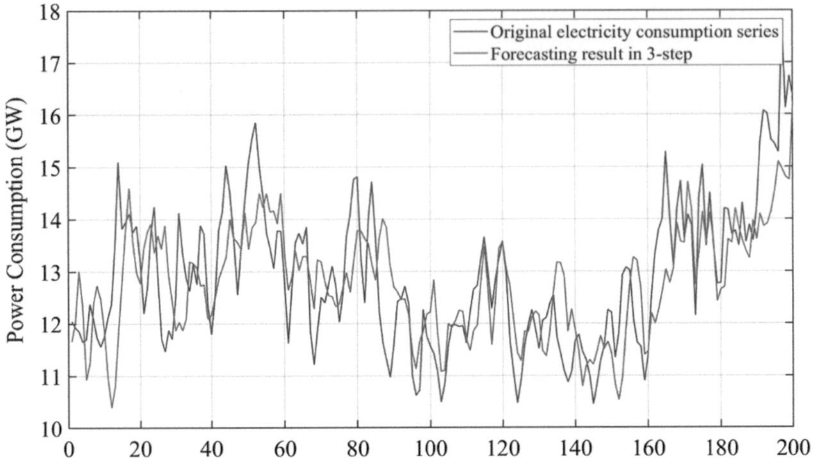

Fig. 3.34 Long-term series forecasting results of the SARIMA models in 3-step

sorted locally. The ReduceTask is called for logical operations, and the operations are collected to obtain the final prediction results.

3.7 Comparative Analysis of Forecasting Performance

In order to comprehensively compare and analyze the predictive performance of the two models of SARIMA and ARIMA for short-term electricity consumption series and long-term electricity consumption series, this chapter uses ARIMA as a comparative model to calculate the four types of precision comparison indicators $P_{MAE}(\%)$, $P_{MAPE}(\%)$, $P_{RMSE}(\%)$, and $P_{SDE}(\%)$, comparing the analysis accuracy of the ARIMA model before and after the addition of seasonal items. The comparison results are shown in Tables 3.12 and 3.13. By analyzing the results of various comparison indicators in the tables, the following conclusions can be drawn.

(a) Among the two types of electricity consumption series, the SARIMA model has a considerable degree of optimization for the short-term electricity consumption series with relatively high-frequency stability. Compared with the ARIMA model, the four evaluation indicators of the SARIMA model are all above 40%. It is indicated that the seasonal items in the selected SARIMA model can better exploit the intrinsic periodic characteristics of the short-term electricity consumption series. The model parameters selected by the BIC criterion can better optimize the model. At the same time, as the number of advance prediction steps increases, it can be clearly seen that the relative optimization range of the SARIMA model also increases. The results of $P_{MAE}(\%)$, $P_{MAPE}(\%)$, $P_{RMSE}(\%)$, and $P_{SDE}(\%)$ are 40.3621%, 41.2197%, 40.6127%, and 40.6083%, respectively. In the first three steps, $P_{MAE}(\%)$, $P_{MAPE}(\%)$, $P_{RMSE}(\%)$, and $P_{SDE}(\%)$ results are

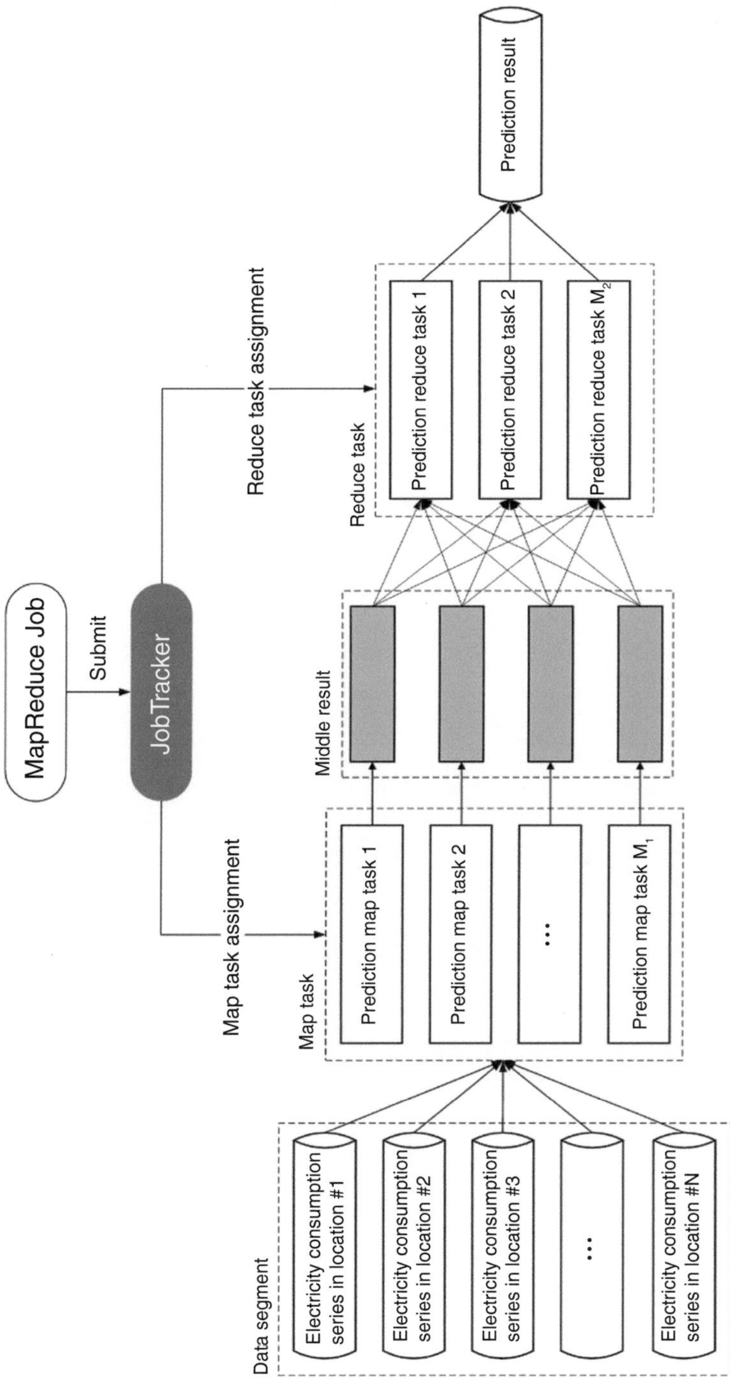

Fig. 3.35 The MapReduce prediction architecture of electricity consumption

56.7186%, 56.9429%, 54.2761%, and 54.2764%, respectively, and the relative improvement gradually increased.

(b) Compared with the short-term electricity consumption series, the long-term electricity consumption series with high-frequency fluctuation and long sequence period is not sufficiently accurate when using the SARIMA model. When the prediction is advanced, the results of $P_{MAE}(\%)$, $P_{MAPE}(\%)$, $P_{RMSE}(\%)$, and $P_{SDE}(\%)$ are 5.8663%, 5.9919%, 6.2190%, and 6.2191%, respectively. When the three-step prediction is made, the results of $P_{MAE}(\%)$, $P_{MAPE}(\%)$, $P_{RMSE}(\%)$, and $P_{SDE}(\%)$ are 10.7233%, 10.4639%, 10.5326%, and 10.5338%, respectively. As the number of advance prediction steps increases, the accuracy improvement gradually increases, but the improvement effect is not obvious enough. At the same time, by analyzing the prediction results of the ARIMA and SARIMA models for the long-term electricity consumption series, it can be found that the prediction effects of the two types of models are poor when the multi-step prediction is advanced. This is mainly due to the fact that the long-term electricity consumption series has a large fluctuation period, and the steps in Sect. 3.5.2 fail to accurately find the true period and the corresponding optimal parameters. To improve the above problems, it is necessary to increase the training sample size and expand the cycle search range.

3.8 Conclusion

This chapter deeply studies the short-term electricity consumption series and long-term electricity consumption series prediction methods based on two statistical models of ARIMA and SARIMA. At the same time, the influence of the two types of models on the prediction accuracy of electricity consumption series in different time dimensions is compared and analyzed. This chapter also uses the ARCH conditional heteroscedastic model to fit the prediction residual of the ARIMA model, so as to realize the uncertainty interval prediction of electricity consumption series and

Table 3.12 The forecasting performance comparison of different model for series 1

Model advance step	Advance step	$P_{MAPE}(\%)$	$P_{MAE}(\%)$	$P_{RMSE}(\%)$	$P_{SDE}(\%)$
SARIMA vs ARIMA	1-step	40.3621	41.2197	40.6127	40.6083
	2-step	53.6741	54.2120	51.8558	51.8549
	3-step	56.7186	56.9429	54.2761	54.2764

Table 3.13 The forecasting performance comparison of different model for series 2

Model advance step	Advance step	$P_{MAPE}(\%)$	$P_{MAE}(\%)$	$P_{RMSE}(\%)$	$P_{SDE}(\%)$
SARIMA vs ARIMA	1-step	5.8663	5.9179	6.2190	6.2191
	2-step	9.7576	9.6799	9.2969	9.2982
	3-step	10.7233	10.4639	10.5326	10.5338

use three precision indicators to evaluate the prediction accuracy of various model intervals. The conclusions of this chapter are summarized as follows:

(a) The long-term electricity consumption series in different regions is highly correlated. The characteristics of the long-term electricity consumption series of the period, peak, and trough of the overall fluctuation trend satisfy the general law. The overall fluctuation trend of each series is the same, with similar peaks and troughs. This shows that users in each region have similar power usage habits as a whole.
(b) The ARIMA model shows a high difference in the prediction results of the two types of electricity consumption series. The short-term electricity consumption series uses the ARIMA model for predictive effects with high prediction accuracy. However, the performance of long-term electricity consumption series using the ARIMA model is not satisfactory. When the number of advanced prediction steps is 1, it can meet the requirements, and there is a high linear correlation between measured data and predicted data. When the number of advanced prediction steps is high, the linear correlation is greatly reduced, the errors are significantly increasing, and the misalignment and fluctuation are obvious. At the same time, the prediction accuracy of the ARIMA model of the two types of electricity consumption series is significantly reduced with the increase of the number of advance prediction steps.
(c) Whether it is deterministic or uncertain prediction, short-term electricity consumption series has better predictive performance than long-term electricity consumption series. And the interval coverage probability of the two is generally lower than the pre-set reliability. The short-term electricity consumption series has the best overall performance in the lead-step interval prediction, while the comprehensive performance in the first two-step interval prediction is relatively the worst. In contrast, for a long-term electricity consumption series with a large relative deviation between the predicted mean and the true value, the wide prediction interval brought by the low significance level can procedurally improve the prediction accuracy of the interval model. But with it, the practicality of the interval prediction model is insufficient.
(d) Like the ARIMA model, the SARIMA model has higher prediction accuracy for the short-term electricity consumption series than the long-term electricity consumption series. For the short-term electricity consumption series test set, the predicted values of the SARIMA model fit well with the true values and perform well in trend follow-up and extreme point deviation. Moreover, even in the prediction of the 3-step, the prediction results still have higher prediction accuracy.
(e) Among the two types of electricity consumption series, the SARIMA model has a considerable degree of optimization for the short-term electricity consumption series with relatively high-frequency stability. Compared with the short-term electricity consumption series, the long-term electricity consumption series with high-frequency fluctuation and long sequence period is not sufficiently accurate

when using the SARIMA model. This is mainly because the real fluctuation period of the long-term electricity consumption series is too large, and the preset period search range fails to accurately find the real period and the corresponding optimal parameters. It needs to increase the training sample size and expand the cycle search range.

References

Bianchi FM, Maiorino E, Kampffmeyer MC, et al (2017) An overview and comparative analysis of recurrent neural networks for short term load forecasting [J]. arXiv preprint arXiv: 170504378

Brockwell PJ, Davis RA (2009) Introduction to time series and forecasting. Springer, Berlin

Chhay L, Reyad MAH, Suy R, Islam MR, Mian MM (2018) Municipal solid waste generation in China: Influencing factor analysis and multi-model forecasting. J Mater Cycles Waste Manag 20(3):1–10

Hilbert S (2017) Correlation coefficient. Springer, Berlin

Jaffel I, Taouali O, Harkat MF, Messaoud H (2017) Kernel principal component analysis with reduced complexity for nonlinear dynamic process monitoring. Int J Adv Manuf Technol 88(9–12):3265–3279

Kavaklioglu K (2011) Modeling and prediction of Turkey's electricity consumption using support vector regression. Appl Energy 88(1):368–375

Mu Y, Liu X, Wang L (2018) A Pearson's correlation coefficient based decision tree and its parallel implementation. Inf Sci 435:40–58

Oprea SV, Pîrjan A, Căruţaşu G, Petroşanu DM, Bâra A, Stănică JL et al (2018) Developing a mixed neural network approach to forecast the residential electricity consumption based on sensor recorded data. Sensors 18(5):1443

Persio LD, Honchar O (2017) Analysis of recurrent neural networks for short-term energy load forecasting. In: American Institute of Physics Conference Series. AIP, College Park

Rahman S, Shrestha G (1991) A priority vector based technique for load forecasting. IEEE Trans Power Syst 6(4):1459–1465

Sadaei HJ, Guimarães FG, Silva CJD, Lee MH, Eslami T (2017) Short-term load forecasting method based on fuzzy time series, seasonality and long memory process. Int J Approx Reason 83(C):196–217

Saeed Madani S (1991) Electric load forecasting using an artificial neural network. IEEE Trans Power Syst 6(2):442–449

Shi H, Xu M, Ran L (2017) Deep learning for household load forecasting—a novel pooling deep RNN. IEEE Trans Smart Grid 99:1–1

Tamura Y, Sato T, Ooe M, Ishiguro M (2010) A procedure for tidal analysis with a Bayesian information criterion. Geophys J Int 104(3):507–516

Wu L, Gao X, Xiao Y, Yang Y, Chen X (2018) Using a novel multi-variable grey model to forecast the electricity consumption of Shandong Province in China. Energy 157:327–335

Zhou X, Guo T, Jiao J (2012) Application of trend extrapolation method to spectrum analysis of microtremor signal. Springer, Berlin, pp 653–658

Chapter 4
Prediction Models of Energy Consumption in Smart Urban Buildings

Abstract In smart cities, the construction is a complex system, which integrates the concepts of the power grid, environment, and other intelligent systems. The development level of intelligent buildings is closely related to the development level of the intelligent city, and it is an indispensable factor in the city. However, the building is also an issue that consumes a lot of resources and energy, and the development of intelligent building has become an urgent task for the construction industry. This chapter describes, which is used to establish two-dimensional and three-dimensional models and obtain data to support building simulation prediction and performance optimization. In this chapter, the support vector machine is used to predict the dataset in the simulation results of DeST software, and an intelligent prediction method for energy consumption of a building in Changsha is proposed. Similarly, the big data computing framework is constructed according to the proposed prediction model to provide support for the design and operation of intelligent buildings in smart cities.

4.1 Introduction

More than 50% of the global population are in cities and in buildings of inefficiency if the energy consumption has been taken into consideration (Joshi et al. 2016). With this assumption, the designers should take technologies and innovations to process development of society and improve people's life (Battista et al. 2014).

The world has entered the era of smart cities. Countries around the world are building more smart cities. In smart cities the construction is a complex and interacted system and the smart city is considered to be collection of connected networks (Barabási 2007), such as the network of transportation, communication, lighting and buildings, the power and water supply, garbage recycling, etc. The idea of smart cities includes an important component, the smart environment, which contains the

H. Liu, *Smart Cities: Big Data Prediction Methods and Applications*,
https://doi.org/10.1007/978-981-15-2837-8_4

efficient improvement of old buildings and construction of the new ones (Battista et al. 2014).

The building has a special position in urban development. The construction of smart city starts with the planning of the city, which must include smart buildings. Therefore, smart building is fundamental, because the developing level of smart buildings is deeply related to the developing level of smart cities (Nam and Pardo 2011).

For the future of city, the implementation "smart city" strategy has far-reaching significance. In addition to the intelligence and facilitation of urban daily life, the development of smart cities will become an important measure to promote the urbanization process of China (Wang et al. 2014) and other developing countries. By the intellectualized reconstruction of architecture and efficient operation in building energy consumption and environmental protection can be achieved and guaranteed.

Architecture is the basic space for human activities and an indispensable factor in cities (Feng et al. 2015). The high level of architecture represents a significant increase in people's living standards and social productivity levels. However, architecture is also an issue that consumes a large sum of resources and energy. Energy consumption of buildings accounts for more than one-third of whole energy consumption (Shaikh et al. 2014). So efficient design, construction, maintenance, and control of building energy are important for increasing green economy (Shaikh et al. 2014). With the continuous development of science and technology, smart buildings or energy efficient buildings, as modern building technology, have significant advantages over traditional buildings and they are turning to the trend of future buildings.

Facing the energy shortages, according to the current economic development and energy consumption levels, the Chinese government has taken measures to implement the coal consumption cap (Shang et al. 2018). In addition, under the high-consumption and high-emission developing mode, air pollution will affect people's health and energy shortages and pollution of the living environment will hinder the process of urbanization in China (Zhou and Zhao 2013). The development of smart cities must be adjusted to sustainable development of society and the environment with the low-carbon developing route. As a consequence, the development of smart buildings and the promotion of environment-friendly and energy-saving building models have become the top priority of the construction industry. Figure 4.1 shows the connection between smart city and smart buildings.

This chapter introduces the DeST (designer's simulation toolkit) software, which establishes the 2D and 3D model and uses it (DeST-c) to achieve data by building thermal analysis, system scheme analysis, AHU solution analysis, duct network simulation, etc., aiming at supporting building simulation prediction and performance optimization (Yan and Jiang 2005).

SVM (support vector machine) is implemented to conduct prediction experiments on the datasets from the DeST software simulation results in this chapter. It presents SVM, a machine learning method to predict energy consumption of a building in Changsha, China. Different parameters of the building model are

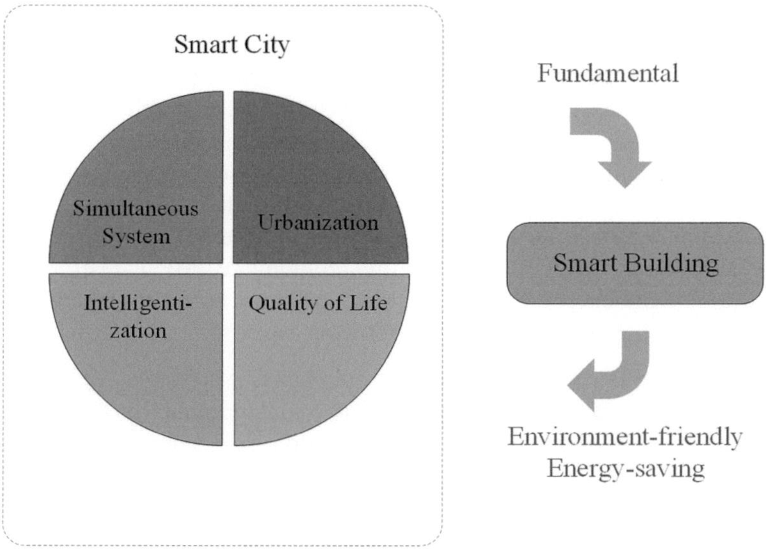

Fig. 4.1 Connection diagram between smart city and smart building

selected as the input of the experimental forecasting process. With a large amount of data in complex structures and forms from the result, big data technology helps to solve the problem of storage, analysis, and data mining to support the design and operation of smart buildings in smart cities.

4.2 Establishment of Building Simulating Model

4.2.1 Description and Analysis of the BEMPs

The fundamental purpose of the building is to provide a comfortable and healthy environment. Building energy consumption is now increasingly recognized as a serious problem (Cao et al. 2016), and a large part of building energy is consumed by heating/ventilation and air-conditioning systems.

The indoor environment is mainly constrained by three main factors: the structure and performance of the building, the composition and operation of the heating system and air-conditioning system, and also the behavior of people in the building (Zhang et al. 2008). These factors influence the indoor environment and energy consumption through different aspects in many ways. At the same time, because the heat and humidity of buildings are easily affected by outdoor meteorological conditions, such mutual influences and constraints vary with meteorological conditions and vary from season to season (Zhang et al. 2008).

Such a complex process is difficult to accurately quantify by conventional methods, which makes it more difficult to evaluate and optimize the design and operation. Therefore, computer simulation and analysis methods have been applied to this field to solve the above quantitative analysis, prediction, and optimization (Yan and Jiang 2005).

Building simulation technology has been developed over more than 40 years and it has been widely used in architecture environment, including design, construction, operation, maintenance, and management of the whole life cycle of the building (Xin et al. 2014). Building energy modeling programs (BEMPs) are efficient methods to evaluate the energy consumption and optimize building construction (Xin et al. 2014). It is mainly applied in the following aspects:

(a) Calculation of building load to select equipment;
(b) Energy consumption analysis by design or maintenance of the building;
(c) Management of building energy consumption to develop the energy-saving ability;
(d) Combined with various standard specifications for the local requirement;

Building energy modeling programs (BEMPs) can be divided into the following two groups.

Air-conditioning system simulation software like TRNSYS, SPRAK, and HVACSIM are mainly for simulation of the control process of components in air-conditioning system. The main target of this group is the characteristics of the system. They use a model of simple room and complex systems for the simulation analysis. Energy consumption simulation software is used for the simulation and analysis of buildings and systems, such as DOE, EnergyPlus, and ESP-r. The main targets of this group are long-term thermal characteristics of buildings and systems. They use a model of complete room and simple system.

The first group of software is flexible to simulate many forms of systems. However, this can also lead to a large calculated amount and complex results. Compared to the first group of software, the second group starts with the building instead of system. Such software starts with the building structure and can be handled flexibly. But it is difficult to form various systems.

After comparing the different characters of the two simulation objects of the building and air-conditioning system, the conclusion can be drawn that simulation software should have both structural flexibility and control ability.

Tsinghua University has started to develop a building simulation software DeST in the early 1980s, aiming at simulating and analyzing both building energy consumption and HVAC (heating, ventilation, and air conditioning) systems (Yan and Jiang 2005). It has been designed to achieve an efficient connection between thermal output of buildings and dynamic result of HVAC systems to improve the system design, to ensure the quality, and to reduce energy consumption (Yan et al. 2008). The flowchart of the BEMPs is shown in Fig. 4.2.

The chapter introduces the main features, structures, and programs of DeST and finds the ideal prediction of the system and compares the selected design conditions in the following example.

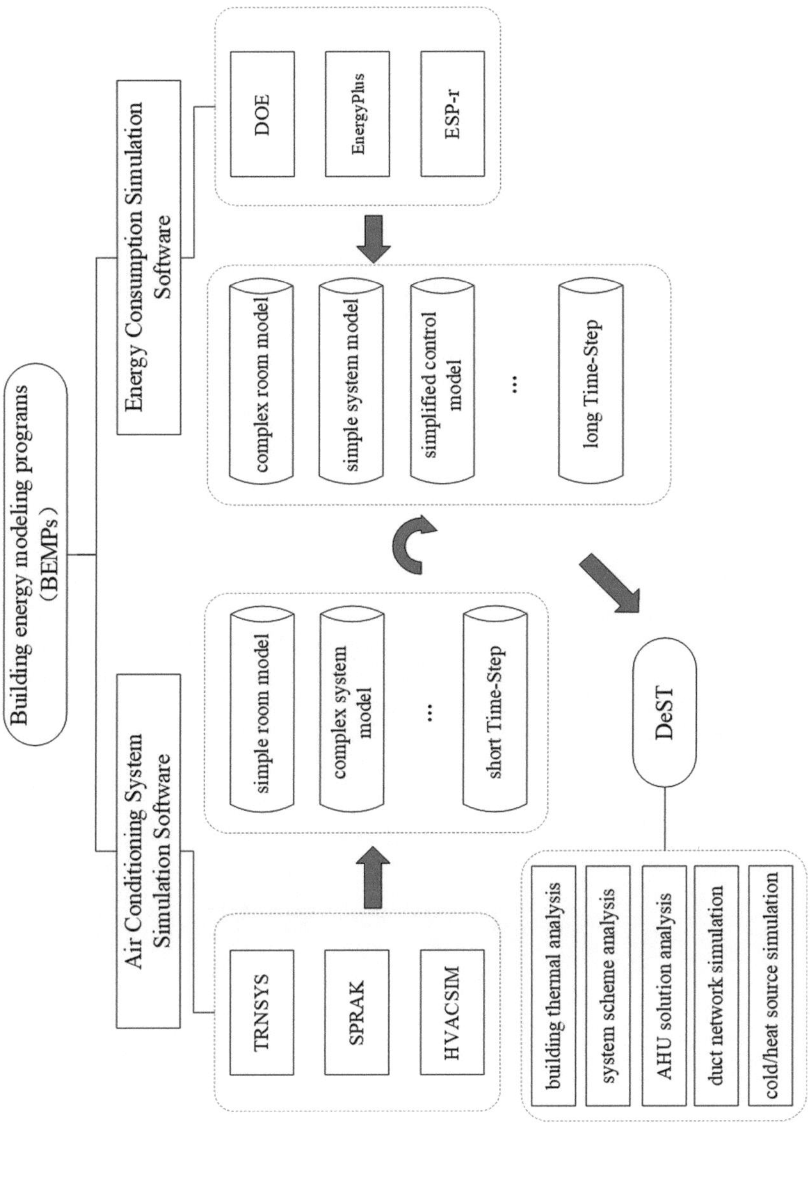

Fig. 4.2 Building energy modeling programs

4.2.2 *Main Characters of DeST Software*

In order to realize the simulation in stages, DeST highlights its own characters in the design and development process (Yan and Jiang 2005):

4.2.2.1 Combine Buildings and Environmental Control Systems with Base Temperature

Base temperature is the building temperature without the HVAC system and the building has both effects from outdoor meteorological conditions and indoor heat sources. This temperature fully reflects the performance of the building and the effects of various thermal disturbances (Yan and Jiang 2005). Figure 4.3 is the distribution of indoor and outdoor temperatures of 10 rooms with the numbers from model of buildings.

The base temperature of all rooms can be obtained by analysis of the thermal results in the building and accurate building model. On the research of HVAC systems, the base temperature and other building parameters can be integrated to form a building module. System modules are also collected with base temperatures modules to form a system model (Yan and Jiang 2005). This is the basic coupling method of DeST.

4.2.2.2 Design and Simulation in Stages

The practical design process includes various steps, while the design goals of each step are different. As the design from initial to detail, the amount of information increases but the flexibility decreases. The uncertain factors of the previous step become defined in the current step.

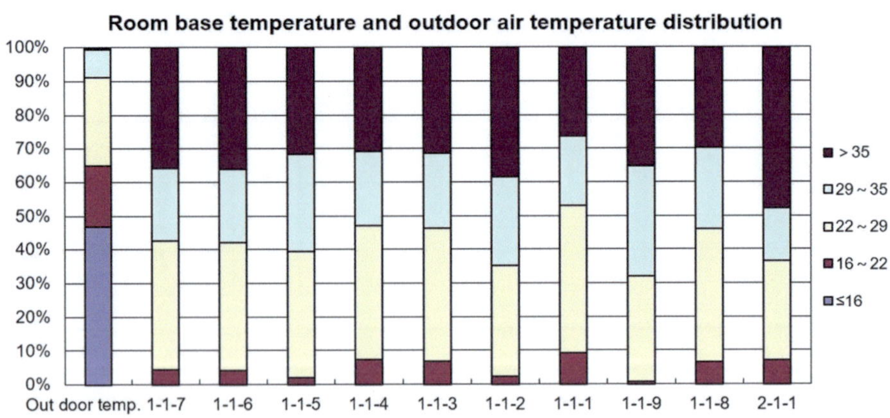

Fig. 4.3 Annual base temperature chart of 10 rooms

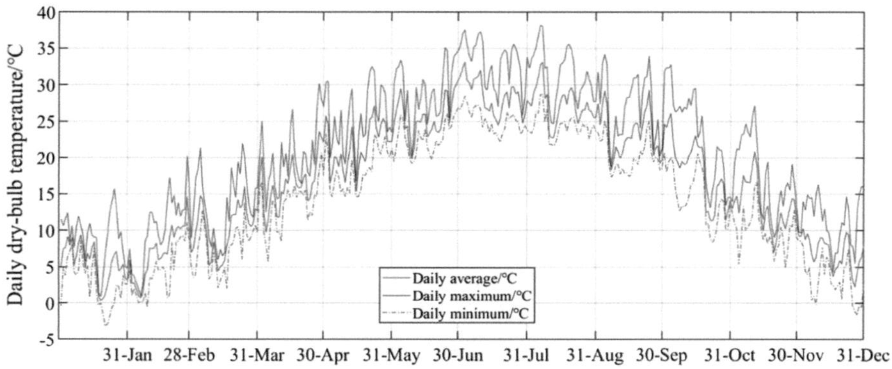

Fig. 4.4 Daily dry-bulb temperature

For example, in the preliminary design step, indoor heating load and external meteorological parameters are certain conditions. Figure 4.4 shows the daily dry-bulb temperature in Changsha, which belongs to the meteorological parameters. Under these parameters, the thermal characters of the building are uncertain. On the contrary, the thermal characters of the building become defined factors at the step. So the designer can use these to compare and select the best plan for HVAC systems (Zhu and Jiang 2003).

DeST combines the phased features of the actual design process in the developing process and divides the simulation into steps to provide accurate and practical analysis results such as building thermal analysis, system scheme analysis, AHU solution analysis, etc. (Yan and Jiang 2005).

4.2.2.3 Graphical Interface

DeST develops a graphical interface for all simulation steps on the user interface based on the AutoCAD, and the interface runs under the Windows operating system. Various data related to the building (material, geometric size, internal disturbance, etc.) are connected to the user interface through the database interface (Yan and Jiang 2005). The simulation module is very convenient to draft for the users in 2-D and 3-D display (Zhu and Jiang 2003). The results of the simulation calculations are in the form of Excel tables. The 2-D floor model diagram and 3-D display model diagram are shown in Figs. 4.5 and 4.6, respectively.

4.2.3 Process of DeST Modeling

DeST combines the periodical characters of the design process in the developing process and provides accurate analysis results for different steps of building design. DeST uses AutoCAD as the design platform, designers can complete the building

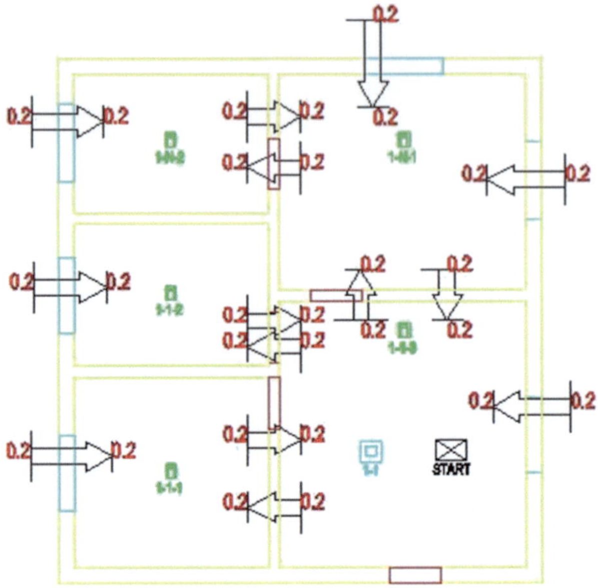

Fig. 4.5 2-D floor plan of 6 story building

modeling and simulation of energy systems in the AutoCAD interface (Hu and Wu 2015). In the chapter, a building model will be established gradually and output the data after calculation.

4.2.3.1 Start Establishing Architectural Models

An office building has been selected as the research object of the simulation. The office is public buildings in the university, which is made of a single frame structure with three floors. The height is 3.6 m for each floor and total construction area of air conditioning is 1728 m^2. It is for meetings and office work.

4.2.3.2 Set the Floors of the Building

Users should draw the draft before using DeST for building energy simulation. In the AutoCAD, it can make the model work with DeST. After editing the outline of each floor along the centerline of the wall, DeST expresses windows, walls, and doors in different colors with different layers, which is quite explicit on the board. The standard layer can be just one floor and other layers can copy the basic floor or set another with different structures (Hu and Wu 2015).

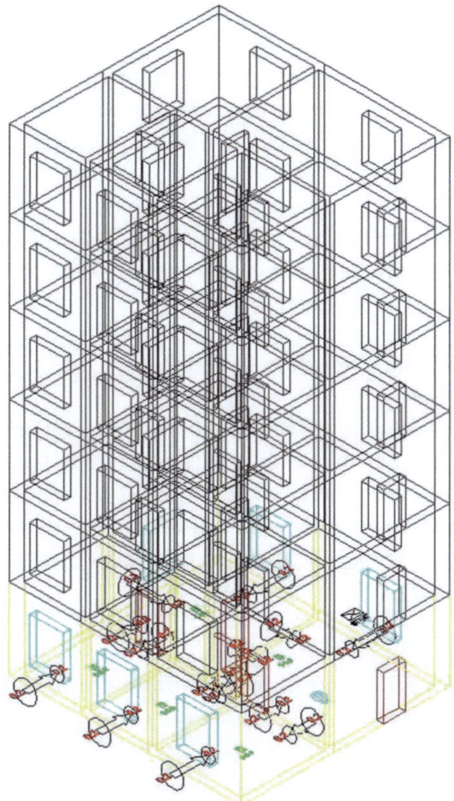

Fig. 4.6 3-D display of 6 story building

4.2.3.3 Set the Windows and Doors

The doors are on vertical walls with length and width according to the location of doors. Windows are also on the vertical walls by given dimension, edge of wall, or window–wall ratio. The material of walls, windows, and doors as enclosure structures should be changed for energy simulation. The process and results will be presented in Sect. 4.3.4.

4.2.3.4 Pre-processing of Construction

The instruction of pre-processing is the first implementing of the whole simulation, in which the whole building can be inspected to determine the integrity of the building envelope and the rationality of the door and window layout, and automatically remove some divergent drawings and add identification and thermal interference automatically. The pre-processing checks the building enclosure structure and guarantee the correctness (Hu and Wu 2015).

4.2.3.5 Set Wind Pressure

Wind pressure is the main driving force of natural ventilation (Gan 2010). To obtain an accurate wind pressure coefficient is the key to calculate the natural ventilation of a building. The distribution of wind pressure on the building surface relates to factors such as wind direction, surrounding environment, and orientation of the building. In practical applications, the wind pressure coefficient is usually to define the wind pressure on surface, and it is also used to characterize the ratio of the dynamic pressure of flowing wind to the wind pressure on the building.

4.2.3.6 Set Indoor Ventilation

The heat transfer of the building is not separated. So ventilation symbols should be added at the prompt between rooms and outside for building description and the following calculation. To fulfill all ventilation requirements users can also add the ventilation manually. After adding all rooms that need ventilation after the pre-processing, the architectural model plan is finished. Figure 4.7 is the architectural model plan for this chapter.

4.2.3.7 Condition of Building Simulation

With the multiplicity and complexity of the parameters, computer simulation is more convenient and quick for experimental testing. Annual simulations can be realized by DeST software. By researching the energy-saving potential of the structure, considering the local temperature characters and that the regular heating season lasts from October to March and the air-conditioning season lasts from May to September, the room temperature setting is between 18 °C and 26 °C.

Individual analysis, combination analysis, and other research methods are used to analyze the potential of energy saving. The impacts on building energy consumption can be analyzed while simulating building energy consumption under various conditions to research the energy-saving potential of the whole building (Hu and Wu 2015).

4.2.3.8 Simulation Process

Based on the establishment of a complete building model, describing the building and the main factors of the building, the simulation process starts after setting the building, rooms, location, directions, and heat conduction. The option of location is very crucial because weather conditions are quite different in cities and the annual hourly weather parameters can be calculated from the local meteorological parameter according to the selected location.

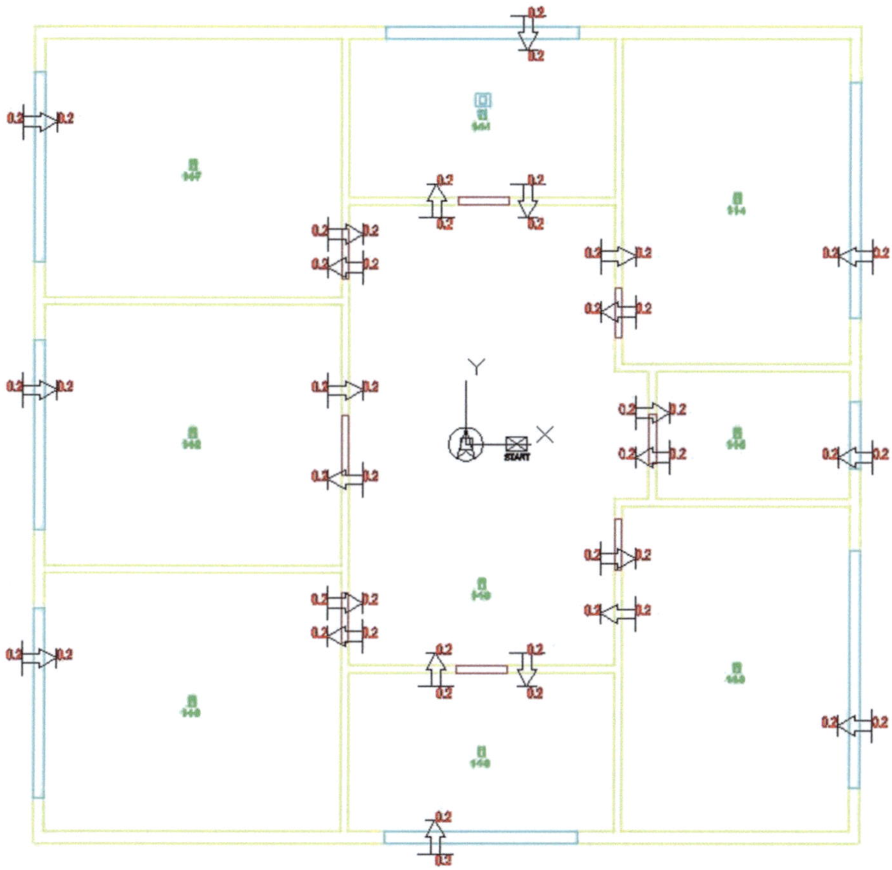

Fig. 4.7 Building model plan

In commercial buildings, room characters are different by internal persons, lighting, equipment, and system conditions, which will affect the final load of the building, so the type of room is very important. The total office version can be used to build more rooms, shopping centers, toilets, hotels, office, and others (Hu and Wu 2015).

4.2.3.9 Global Building Settings

Total settings include global building, air conditioning, AHU, and global heat/cold settings, etc. Global building setting involves every parameter of the building, such as location, room type, settings of ventilation, windows and walls, and other building parameters (Hu and Wu 2015). The process runs to the complete result only when global setting is correct.

4.2.3.10 Calculation of Shadow and Lighting

In the dynamic simulation of the built environment, solar radiation is an important factor affecting the indoor thermal environment of the building. It directly affects the building base temperature, load calculation, natural lighting, the building's air-conditioning energy consumption, and lighting power consumption.

The architectural solar path module BShadows in DeST and the building lighting module calculate building shadow and daylight. These results of calculation are also for the input conditions for room temperature and load calculations. Shadow calculation helps to analyze the shelter between buildings, the shelter of the building itself, and the shelter of sunshade components (Yan et al. 2008).

Figure 4.8 shows the interface of BShadows with the 3-D building model. The following impacts are included:

(a) Distribution of building shadow in different locations of the year;
(b) Detailed geometric information of the shadow;
(c) Spot distribution in special types of buildings (Yan et al. 2008).

In addition, BShadows also provides many auxiliary analysis functions. Based on the simulation results, building components can be optimized to minimize building energy consumption.

4.2.3.11 Output of Simulation

The main function of the DeST is to simulate and calculate building loads, natural light effects, loads of air conditioning, cold/heat sources, and energy consumption of systems with output in EXCEL.

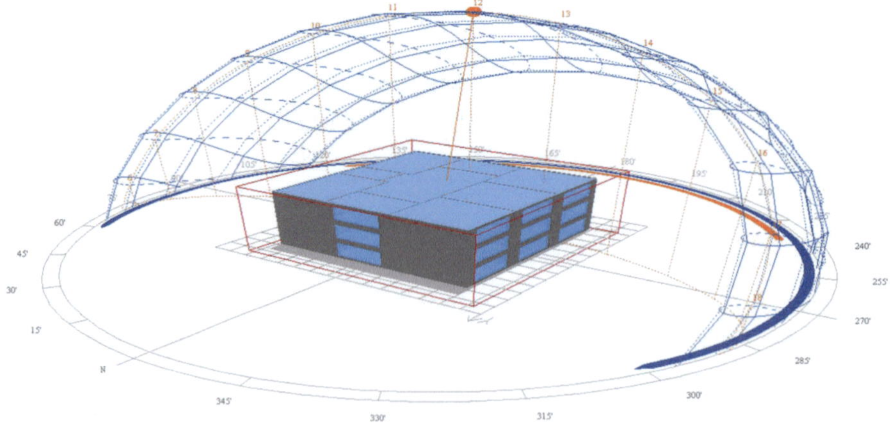

Fig. 4.8 Shadow analysis of the building model from daily and annual solar path

The steps of DeST Modeling are shown in Fig. 4.9 and the duration of process can be variable depending on the complexity of the building.

4.3 Analysis and Comparison of Different Parameters

4.3.1 Introduction of the Research

With the calculation results in Sect. 4.2, there are different reports of the building model, which contains various information about different parameters. The DeST model in Fig. 4.7 represents a 3-floor building in Changsha. All the parameters may lead to changes for energy consumption. The research needs the most influential parameters through analysis and comparison in the process. So meteorological parameters, enclosure structure, indoor thermal perturbation, and indoor design parameters have been chosen to begin data processing and to observe the results

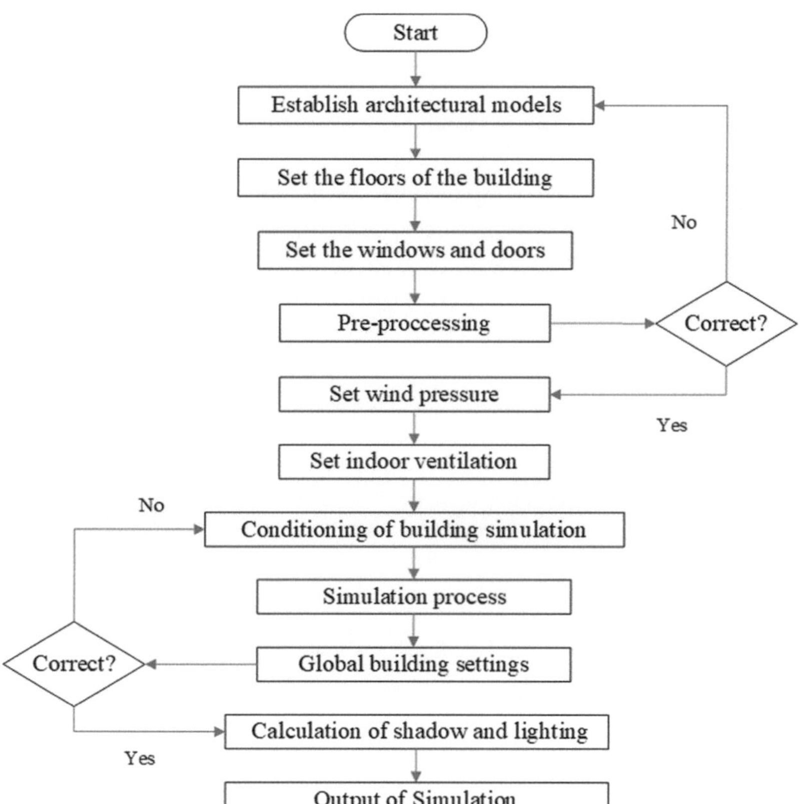

Fig. 4.9 The steps of modeling

under designated circumstances. The research aims to reduce energy consumption, which in summer mainly reflects in building load, especially the cold load. Figure 4.10 shows the annual hourly load of the building model and Fig. 4.11 shows the annual hourly load per area of the building model.

From Figs. 4.10 and 4.11 it can be concluded that the cooling load of building in summer is relative high, resulting in huge energy consumption in the city. Hence, the data from July to August are suitable for the research.

4.3.2 Meteorological Parameters

The basic meteorological data are from the measured daily data (including temperature, humidity, solar radiation, wind speed and direction, hours of sunshine, and atmospheric pressure) of 270 cities in different parts of China in the past 40 years (Yan et al. 2008). According to the typical year selection method in air-conditioning load calculation, the software first selects a representative year. Then the daily variations of each meteorological parameter are for simulating the hourly meteorological data (such as air temperature, humidity, direct solar radiation, and wind speed) (Yan et al. 2008).

From the output of the weather data there are hourly weather data, daily dry-bulb temperature, monthly average dry-bulb temperature (Fig. 4.4), hottest month (July) temperature (Fig. 4.12), coolest month (January) temperature (Fig. 4.13), annual temperature and direct solar radiation (Fig. 4.14)

So when the meteorological data has changed its completeness and accuracy will also affect the energy performance of the building (Qiang et al. 2015). In order to evaluate the impact of this change, it should be compared with the original data of July and August in Fig. 4.15.

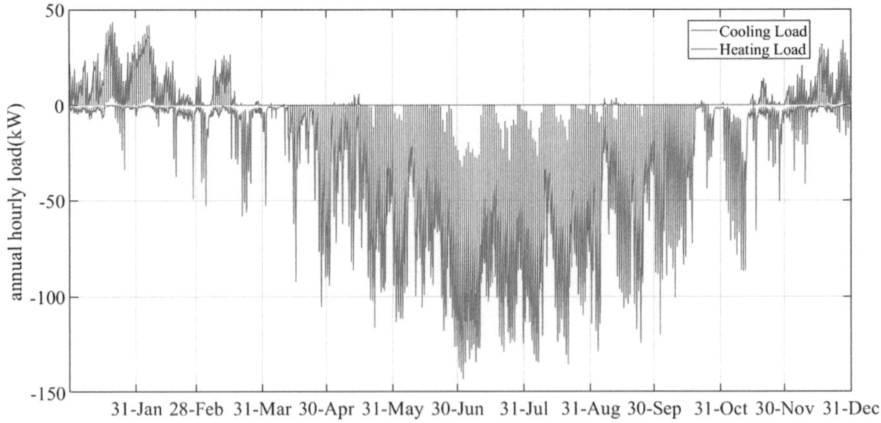

Fig. 4.10 Annual hourly load in the building

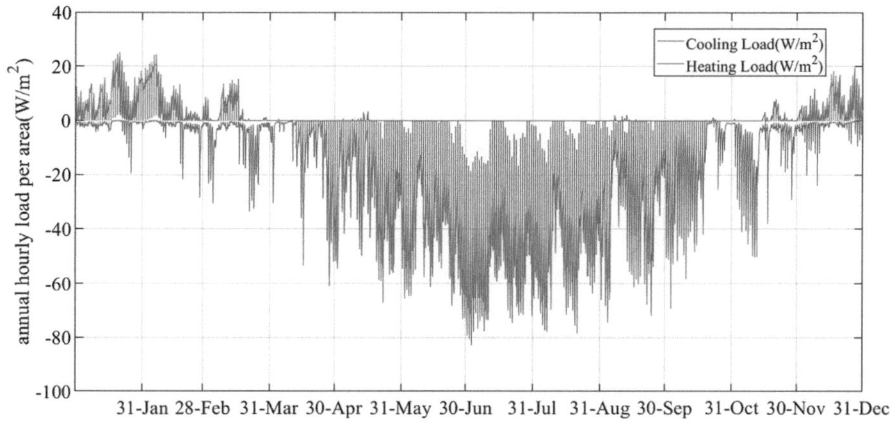

Fig. 4.11 Annual hourly load per area in the building

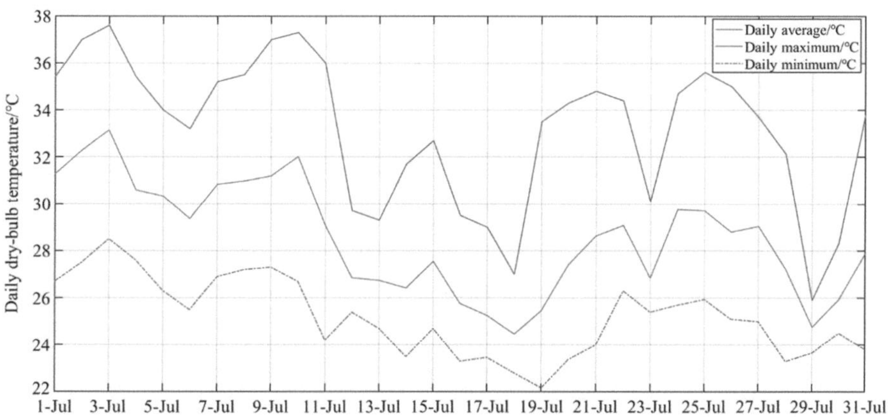

Fig. 4.12 The temperature of the hottest month

The heating load is zero in summer, so only the cooling load has been taken into account. Figure 4.10 clearly shows the high building load by continuous high temperature in summer. If the meteorological data like solar radiation, solar duration, humidity, or wind pressure changes, which can be achieved in DeST, the situation is surely quite different, as shown in Fig. 4.16, which has almost the same trend in Fig. 4.15, but energy consumption has an obvious rise.

4.3.3 Indoor Thermal Perturbation

In commercial buildings, indoor heat gains are one of the most crucial impacts that influence the building thermal environment (Yan et al. 2008). Because the inner heat gains run in an uncertain way in time serial and space, building thermal process model in DeST is especially for simulation of the indoor thermal perturbation in the buildings.

The indoor thermal perturbation contains the person perturbation, lighting perturbation, equipment perturbation, and air condition perturbation. At first, the function of the room is changeable, such as living room, study room, meeting room, restaurant, hotel, hospital, etc. Each has quite different thermal perturbations, together with equipment, lighting, and air condition, as the example shown in Fig. 4.17, different colors represent different working status. The date and duration are alterable, which lead to changes in total thermal perturbation.

Therefore, the function of room for indoor thermal perturbation changes just by doubling the persons and the duration. The load of the building has increased significantly, which is more than double or even triple the original load as following Fig. 4.18.

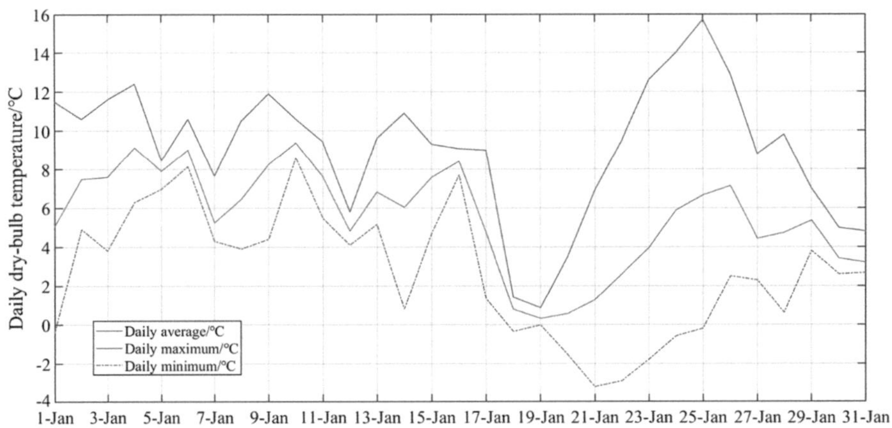

Fig. 4.13 The temperature of the coldest month

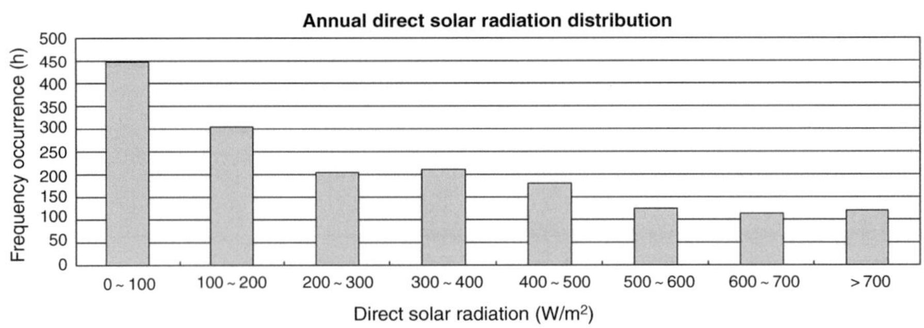

Fig. 4.14 Annual direct solar radiation

4.3.4 Enclosure Structure and Material Performance

Enclosure structures are essential parts of a building like walls, doors, and windows, which physically separate the external from the interior environment and affect the heat convection and conduction. The enclosure structure for energy saving is selected to improve the efficiency of heating and cooling and to reduce energy consumption while ensuring the building's performance and comfort for internal environment. To research and evaluate the exact role this parameter plays for the energy consumption in the building, the simulation process runs after changing the material

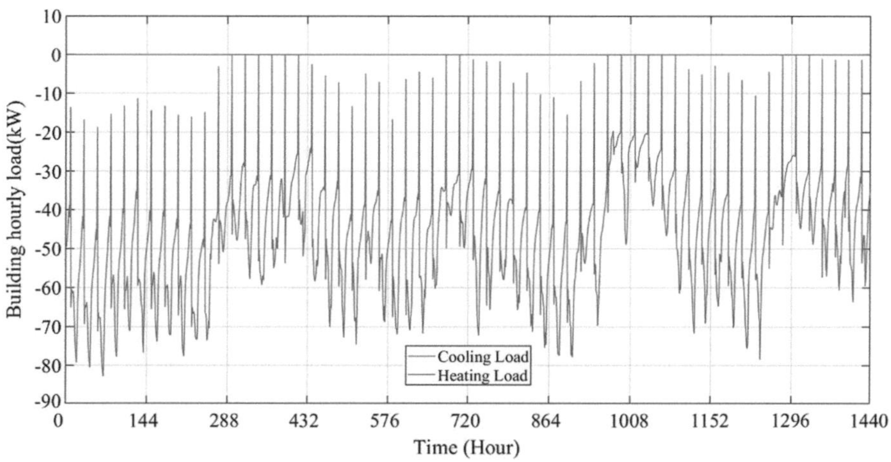

Fig. 4.15 The original data of 60 days load

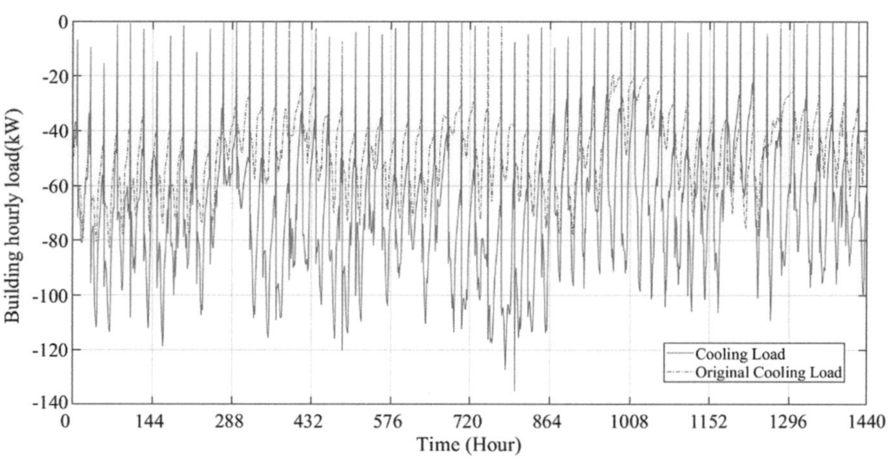

Fig. 4.16 The temperature of 60 days after adjustment of meteorological parameter

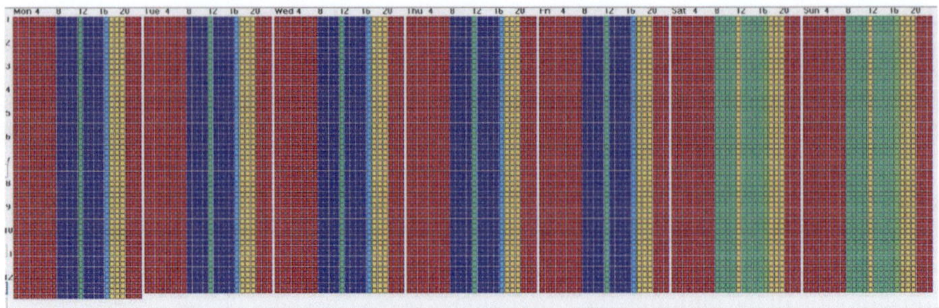

Fig. 4.17 Indoor thermal perturbation setting of a room in a week

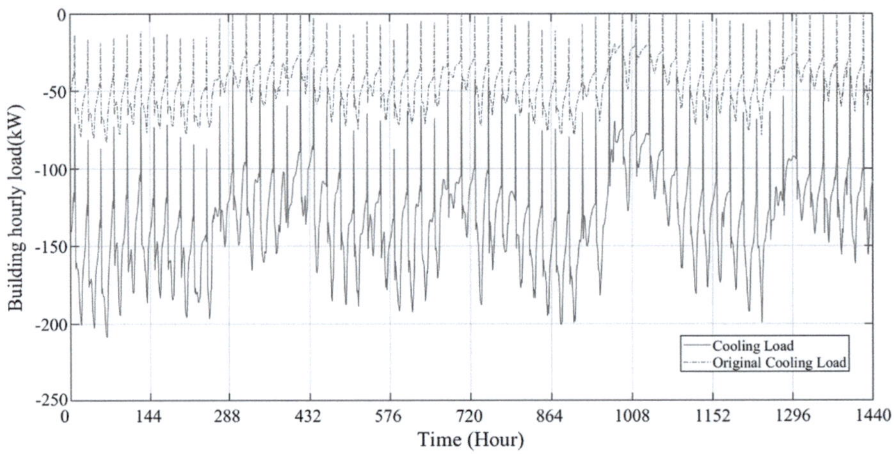

Fig. 4.18 The temperature of 60 days after adjustment of indoor thermal perturbation

of structure and compares the result with the original data, and the result is shown in Fig. 4.19.

Table 4.1 shows the changes of different parts and from Fig. 4.20 the building load also keeps the same trend with more costs. The results show that the original material is better for summer while the new materials may be suitable for keep warm in winter but increase the temperature in summer.

4.3.5 *Indoor Design Parameters*

The building layout generally has a direct influence on the quality of indoor ventilation. To obtain the efficient natural ventilation effects, the direction and rational structure in the overall layout are the foremost factors, so that the natural wind can flow through the whole building to reduce the indoor heat quickly (Hui et al. 2012).

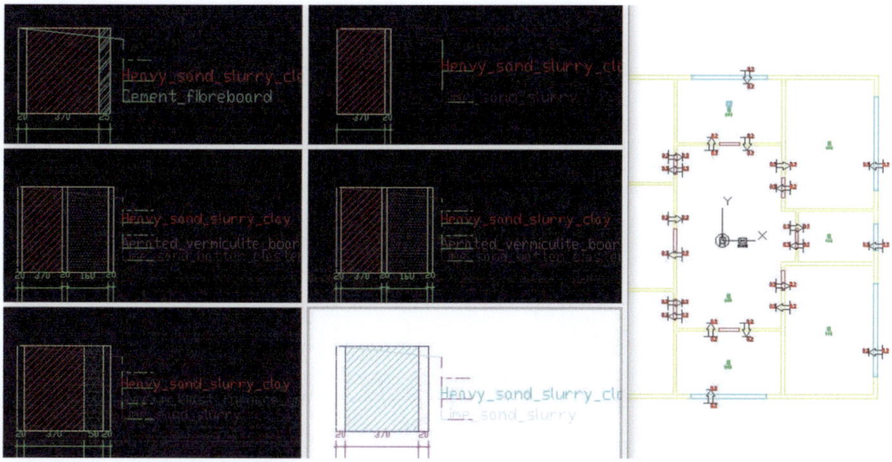

Fig. 4.19 The material change of the exterior wall

Table 4.1 The change of structure materials

Parts	Original	After changing
Exterior wall	240 mm-depth heavy sand slurry clay + 10 mm-depth plasterboard + 60 mm-depth polystyrene foam + 8 mm-depth plasterboard	20 mm-depth cement sand slurry + 240 mm-depth heavy sand slurry clay + 20 mm-depth lime sand slurry
Interior wall	20 mm-depth cement slurry clay + 120 mm-depth heavy sand slurry clay + 20 mm-depth cement slurry clay	20 mm-depth oak + 50 mm-depth cement slurry clay + 120 mm-depth heavy sand slurry clay + 50 mm-depth cement slurry clay + 20 mm-depth oak
Door	Single layer of wooden exterior door	Glass door in shopping mall
Window	6 mm-depth single layer glass	Ordinary insulating glass

There are various designs for the structure of modern buildings, but many of them are also energy consumers. Compared to the traditional 3-floor house in the chapter, the building with the elevator may lead to other energy data. This 2-D plan in Fig. 4.21 is from the 3-floor building, the height is also 3.6 m for each floor and the length and width are the same as the original plan. The layout is clearly that there are few windows and doors with the elevator right in the middle, which hinders the ventilation. After the simulation the data in Fig. 4.22 demonstrates the construction of elevator in building design is obviously costing more energy.

The research results indicate that meteorological parameters, indoor thermal perturbation, enclosure structure or material and indoor design affect energy consumption. However, the effect of indoor thermal perturbation is far more significant to others.

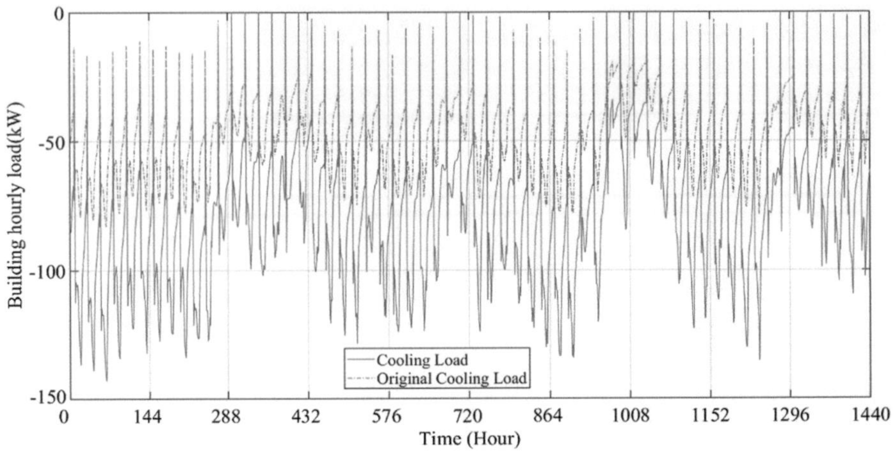

Fig. 4.20 The temperature of 60 days after adjustment of enclosure structure and material

4.4 Data Acquisition of Building Model

4.4.1 Data After Modeling

Due to the large amount of data to be calculated, DeST provides a number of different options in order to meet the needs so that users can selectively extract data and continue to process the data according to their needs.

The data after modeling without calculation are shown in Table 4.2.

4.4.2 Calculation of Room Temperature and Load

Room base temperature refers to the indoor air temperature caused by the combination of outdoor meteorological conditions and various indoor thermal perturbation by a heating or air-conditioning system. It fully reflects the performance of the building and the impact of passive thermal perturbation in the building. The related data are shown in Tables 4.3 and 4.4, respectively.

4.4.3 Calculation of Shadow and Light

It can save a lot of lighting power and effectively reduce the annual energy consumption of air conditioners by making full use of natural light. The shadow and light calculations of building base on the hourly illuminance of each area of the building under natural lighting conditions, and optimize the lighting system design

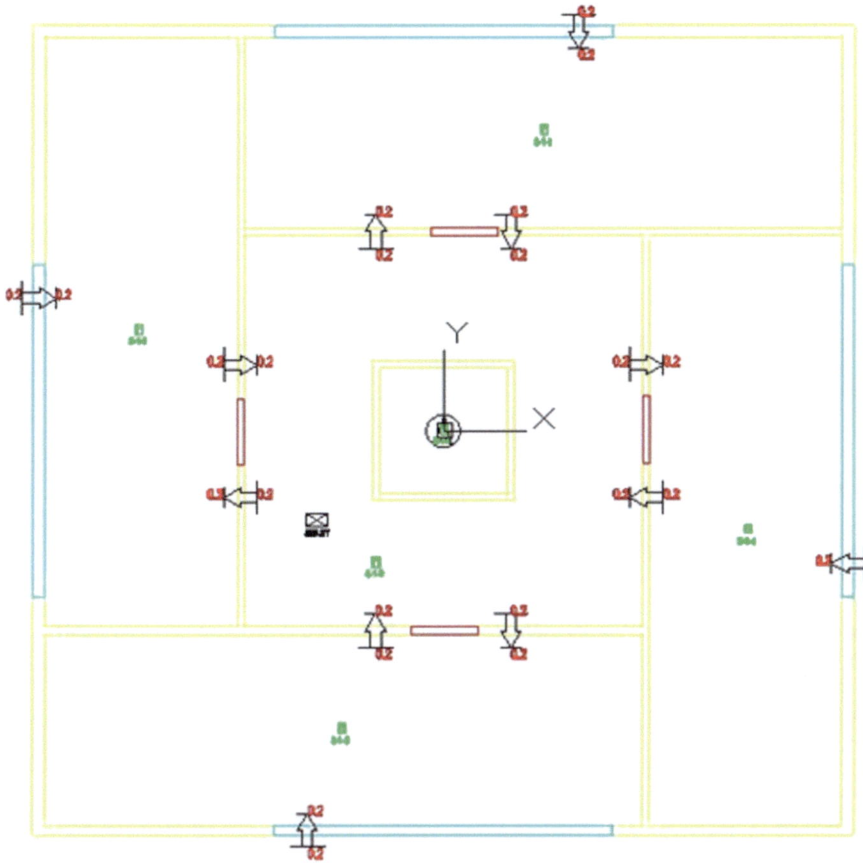

Fig. 4.21 The construction of a 3-floor building with elevator

or control the lighting system according to the calculation results. The related data are shown in Tables 4.5 and 4.6, respectively.

4.4.4 Calculation of Natural Ventilation

The natural ventilation can reduce the energy consumption of building air conditioners, and it also contributes to reduce the indoor pollutants and carbon dioxide. So accurate simulation of natural ventilation is one of the key factors affecting the simulation results. The data after calculation of natural ventilation is shown in Table 4.7.

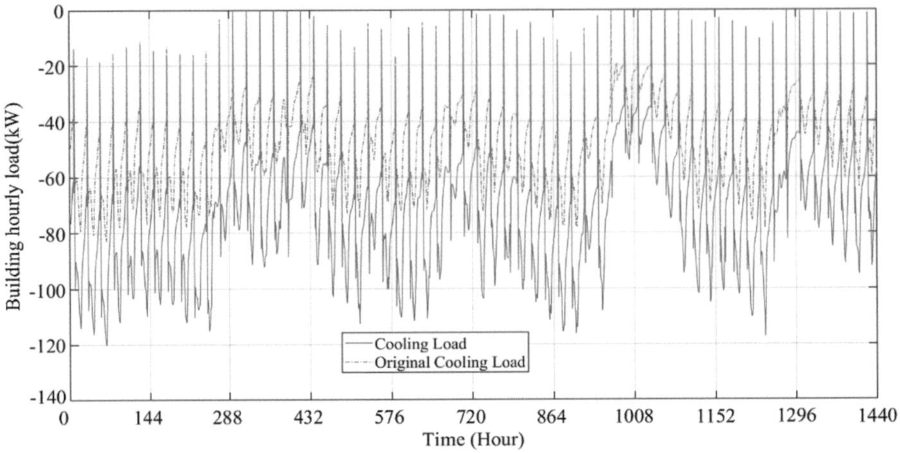

Fig. 4.22 The temperature of 60 days after adjustment of building construction

Table 4.2 Output after modeling

Weather	Hourly weather data of meteorological parameters
Enclosure report	Statistics on the area of the enclosure structures, internal and external walls, doors, windows and information of building component with costs
Room function	Summary of the thermal perturbation from room functions

4.4.5 Simulation of the Air-Conditioning System

The air-conditioning equipment is one of the main facilities for cooling, heating, dehumidifying, humidifying, etc. The air-conditioning system consumes a large amount of energy, so the design of the air-conditioning solution is a very important part of the building design process. The related data are shown in Tables 4.8, 4.9, and 4.10, respectively.

4.5 SVM Prediction Model for Urban Building Energy Consumption

4.5.1 The Theoretical Basis of the SVM

Support vector machine (SVM) is a type of learning machine. It is from the principle of structural risk minimization to reduce the universalization error from training error (Dong et al. 2005). SVM is very stubborn to the overfitting problem, ultimately getting better universalization ability (Huang et al. 2005) than the traditional methods to solve

Table 4.3 Output after room temperature calculation

Room base temperature	Base and ambient temperature, degree hourly summary, annual base temperature

Table 4.4 Output after room load calculation

Room temperature	Air condition temperature
Room load	Room hourly cooling load, room hourly heating load, dehumidification load, humidification rate, room latent load
Building load	Building hourly load, building hourly load per area
Load report	Building load, system heating load, system cooling load, room heating load, room cooling load

various problems (Dong et al. 2005). SVM has been utilized to implement text classification (Joachims 1998), handwriting recognition, image recognition, gene classification (Brown et al. 2000), and real forecasting computation in time series (Liu et al. 2015).

The basic steps of SVM applying for classification are as follows: The input data should be taken into a feature space (with higher dimension if possible). Then an optimized linear division is accomplished with the feature space to form a plane to separate the minimum error and maximum margin (Cai et al. 2001).

The SVM algorithm constructs the following iteration formula (Ma et al. 2018): input a set of training data $(x_1, d_1)\ldots(x_j, d_j)$ (x_i is the input data, d_j is the corresponding outputs) is independently and randomly produced. SVM will approximate the regression with the following steps (Ma et al. 2018):

$$f(x) = \omega \cdot \psi(x) + b \tag{4.1}$$

where, $\psi(x)$ is for the feature spaces, ω represents weight vector; b is the coefficient.

$$R = \frac{1}{2} \| \omega \|^2 + E\frac{1}{N}\sum_{i=1}^{N} L_\varepsilon(d_i, y_i) \tag{4.2}$$

where, E is the regularized constant and ε is the threshold of SVM support vector machine, L_ε is loss function (Ma et al. 2018) as follows:

$$L_\varepsilon(d_i, f(x_i)) = \begin{cases} |d_i - y(x_i)| - \varepsilon, |d_i - y(x_i)| \geq \varepsilon \\ 0 \end{cases} \tag{4.3}$$

Equation (4.3) is changed to the function with the variables η_i^*.

$$R = \frac{1}{2} \| \omega \|^2 + E\frac{1}{N}\sum_{i=1}^{N} (\eta_i + \eta_i^*) \tag{4.4}$$

Table 4.5 Output after calculation of building daily shadow

Surface shadow	The proportion of illumination on the surface under different solar elevations and azimuths

Table 4.6 Output after calculation of building illumination

Room daylighting	Room hourly illumination

The Eq. (4.2) can be obtained to the solution:

$$y(x)=\sum_{i=1}^{n}\left(\alpha_i-\alpha_i^*\right)\psi\left(x_i\right)\cdot\psi\left(x\right)+b \tag{4.5}$$

where, α_i and α_i^* are Lagrange multipliers with function $K(x_i, x)$, Eq. (4.2) is changed as follows with RBF kernel (Niu et al. 2010):

$$f(x)=\sum_{i=1}^{n}\left(\alpha_i-\alpha_i^*\right)\cdot K\left(x_i,x\right)+b \tag{4.6}$$

$$K\left(x_i,x\right)=\exp\left\{-\frac{\| x_i-x \|^2}{2\sigma^2}\right\} \tag{4.7}$$

4.5.2 *Steps of Modeling*

According to the method in Sect. 4.5.1, the steps of using the SVM model to predict energy consumption are as follows:

4.5.2.1 Data Selection

For the prediction of energy consumption, various parameters may lead to different influences. According to the comprehensive analysis in Chap. 2, there are nine selected parameters, which are the elements of indoor thermal disturbance, system load and meteorological factors, etc. After cross-correlation analysis the conclusion can be drawn that the cooling load $\{x_{\mathrm{cl}}\}$, the total solar radiation $\{x_{\mathrm{tsr}}\}$, and the personnel calorific value $\{x_{\mathrm{pca}}\}$ are the three groups of parameters with the strongest correlation with the energy consumption. So the data from the three groups are given as input *in* R^n, and the cold loads also are given as the corresponding outputs in R.

Table 4.7 Output after calculation of natural ventilation

Room temperature	Temperature by natural ventilation
Room load	Room hourly cooling and heating load, dehumidification load, humidification rate, and room latent load by natural ventilation
Building load	Building hourly load and building hourly load per area by natural ventilation
Load report	Building load, system heating and cooling load, room heating and cooling load by natural ventilation
Room ventilation	Building hourly fresh air volume by natural ventilation

Table 4.8 Output after calculation of air-conditioning scheme

Scheme result	Results of air-conditioning system simulation

Table 4.9 Output after calculation of wind network

Fan result	Wind pressure and power consumption of Fan
Duct result	Wind system power consumption, and wind system hourly work status

Table 4.10 Output after calculation of AHC simulation

AHU result	Results of AHC system simulation

4.5.2.2 Samples Setting

The data for training sample set and test sample set are chosen randomly in a role from the results in DeST. There are 500 as the training samples and 100 as the test samples. In this step, the training set and test set of three groups of parameters, respectively, are ready for the next step.

4.5.2.3 Initialization and Training of the SVM Model

This step works to set the value of the variable, such as initial value x_1, precision threshold b, and other important parameters of the SVM model. And this step sets normalization, feature selection, and regression prediction. In order to achieve accurate results, the 1-step, 2-step, and 3-step strategies of forecast strategy are used, so the SVM model needs to go through three times, end up with three separately predicted diagrams. The final training also needs to go through several iterations.

4.5.2.4 Testing the Trained SVM Model

The divided test set is used to conduct prediction experiments on the trained SVM model. The indices of MAE, MAPE, RMSE, SDE, and *R* are used to compare and analyze the forecasting results of different parameter datasets.

4.5.3 Forecasting Results

The test samples set input into the trained SVM to get the forecasting results. Figures 4.23, 4.24, and 4.25 show the prediction results of the SVM model from different forecasting strategies. From the figures, it can be concluded that:

(a) The SVM model has a good prediction effect on test samples. The fitting degree of both the main trend component and the peak and trough of the wave is close to the real value of the test set. From those figures, it shows that all the prediction curves well fit the main trend of the original load data.
(b) The prediction accuracy of 1-step strategy is better than that of 2-step and 3-step strategy. In Figs. 4.24 and 4.25, the datasets of 2-step and 3-step strategy fluctuate violently. Therefore, the errors of the prediction curve are mainly concentrated around the peaks and troughs, and the results have a certain delay. In the practical application, time compensation can be added to obtain the prediction results with higher accuracy.
(c) From the figures, it seems that 2-step and 3-step forecasting strategies of the SVM model do not have a high prediction accuracy for the situation. The prediction error of the model is mainly caused by the error accumulation in the process, while multi-step forecasting leads to more errors, resulting in decreased prediction accuracy.

In order to make a more accurate evaluation of model prediction performance, Table 4.11 gives some statistical indices of the prediction results of the SVM model for 1-step, 2-step, and 3-step forecasting strategies of datasets. According to Table 4.11, it can draw the following conclusions:

(a) The SVM model 1-step forecasting strategy has the best prediction effect for cold load dataset. The MAE, MAPE, RMSE, SDE, and *R* of forecasting results for 1-step, 2-step, and 3-step forecasting strategies are shown in Table 4.11. The statistical indices of forecasting results of dataset of 1-step forecasting strategy in SVM model are smaller than that of other two advanced steps forecasting. This is consistent with the conclusion acquired from the figures, which reflects the result of the experiment.
(b) From the data in the tables, it can be seen that 1-step forecasting strategy in the SVM model has a better prediction effect than that on the test dataset of 2-step and 3-step forecasting strategies. But 1-step forecasting strategy still has potential to improve and increase accuracy by other methods like hybrid models or recursive hybrid strategies, which may be the research emphasis in the future.

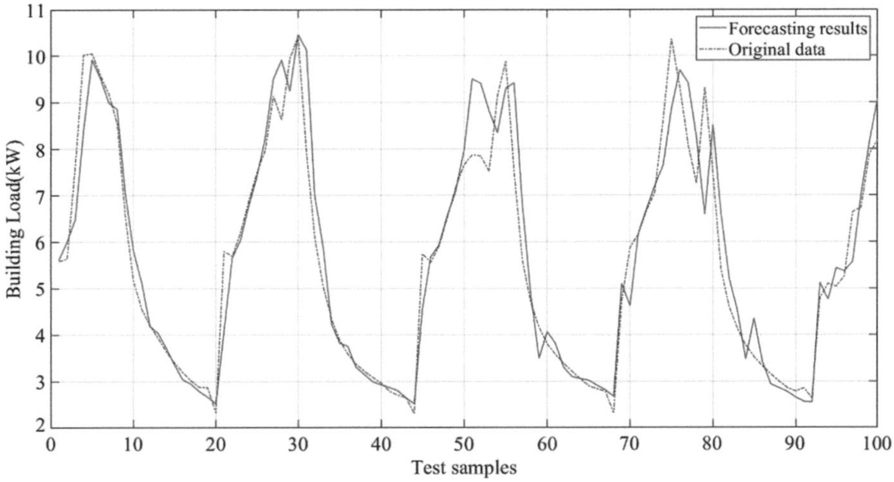

Fig. 4.23 Forecasting results of 1-step strategy

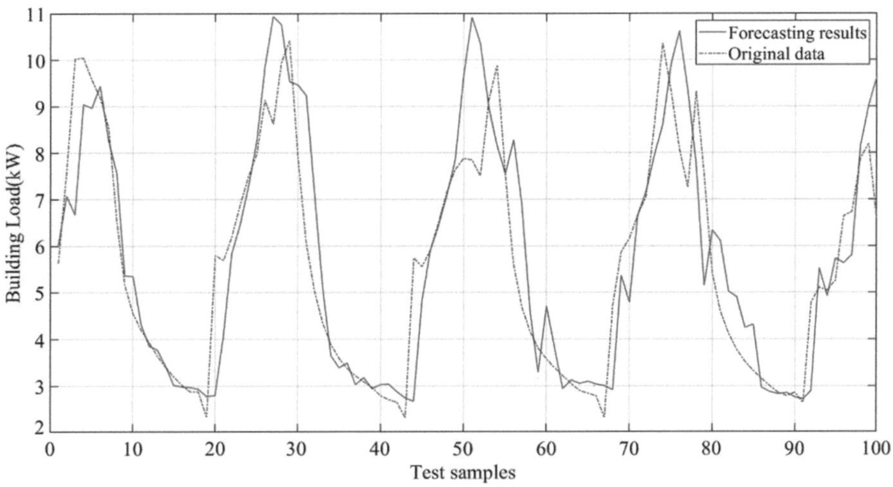

Fig. 4.24 Forecasting results of 2-step strategy

4.6 Big Data Prediction of Energy Consumption in Urban Building

With the rapid development of sensor technology and the urbanization process, the energy consumption in the urban buildings increases remarkably and represents a large part of the energy consumption worldwide (Amasyali and El-Gohary 2018). With efficient sensors and simulation software like DeST, large amounts of data can be accumulated in almost every aspect of modern buildings. The data contain such

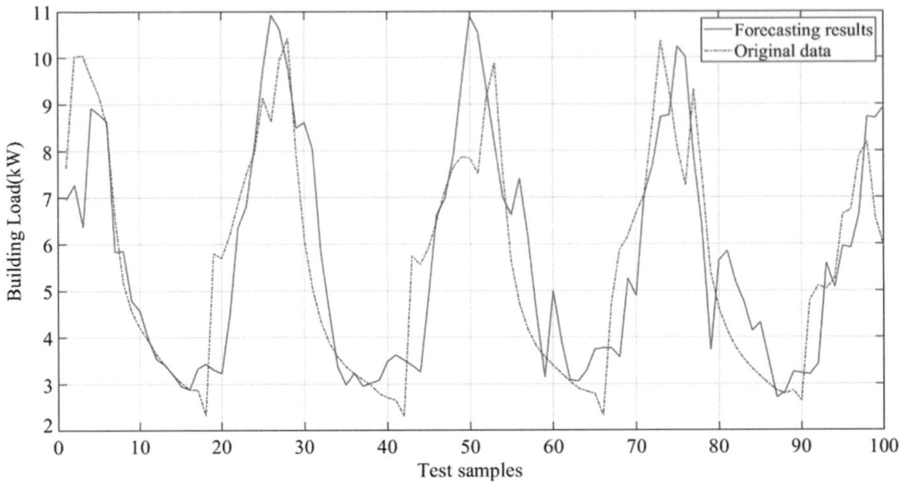

Fig. 4.25 Forecasting results of 3-step strategy

Table 4.11 The forecasting performance indices of the SVM model

Model	Step	MAE (kW)	MAPE (%)	RMSE (kW)	SDE (kW)	*R*
SVM	1	0.4793	7.8528	0.7347	0.7296	0.9541
	2	0.8371	14.4586	1.2301	1.2208	0.8787
	3	1.0486	19.8617	1.3775	1.3725	0.8334

wide range of building parameter data types that there are various useful information from the original data, and that means, with a large number of collected data the prediction of energy consumption in building will play an important role in the utilization process through the data-driven design and management mode.

However, the volume of data in urban buildings is growing fast with increasingly complex structures and forms, which brings more difficulties than ever to realize accurate predictions of energy consumption. Specifically the amount of the data is extremely huge, as shown in Sect. 4.3. Just the meteorological data, thermal perturbation data, and enclosure structure data take a large part of the analysis works. Comparing those with buildings in different locations even different countries, the data will be enormous. So it is facing the unavoidable challenges that how to effectively collect, store, and manage the data and how to efficiently analyze and mine the data. Only by the breakthrough of technology users are able to find the solution to support more effective and efficient decision of building design and prevent risks and troubles in the process.

Scientific research and industrial production of big data-driven by technology have developed rapidly in recent years. The big data have created opportunities to develop new methods to analyze energy consumption, improve energy-saving potential, and increase energy efficiency (Grolinger et al. 2016). With the function like energy control, data collection, and analysis, big data technology plays important role to support the design and operation of smart buildings in smart cities.

4.6.1 Big Data Framework for Energy Consumption

In this chapter, the Hadoop big data platform is chosen to build SVM prediction model of energy consumption in smart cities. Big data is too huge and complex to process with traditional database system tools (Lee et al. 2014), so it requires the suitable ones which are able to handle huge amount of data. Hadoop is the combination of Google GFS and MapReduce for storing and processing huge amounts of data (Dede et al. 2013). Hadoop framework contains a lot of components and one of the best components is the HDFS, which contains a quantity of data (Idar et al. 2018). That precisely fulfills the requirement of building information, which represents various forms and quantities.

MapReduce separates programs and it is also core framework to develop Hadoop-based data analysis. A simple method to improve SVM training efficiency is by separating training data and evaluating respective MapReduce functions. For the Hadoop MapReduce, the data separating strategy can be finished by the number of MapReduce tasks (Caruana et al. 2011). Figure 4.26 shows the framework of Hadoop-SVM model.

4.6.2 Big Data Storage and Analysis for Energy Consumption

HDFS (Hadoop Distributed File System) is developed by Java for Hadoop to store huge amount of files in the mode of streaming data access and runs on commercial hardware cluster. HDFS uses the NameNode server to set metadata and DataNodes servers to set application data. HDFS is aiming to store huge amounts of data safely and to send the data to user requirements (Kiran et al. 2013).

Hive is a database, which is deployed on a Hadoop cluster. Its data are stored on HDFS. The table built by Hive corresponds to a folder on HDFS. It can store huge sum of data and analyze stored data. It can also handle a huge sum of data with Hadoop MapReduce and take Zookeeper as its collaborative service.

HDFS provides user with a robust, reliable, and convenient file system. It can detect and quickly respond to hardware errors with simplified models and high fault tolerance to run on clusters of commercial hardware.

4.6.3 Big Data Mining for Energy Consumption

The data mining and modeling platforms for Hadoop are CDH, HDP, TipDM-HB, etc.

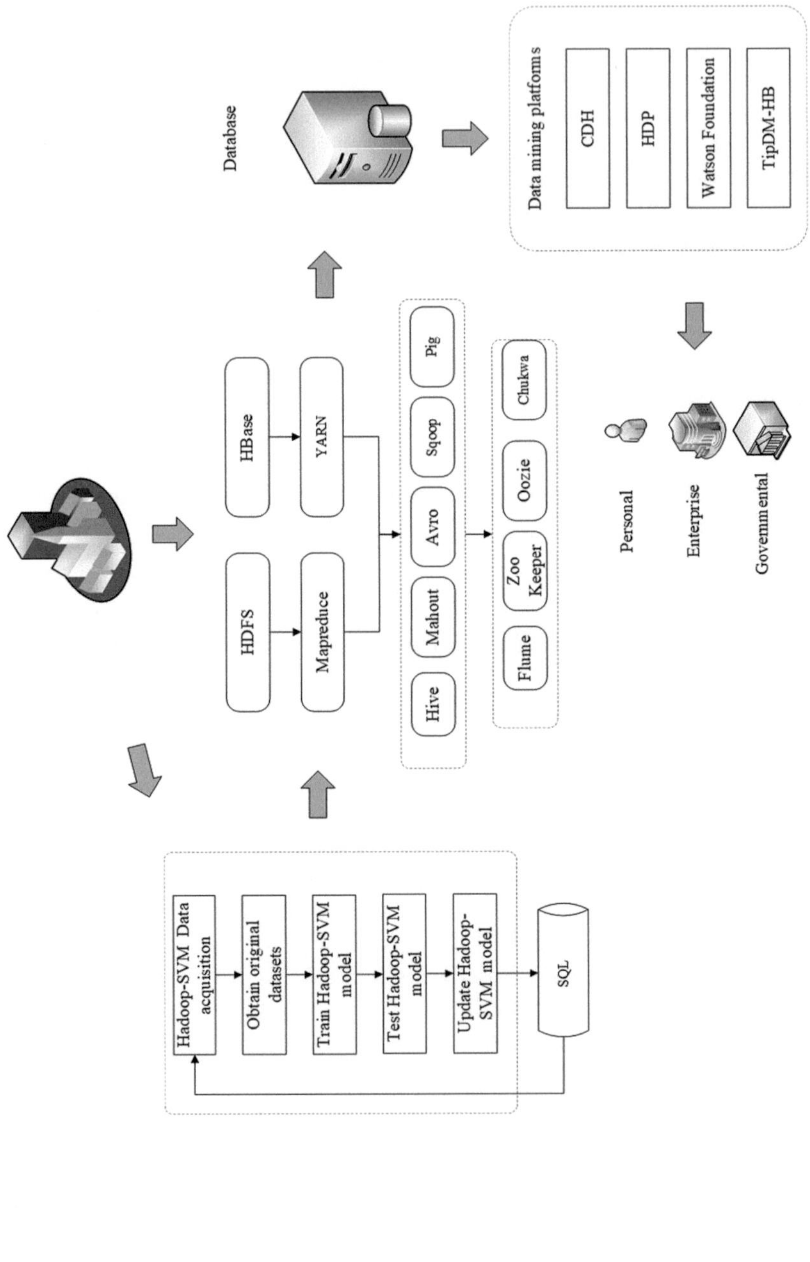

Fig. 4.26 The framework of Hadoop-SVM model

CDH (Cloudera's Distribution Including Apache Hadoop) is the open resource project of Hadoop from Cloudera company. It has integrated SQL, coactive search, and individual access control.

HDP (Horton Works Data Platform) is a big data platform from Horton Company. It is based on YARN (Yet Another Resource Negotiator) and HDFS with total open resources.

The platform TipDM-HB is a set of customizable data mining modeling platforms based on the Hadoop structure. Through the application model of enterprise data mining, data application development is faster with lower costs to make enterprise data mining easier. The common algorithms include Naive Bayes, Random Forests, k-nearest neighbor, etc.

After exploration and pre-processing, the data can be directly modeled. According to the mining target and data form, the results are models of classification, prediction, association rules, intelligent recommendation, and other models. Finally, the results will be output through the platforms.

4.7 Conclusion

In this chapter, the DeST software is used to establish the model of a 3-floor building for energy consumption research. DeST combines the characters of the actual design process in the development process and provides various analysis results for different stages of building design. The data of cold load from July to August and meteorological parameters, enclosure structure, indoor thermal perturbation, and indoor design parameters have been chosen for the analysis and comparison. At last the effect of indoor thermal perturbation is far more significant to others.

After the steps, the simulation and calculation process run to achieve a large amount of data, which are related to building, system, and environment. SVM is the forecasting model in this chapter, it takes three parameters with the strongest correlation with the energy consumption after cross-correlation analysis to set the modeling and obtains the conclusions.

The forecasting accuracy of 1-step strategy is better than that of 2-step and 3-step strategies and the SVM model has good forecasting effect on test samples. But the strategy still has potential to be improved by integrating other methods. More experiments should be implemented in the future to increase accuracy.

At last, for the huge amount of data with various structures the big data technology could create opportunities to develop new methods to analyze energy consumption and improve energy-saving potential. The big data platform of Hadoop helps to build the SVM forecasting model of energy consumption in smart cities with storage, analysis, and data mining.

References

Amasyali K, El-Gohary NM (2018) A review of data-driven building energy consumption prediction studies. Renewable Sustainable Energy Reviews 81:1192–1205

Barabási A-L (2007) The architecture of complexity. IEEE Control Systems Magazine 27(4):33–42

Battista G, Evangelisti L, Guattari C, Basilicata C, de Lieto Vollaro R (2014) Buildings energy efficiency: Interventions analysis under a smart cities approach. Sustainability 6(8):4694–4705

Brown MP, Grundy WN, Lin D, Cristianini N, Sugnet CW, Furey TS et al (2000) Knowledge-based analysis of microarray gene expression data by using support vector machines. Proceedings of the National Academy of Sciences 97(1):262–267

Cai Y, Liu X, Xu X, Zhou G (2001) Support vector machines for predicting protein structural class. BMC Bioinformatics 2(1):3

Cao X, Dai X, Liu J (2016) Building energy-consumption status worldwide and the state-of-the-art technologies for zero-energy buildings during the past decade. Energy Buildings 128:198–213

Caruana, G., Li, M., & Qi, M. (2011). A MapReduce based parallel SVM for large scale spam filtering. In *2011 eighth international conference on fuzzy systems and knowledge discovery (fskd)* (Vol. 4, pp. 2659–2662). IEEE.

Dede, E., Govindaraju, M., Gunter, D., Canon, R. S., & Ramakrishnan, L. (2013). Performance evaluation of a MongoDB and Hadoop platform for scientific data analysis. In *Proceedings of the 4th ACM workshop on Scientific cloud computing* (pp. 13–20). ACM.

Dong B, Cao C, Lee SE (2005) Applying support vector machines to predict building energy consumption in tropical region. Energy Buildings 37(5):545–553

Feng X, Yan D, Hong T (2015) Simulation of occupancy in buildings. Energy Buildings 87:348–359

Gan G (2010) Interaction between wind and buoyancy effects in natural ventilation of buildings. The Open Construction Building Technology Journal 4(1)

Grolinger, K., Capretz, M. A., & Seewald, L. (2016). Energy consumption prediction with big data: Balancing prediction accuracy and computational resources. In *2016 IEEE International Congress on Big Data (BigData Congress)* (pp. 157–164). IEEE.

Hu J, Wu J (2015) Analysis on the influence of building envelope to public buildings energy consumption based on DeST simulation. Procedia Engineering 121:1620–1627

Huang W, Nakamori Y, Wang S (2005) Forecasting stock market movement direction with support vector machine. Computers Operations Research 32(10):2513–2522

Hui, Z., Hao, X., Wei, W., & Jie, L. S. (2012). Research on natural ventilation energy-saving design of traditional residential building. In *International Conference on Green Buildings and Optimization Design (GBOD 2012)*. ASME Press.

Idar, H. A., Aissaoui, K., Belhadaoui, H., & Hilali, R. F. (2018). Dynamic data sensitivity access control in Hadoop platform. In *2018 IEEE 5th International Congress on Information Science and Technology (CiSt)* (pp. 105–109). IEEE.

Joachims, T. (1998). Text categorization with support vector machines: Learning with many relevant features. In *European conference on machine learning* (pp. 137–142). Springer.

Joshi S, Saxena S, Godbole T (2016) Developing smart cities: An integrated framework. Procedia Computer Science 93:902–909

Kiran M, Kumar A, Mukherjee S, Prakash RG (2013) Verification and validation of MapReduce program model for parallel support vector machine algorithm on Hadoop cluster. International Journal of Computer Science Issues 10(3):317

Lee, W., On, B.-W., Lee, I., & Choi, J. (2014). A big data management system for energy consumption prediction models. In *Ninth International Conference on Digital Information Management (ICDIM 2014)* (pp. 156–161). IEEE.

Liu H, Tian H, Li Y, Zhang L (2015) Comparison of four AdaBoost algorithm based artificial neural networks in wind speed predictions. Energy Conversion Management 92:67–81

Ma Z, Ye C, Li H, Ma W (2018) Applying support vector machines to predict building energy consumption in China. Energy Procedia 152:780–786

Nam, T., & Pardo, T. A. (2011). Conceptualizing smart city with dimensions of technology, people, and institutions. In *Proceedings of the 12th annual international digital government research conference: Digital government innovation in challenging times* (pp. 282–291). ACM.

Niu D, Wang Y, Wu DD (2010) Power load forecasting using support vector machine and ant colony optimization. Expert Systems with Applications 37(3):2531–2539

Qiang G, Zhe T, Yan D, Neng Z (2015) An improved office building cooling load prediction model based on multivariable linear regression. Energy Buildings 107:445–455

Shaikh PH, Nor NBM, Nallagownden P, Elamvazuthi I, Ibrahim T (2014) A review on optimized control systems for building energy and comfort management of smart sustainable buildings. Renewable Sustainable Energy Reviews 34:409–429

Shang Y, Hei P, Lu S, Shang L, Li X, Wei Y et al (2018) China's energy-water nexus: Assessing water conservation synergies of the total coal consumption cap strategy until 2050. Applied Energy 210:643–660

Wang, Q. K., Zuo, W. W., & Xiao, Y. P. (2014). The development potential and countermeasures of China's urbanization based on energy security. In *Advanced Materials Research* (Vol. 971, pp. 2223–2226). Trans Tech Publ.

Xin Z, Hong T, Da Y (2014) Comparison of HVAC system modeling in EnergyPlus, DeST and DOE-2.1E. Building Simulation 7(1):21–33

Yan D, Jiang Y (2005) An overview of an integrated building simulation tool-designer's simulation toolkit (DeST). Building Simulation:1393–1400

Yan D, Xia J, Tang W, Song F, Zhang X, Jiang Y (2008) DeST—An integrated building simulation toolkit Part I: Fundamentals. Building Simulation 1(2):95–110

Zhang X, Xia J, Jiang Z, Huang J, Qin R, Zhang Y et al (2008) DeST—An integrated building simulation toolkit Part II: Applications. Building Simulation 1(3):193–209

Zhou N, Zhao S (2013) Urbanization process and induced environmental geological hazards in China. Natural Hazards 67(2):797–810

Zhu, Y., & Jiang, Y. (2003). DeST-A simulation tool in HVAC commissioning. In *Document prepared for IEA BCS Annex 40 Workshop* (Vol. 8).

Part III
Smart Traffic Systems

Smart city refers to using various information technologies or innovative concepts to open up and integrate urban systems and services, improving the efficiency of resource utilization, optimizing urban management and services, and improving the quality of life of citizens. Traffic construction is the first step in urban construction, transportation is the artery of urban economic development, and intelligent transportation system is an important part of smart city construction. The intelligent transportation system can alleviate traffic congestion, improve urban traffic conditions, maximize the efficiency of urban traffic, and establish a new-generation comprehensive traffic operation coordination system that coordinates the operation of people, vehicles, roads, and environments, and realizes the overall operational efficiency of urban transportation systems. In short, the intelligent transportation system plays a very important role in the construction of smart city.

Smart transportation system takes advantage of the Internet of Things, cloud computing, Internet, artificial intelligence, automatic control, and other technologies in the transportation field. It comprehensively applies theories of traffic science, system methods, artificial intelligence, and data mining to analyze the mechanism of traffic information. Then, a real-time dynamic information service system can be built, which can direct traffic management, transportation, public travel, other aspects of the transportation sector, and the entire process of traffic construction management. Smart transportation system can make the region and even the whole city have the abilities of perception, interconnection, analysis, prediction, and control, so that it can fully guarantee traffic safety and improve the efficiency and management of the transportation system.

With the rapid improvement of economic and the acceleration of transportation infrastructure construction, the contradiction between transportation supply and demand has become increasingly prominent. Traffic congestion has become a bottleneck restricting urban social and economic development. Sensing the change of traffic flow in advance and predicting the driving trajectory of the vehicle are an important part of smart transportation system construction. It can help people understand the traffic status of urban roads in real-time and promptly induce the people to reasonably choose the travel route to reduce the current situation of congestion. This part studies traffic flow prediction and vehicle trajectory prediction and establishes traffic flow prediction model and vehicle trajectory prediction model, respectively. Chapter 5 studies the vehicle trajectory and establishes the vehicle trajectory prediction model. Chapters 6 and 7 study the traffic flow prediction model separately. Chapter 6 focuses on single-data-driven traffic flow prediction and establishes deterministic prediction models and interval prediction models for traffic flows. Chapter 7 focuses on multi-dimensional data-driven traffic flow prediction, establishes the spatiotemporal relationship of traffic flow in different sections, and establishes the traffic flow prediction model.

Chapter 5
Characteristics and Analysis of Urban Traffic Flow in Smart Traffic Systems

Abstract Traffic congestion has always been a problem that is difficult to solve in modern cities. Traffic trajectory prediction is one of the effective means to solve traffic congestion. The problem studied in this chapter is to establish a vehicle trajectory prediction model based on vehicle historical trajectory data. The research idea is to use machine learning algorithms and neural network algorithms to predict future trajectories after pre-processing the vehicle trajectory data. A decomposition algorithm is then used to improve the performance of the prediction model. The methods used include the ELM, BPNN, and EWT. The trajectory data of three different cars during the day is selected for prediction experiments. The experimental results show that the ELM model and the BPNN model have good prediction accuracy and practical useful value in the dataset selected in this chapter. The EWT method can effectively improve the prediction accuracy of individual models.

5.1 Introduction

5.1.1 Overview of Trajectory Prediction of Smart Vehicle

With the development of the Internet and the continuous improvement of information intellectualization, the wide application of global positioning system (GPS), radio frequency identification (BPID), and mobile intelligent devices, the geographic location information of mobile objects is becoming more and more convenient (Xie et al. 2018). Huge amounts of track data can be obtained by collecting signals generated by mobile devices. Rich knowledge is contained in the track data. The action route of a single-source track reflects the individual's living habits and travel characteristics, while the multi-source track of urban groups reflects the overall travel characteristics of the city (Wang et al. 2019). Therefore, by studying trajectory data and mining human activity patterns, it can understand the relationship

H. Liu, *Smart Cities: Big Data Prediction Methods and Applications*,
https://doi.org/10.1007/978-981-15-2837-8_5

between people, people, and the environment, understand urban behavior, grasp the pulse of urban traffic, and build a smarter city (Woo et al. 2017). At present, trajectory data mining has been widely applied in various fields of cities, including user's trajectory prediction, user's dynamic distribution, urban traffic simulation and monitoring, urban functional unit identification, and other fields (Xia et al. 2017). Through track data mining, the hidden content is discovered, the trend of urban traffic development is explored, and major social problems such as urban environment, urban planning, and urban congestion are solved (Kai et al. 2018).

Vehicle trajectory prediction refers to the use of historical trajectory data of vehicles, in the case of unknown destination, a certain prediction model is adopted to give the position of vehicles at the next moment or at several consecutive moments. Trajectory data prediction can alleviate traffic congestion in large cities that both developed and developing countries face today (Zhu et al. 2018). Countries have adopted a variety of measures to ease congestion. Studies have shown that traffic congestion can be reduced effectively by providing drivers with timely and accurate traffic information, which can be used to choose other routes to reach their destinations or change the travel time to avoid congestion. Intelligent navigation systems that can predict traffic jams could potentially help relieve pressure on sections of the road, directing drivers to alternative routes by predicting possible future congestion spots. Modern intelligent navigation systems strive to capture real-time traffic conditions so that they can be used to predict congestion. However, the existing navigation system is still short of prediction function. If technology can predict the next location and destination of vehicles currently on the road, it can predict the coming rush hour.

The research problem faced by this chapter is to establish a vehicle trajectory prediction model based on vehicle historical trajectory data. The research idea is to use a machine learning algorithm and neural network algorithm to predict the future trajectory after pre-processing vehicle trajectory data. Then the decomposition algorithm is used to improve the performance of the prediction model.

5.1.2 The Significance of Trajectory Prediction for Smart City

With the increasing urban car ownership in China, the problem of traffic congestion in urban areas is becoming more and more obvious. How to effectively alleviate urban traffic congestion has become an urgent problem to be solved. Studies show that during rush hours, drivers can choose other routes or travel times in advance if they are informed of traffic conditions in urban areas, thus avoiding possible traffic jams. Vehicle trajectory prediction is the core of regional congestion assessment and the basis of urban traffic congestion (Wang et al. 2016). This chapter studies the historical track data of urban vehicles accumulated in the project, and gives the track prediction of future periods of vehicles through the historical track data information of vehicles, providing basic support for traffic congestion assessment, congestion status prediction, travel planning, and other fields in urban areas.

Meanwhile, with the rapid development of mobile Internet technology, global positioning technology, Internet of things technology, and the gradual popularization of intelligent wearable devices, a large amount of track data are generated (Xie et al. 2018). Trajectory data include traffic trajectory data, human movement data, animal migration data, and trajectory data generated by other movable objects. Typically, trajectory data are characterized by timing, data sampling frequency, and uneven quality (Zhu et al. 2018). By processing the trajectory data and extracting the features, the activity rules of related objects can be mined. For example, the analysis and mining of traffic trajectory data can predict the location and regional distribution of vehicles in future periods. Through the analysis of human activity trajectories, individual behavior patterns, social relations, and activity rules can be mined (Wang et al. 2016).

Trajectory prediction is also one of the key technologies for intelligent vehicles (Granda et al. 2018). Trajectory planning requires an accurate risk assessment of the planning state to ensure the safety of intelligent vehicles in traffic environments. Therefore, it is necessary to make a reasonable prediction of the vehicle trajectory in trajectory planning. As an important part of smart city in the future, smart car makes accurate and effective vehicle trajectory prediction technology become very important.

The research of this chapter is on the practical application of trajectory data for further excavations, making more effective use of trajectory data collected. The trajectory data mining results can provide reference for urban traffic planning and development and help users avoid traffic hot spots in the travel, which have practical value and practical significance.

5.1.3 Overall Framework of Model

In this chapter, the machine learning method and neural network method are used to predict the vehicle trajectory. The research idea is to use the extreme learning machine (ELM) in machine learning algorithm and back propagation neural network (BPNN) in neural network algorithm to predict the future trajectory after preprocessing vehicle trajectory data. Then the decomposition algorithm is used to improve the performance of the prediction model.

The ELM is a kind of machine learning method based on the feedforward neural network. It is suitable for supervised learning and unsupervised learning (Huang et al. 2006). The traditional ELM has a single hidden layer. In comparison with other shallow learning systems, such as single-layer perceptron and support vector machines, it is considered that there may be advantages in learning rate and generalization ability (Huang et al. 2012). The BPNN is one of the used widely feed forward network. But compared to traditional feed forward networks, BPNN adds back propagation algorithm to the structure of feed forward network (Hecht-Nielsen 1992). The above two algorithm models will be used as the basic predictors for the trajectory prediction experiment in this chapter. Single data prediction and multiple data predictions are carried out, respectively. Then the EWT algorithm and two basic predictors are combined to explore the prediction effect of the ELM model and BPNN model under the decomposition framework (Hui et al. 2018). The framework of this chapter can be described below and it is shown in Fig. 5.1.

(a) Dataset: In this chapter, three vehicle trajectory datasets are selected. Each dataset is divided into training set and test set according to the method described in Sect. 5.2.2.
(b) Based-predictor: In this chapter, the ELM model and BPNN model are selected as based-predictors. Based-predictors are the core components of the model framework, which are utilized for training and testing.

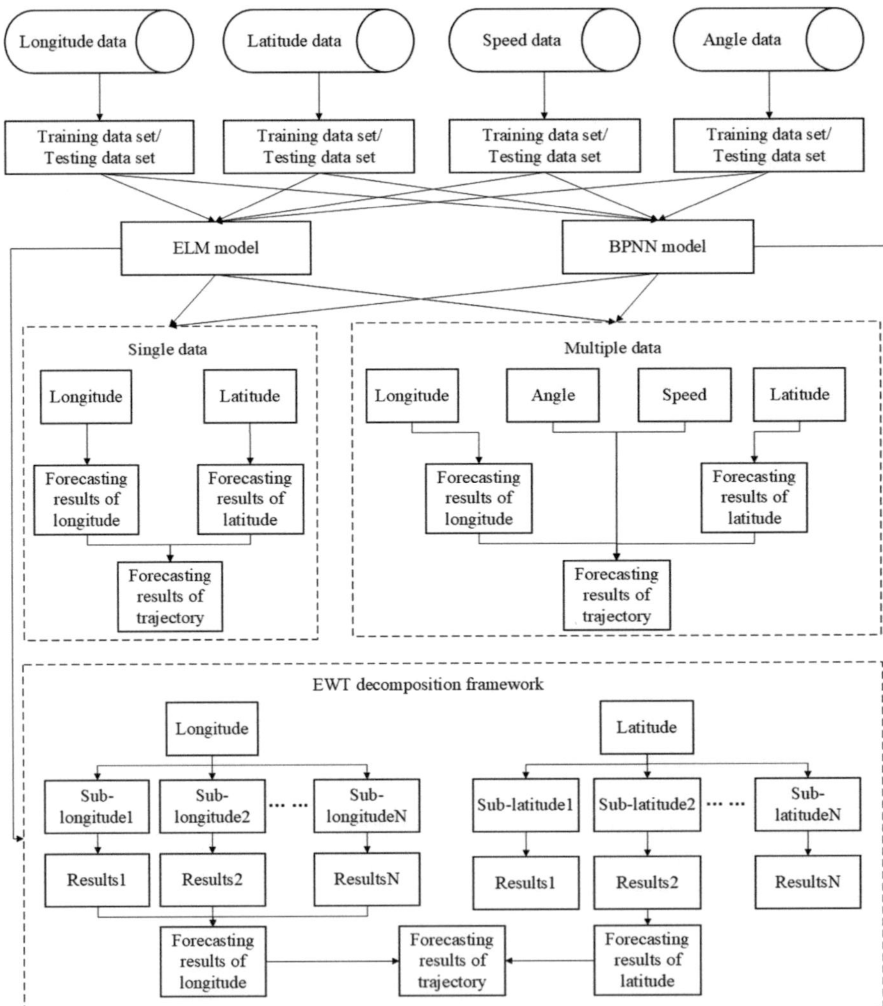

Fig. 5.1 The general framework of the chapter

(c) Decomposition algorithm: In this part, the EWT algorithm is used to preprocess input data. The EWT decomposition algorithm can decompose the complex original time series into a series of more stable sub-sequences, thus reducing the training difficulty of the model. Then, the experimental forecasting results are compared and analyzed to explore the difference in the prediction performance of different models. This part is the summary and sublimation of the prediction model.

5.2 Traffic Flow Time Distribution Characteristics and Analysis

5.2.1 Original Vehicle Trajectory Series

In the chapter, three sets of vehicle trajectory datasets are selected for vehicle trajectory prediction experiments. Each dataset contains four types of data, which is namely longitude data, dimension data, velocity data, and angle data. Moreover, the velocity data here refers to the instantaneous velocity of the car at the corresponding moment. The angle data is the angle from the north in clockwise direction with a unit of 2 degrees, for example, "157" would be 314 degrees from the north in clockwise direction. The sampling interval of each dataset is 1 min, and the missing sample points are interpolated by the method of cubic spline interpolation. In the end, each dataset contains 1440 sample points. Figures 5.2, 5.3, and 5.4 show the data

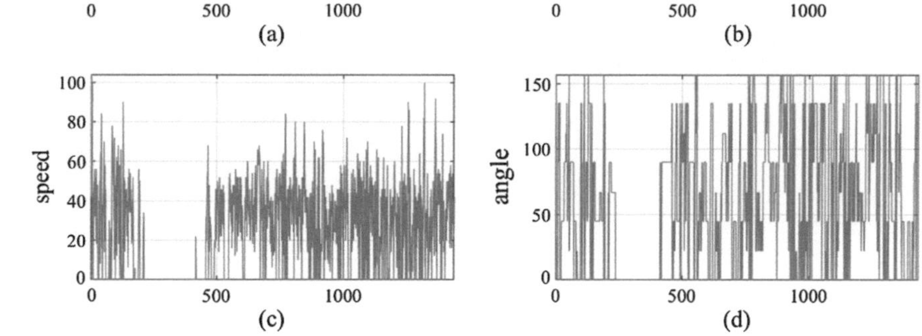

Fig. 5.2 D1: (**a**) Original longitude series; (**b**) Original latitude series; (**c**) Original speed series; (**d**) Original angle series

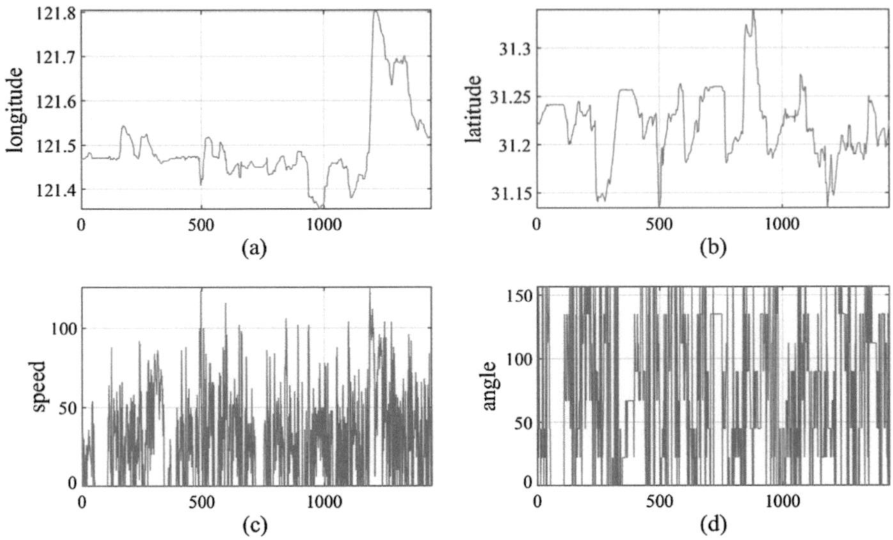

Fig. 5.3 D2: (**a**) Original longitude series; (**b**) Original latitude series; (**c**) Original speed series; (**d**) Original angle series

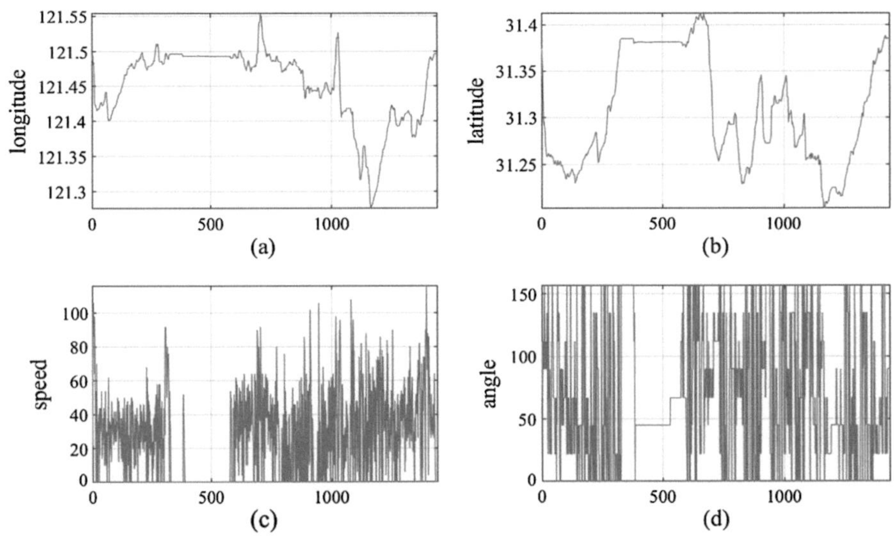

Fig. 5.4 D3: (**a**) Original longitude series; (**b**) Original latitude series; (**c**) Original speed series; (**d**) Original angle series

distribution of the three datasets. From Figs. 5.2, 5.3, and 5.4, it can be seen that the longitude and latitude data of the three sets of data are all of strong non-stationarity, which brings great challenges to subsequent model prediction experiments. Moreover, the cars corresponding to each set of datasets contain the time period for parking, thus increasing the complexity of the dataset. For convince, the first dataset is named as D1, the second dataset is named as D2, and the third dataset is named as D3.

In order to make a more comprehensive analysis of the selected vehicle trajectory datasets, Table 5.1 shows the respective statistical index values of the three datasets, which contain minimum, maximum, mean value, standard deviation, skewness, and kurtosis. Among those, skewness is the third-order ratio of central moment and standard deviation of time series, which reflects the deviation of time series data distribution relative to symmetric distribution. Skewness value is negative, its distribution is skewed to the left when skewness value is positive, its distribution is skewed to the right, and when skewness value is 0, its distribution is symmetrical. Kurtosis is the ratio of the fourth-order central moment of time series to the fourth power of standard deviation, which reflects the outliers of data. The kurtosis value adopted is the kurtosis value relative to the standard normal distribution. When the kurtosis value is positive, it indicates that the dispersion degree of the distribution is greater than the standard normal distribution.

5.2.2 *Separation of Sample*

In order to construct the models and evaluate the models' performance, the vehicle trajectory datasets selected for simulation need to be grouped. The three datasets in this chapter all contain 1440 sample points and the data are divided into a training dataset and testing dataset. The 1st–1200th sample points are used as training samples to train the prediction model, and the 1201st–1440th sample points are used as testing samples to test the obtained model after the model training is completed, so as to obtain the prediction output of the model, calculate the prediction error of the model, and evaluate the prediction performance of the model.

Table 5.1 Statistical characteristics of the original series

Original series	Data types	Minimum (°)	Maximum (°)	Mean value (°)	Standard deviation (°)	Skewness (°)	Kurtosis (°)
D1	Longitude	121.3845	121.5341	121.4623	0.0283	0.0127	3.5463
	Latitude	31.2246	31.3534	31.3117	0.0275	−1.3107	4.0810
D2	Longitude	121.3548	121.8040	121.4931	0.0865	1.7764	6.0784
	Latitude	31.1336	31.3400	31.2205	0.0351	0.2370	4.0174
D3	Longitude	121.2750	121.4523	121.4523	0.0528	−1.1778	3.9365
	Latitude	31.2030	31.4126	31.3072	0.0595	0.2175	1.6145

5.3 The Spatial Distribution Characteristics and Analysis of Traffic Flow

5.3.1 Trajectory Prediction of Urban Vehicles Based on Single Data

5.3.1.1 Theoretical Basis of the ELM and BPNN

The Theoretical Basis of the ELM

Extreme learning machine (ELM) is a generalized single-hidden layer feedforward neural network (Huang et al. 2004, 2006), which has fast learning speed, good generalization ability, and strong ability to search for global optimal solution (Dong et al. 2017). The extreme learning machine can randomly generate the connection weights of the input layer and the hidden layer and the bias of the hidden layer, and there is no need to adjust in the learning process (Huang and Chen 2007). Given N samples of $X_k = \{x_k, y_k\}$, $k = 1, 2, \ldots, N$, where x_k is the input data, y_k is the true value, activation function is $f(\cdot)$, and hidden layer nodes are n, the solution process of ELM is shown below (Dong et al. 2017).

$$o_j = \sum_{i=1}^{n} \lambda_i f\left(W_i \cdot X_k + b_i\right), \quad k = 1, 2, \ldots, N \tag{5.1}$$

where, o_j is the output value of the ELM, $W_i = \{\omega_{i1}, \omega_{i2}, \ldots, \omega_{in}\}'$ is the connection weight between the input node and the ith hidden layer node, b_i is the bias of the ith node of the hidden layer, λ_i is the connection weight of the ith hidden layer node and the output node.

Therefore, only the number of hidden layer nodes and the activation function of the extreme learning machine need to be set to train it.

The Theoretical Basis of the BPNN

The BP neural network is a classic forward feedback neural network, which fits nonlinear data according to error back propagation and has a good approximation ability of nonlinear data (Hecht-Nielsen 1992). This chapter will use the BP neural network as the basic predictor to predict trajectory. Figure 5.5 shows the basic structure of the BP neural network. It consists of three parts: the input layer, the hidden layer, and the output layer. The input layer is used to receive input data for trajectory prediction and to transmit the input data to the hidden layer. The hidden layer processes and computes the input data using weights

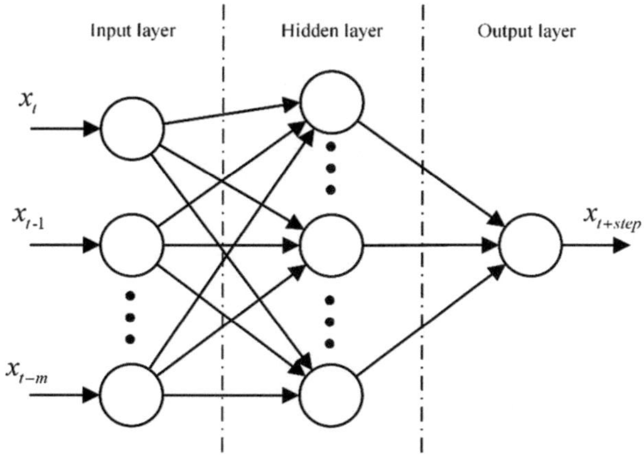

Fig. 5.5 The prediction principle of BP network

and threshold matrices. Finally, the output layer weighted sums the data calculated by the hidden layer and outputs the final result. The calculation process of the BP neural network used in the chapter can be expressed as follows (Zhang et al. 2018):

$$h_j = S\left(\sum_{i=1}^{n_{\text{input}}} w_{ij}x_i + b_j\right) \tag{5.2}$$

$$o = \sum_{j=1}^{n_{\text{hidden}}} w_j' h_j + b' \tag{5.3}$$

where, x_i is the input data of the ith node of the input layer; n_{input} is the number of nodes of the input layer; w_{ij} is the weight between the ith input layer node and the jth hidden layer node; b_j is the threshold of the jth hidden layer node; h_j is the output data of the jth hidden layer node; w_j' is the weight between the jth hidden layer node and the output layer node; b' is the threshold of the output layer node; n_{hidden} is the number of the hidden layer; o is the final output data of the output layer; $S(\cdot)$ is the activation function of the hidden layer. This chapter uses the sigmoid function as the activation function, it is defined as follows:

$$S(x) = \frac{1}{1+e^{-x}} \tag{5.4}$$

Since the output of this chapter is a predicted trajectory value, the number of output layer nodes of the BP neural network is 1.

5.3.1.2 Framework of Model

The model prediction method based on single data input is simple. The number of 10 historical track sample points are used as model input in this part of the experiment. The trajectory sample point at the next time is used as the output. Three datasets are predicted by the rolling prediction strategy. Figure 5.6 shows a block diagram of the method.

5.3.1.3 Steps of Modeling

The steps of using single data to predict trajectory data are as follows:

(a) The training sample set and test sample set are divided according to the method in Sect. 5.2.2. In this step, the training set and test set of the three trajectory datasets will be obtained, respectively;
(b) Setting up the necessary training parameters of the ELM model and BPNN model.
(c) Conducting model training of predictors. Ten historical track sample points are used as the input of the model. Take the trajectory sample points at a future time as output. The rolling prediction training is completed by using the training sample set obtained in the previous step;
(d) Testing model. The model test after training is carried out by using the test sample set. The rolling prediction strategy is also used for testing. Finally, track data for the last 4 h are obtained and compared.

5.3.1.4 Forecasting Results

Forecasting Results of the ELM Based on Single Data

The test sample set is input into the trained ELM model (single data) to get the predicted results. Figures 5.7, 5.8, and 5.9 show the prediction result curves of the ELM model (single data) for three test sample sets. The left two subgraphs of each graph

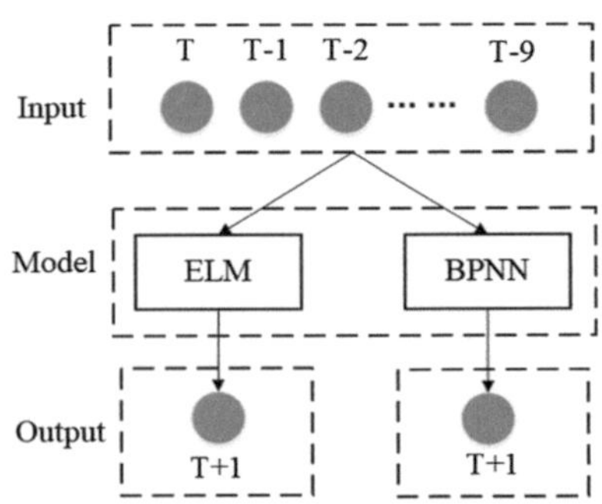

Fig. 5.6 The general framework of this section

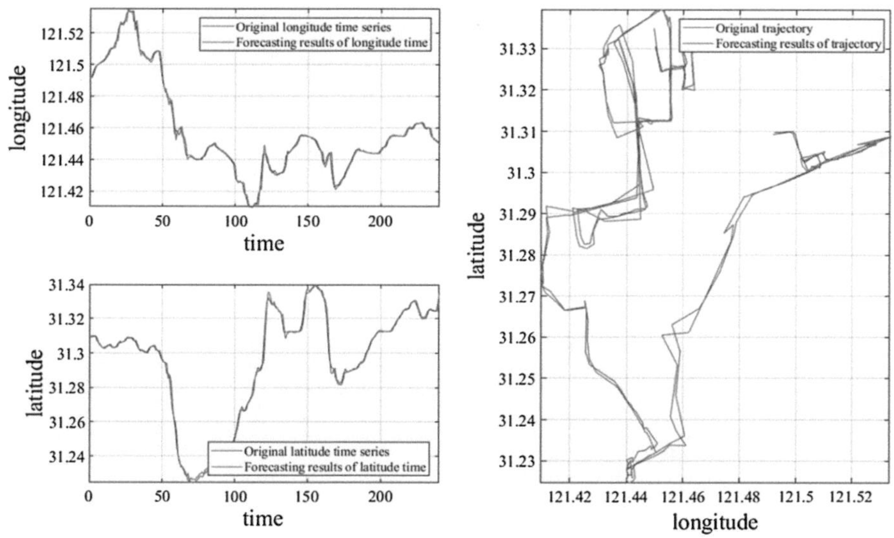

Fig. 5.7 Trajectory prediction results of the ELM model (single data) for D1

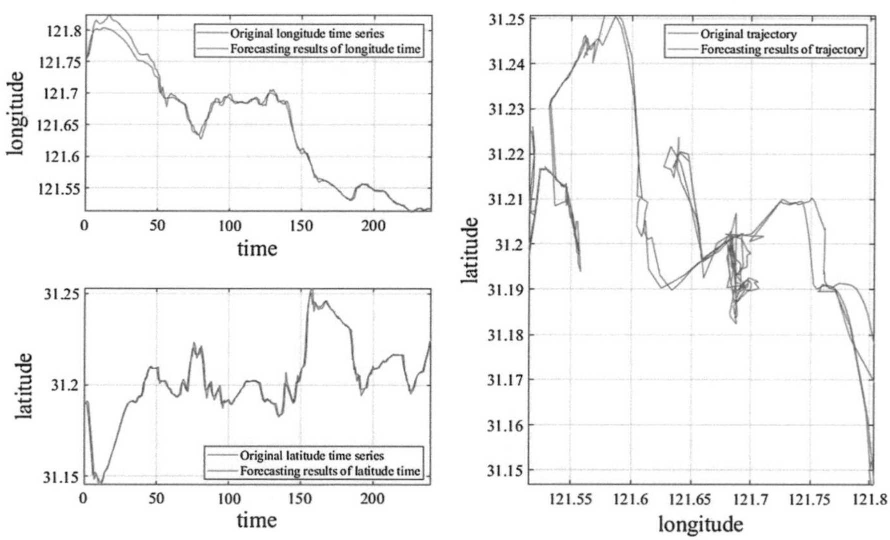

Fig. 5.8 Trajectory prediction results of the ELM model (single data) for D2

represent longitude prediction results and dimension prediction results, respectively, from top to bottom. The subgraph on the right represents the final trajectory prediction results. From the figures the following conclusions can be drawn:

(a) The ELM model (single data) has a good prediction effect on the test sets of three datasets. The fitting degree of both the main trend component and the peak

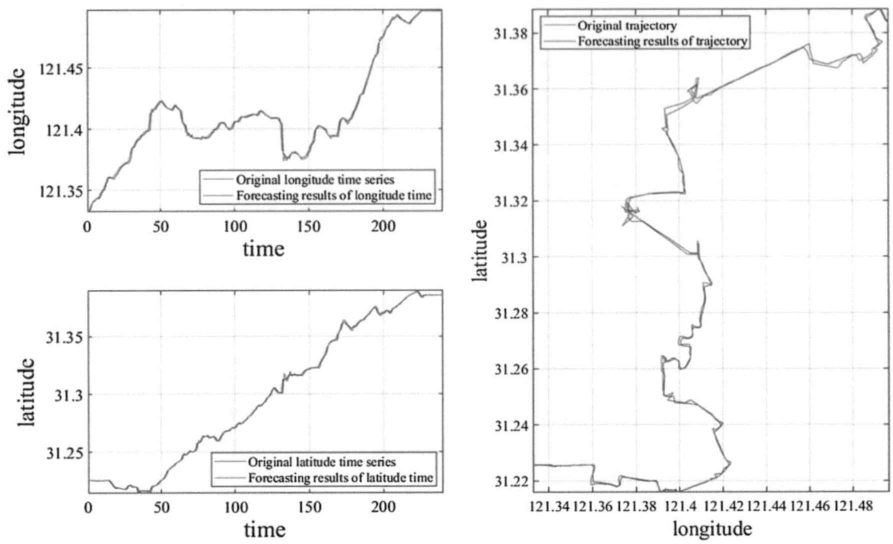

Fig. 5.9 Trajectory prediction results of the ELM model (single data) for D3

and trough of the wave is close to the real value of the test set. It can be concluded that the ELM model (single data) has a good effect on vehicle trajectory prediction.

(b) The prediction effect of longitude and dimension of each dataset is good, but the final trajectory prediction result has a larger prediction error than the longitude and latitude prediction. Because the strategy of separate prediction of latitude and longitude is adopted here, the final trajectory prediction results will show the product effect of errors. Thus, the geometric multiple of the error increases, leading to a significantly larger error.

(c) Graphically, the ELM model (single data) has the best prediction effect on D3, D1 comes second, D2 is the worst predictor. This is because there is a big prediction deviation at the beginning of the prediction data when the longitude of dataset D2 is predicted. Finally, it reflected in the trajectory prediction results leads to a serious prediction deviation. The best prediction effect of D3 may be caused by the relatively flat curve change of the dataset. D3's trajectory curve is not as complex as the other two datasets.

In order to better analyze the prediction performance of the ELM model (single data) for three datasets, Table 5.2 gives some statistical indices of the forecasting results of the ELM model (single data) for three datasets. According to Table 5.2, it can draw the following conclusions:

(a) The ELM model (single data) has the best prediction effect for D3. The MAE, MAPE, RMSE, and SDE of forecasting results for longitude data in D1 are 0.0015°, 0.0012%, 0.0023°, and 0.0023°, respectively. The MAE, MAPE, RMSE, and SDE of forecasting results for latitude data in D1 are 0.0013°,

Table 5.2 The forecasting performance indices of the ELM (single data) model

Model	Dataset	Data types	MAE (°)	MAPE (%) (°)	RMSE (°)	SDE (°)
ELM (single data)	D1	Longitude	0.0015	0.0012	0.0023	0.0023
		Latitude	0.0013	0.0041	0.0021	0.0021
	D2	Longitude	0.0143	0.0118	0.0188	0.0130
		Latitude	0.0017	0.0056	0.0027	0.0027
	D3	Longitude	0.0014	0.0011	0.0023	0.0023
		Latitude	0.0009	0.0030	0.0015	0.0015

0.0041%, 0.0021°, and 0.0021°, respectively. The MAE, MAPE, RMSE, and SDE of forecasting results for longitude data in D2 are 0.0143°, 0.0118%, 0.0188°, and 0.0130°, respectively. The MAE, MAPE, RMSE, and SDE of forecasting results for latitude data in D2 are 0.0017°, 0.0056%, 0.0027° and 0.0027°, respectively. The MAE, MAPE, RMSE, and SDE of forecasting results for longitude data in D3 are 0.0014°, 0.0011%, 0.0023°, and 0.0023°, respectively. The MAE, MAPE, RMSE, and SDE of forecasting results for latitude data in D3 are 0.0009°, 0.0030%, 0.0015°, and 0.0015°, respectively. The statistical indices of forecasting results of dataset D3 by the ELM model (single data) are far smaller than that of the other two datasets, which has the best prediction effect. This is consistent with the conclusion got from the figures, which reflects the rigor of the experiment.

(b) From the data in the tables, it can be seen that the ELM model (single data) has a better prediction effect on the test dataset D3 than on the test dataset D1 and D2. This is because dataset D3 compared to dataset D1 and D2, its data fluctuations are smoother so that the prediction accuracy of the ELM model (single data) is higher.

Forecasting Results of the BPNN Based on Single Data

The test sample set is input into the trained BPNN (single data) to get the predicted results. Figures 5.10, 5.11, and 5.12 show the prediction result curves of the BPNN (single data) model for three test sample sets. From the figures the following conclusions can be drawn:

(a) The prediction results of BPNN (single data) can well fit the motion of the vehicle trajectory data curve. BPNN has stable prediction accuracy for the three datasets selected in this experiment, and no extreme prediction results appear. The BP neural network has the potential of trajectory prediction in real life.

(b) The prediction results of the BPNN (single data) can reflect the overall fluctuation of the original test set. However, overfitting may occur in some details. As can be seen from the figures, when the vehicle trajectory changes are complex, the prediction error is large. The corresponding position is reflected in the

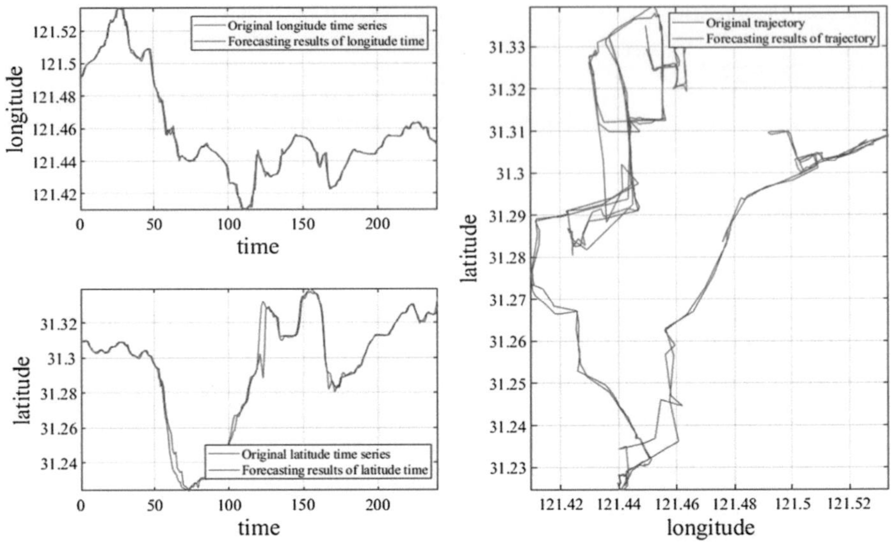

Fig. 5.10 Trajectory prediction results of the BPNN (single data) for D1

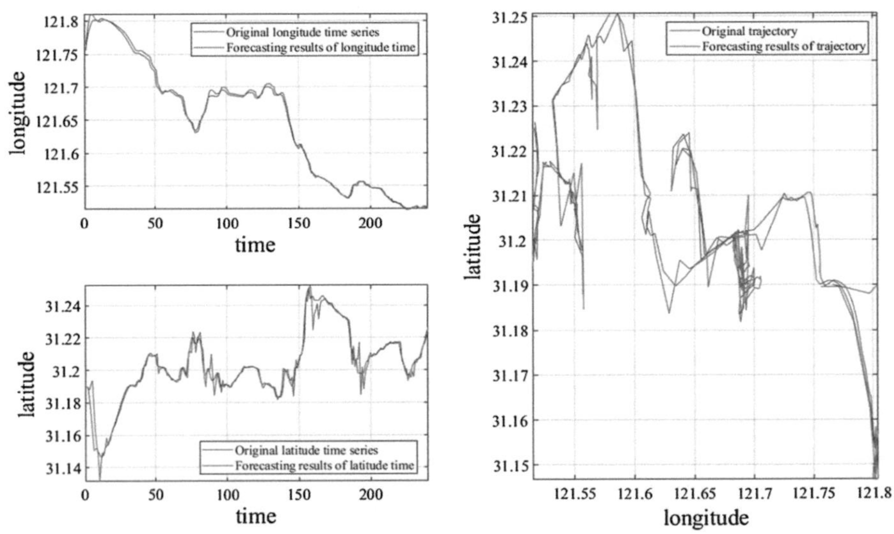

Fig. 5.11 Trajectory prediction results of the BPNN (single data) for D2

longitude prediction curve and latitude prediction curve, where the overfitting situation is intensified.

(c) By comparing the prediction results of three datasets, it can be seen that the BPNN (single data) has the highest prediction accuracy for dataset D3. Dataset D1 has the second-best prediction effect. The BPNN (single data) has the worst

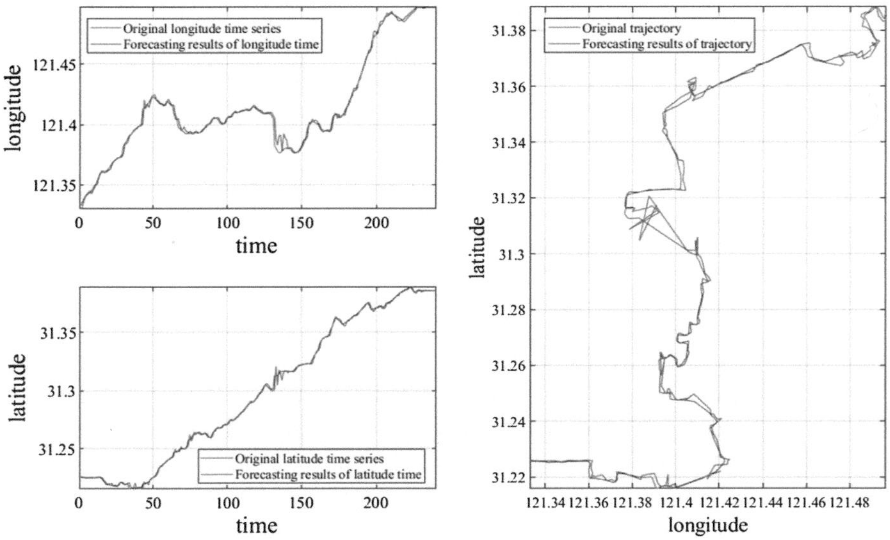

Fig. 5.12 Trajectory prediction results of the BPNN (single data) for D3

Table 5.3 The forecasting performance indices of the BPNN (single data) model

Model	Datasets	Data types	MAE (°)	MAPE (%)	RMSE (°)	SDE (°)
BP	D1	Longitude	0.0013	0.0011	0.0021	0.0021
		Latitude	0.0029	0.0094	0.0063	0.0063
	D2	Longitude	0.0041	0.0034	0.0081	0.0080
		Latitude	0.0023	0.0075	0.0036	0.0036
	D3	Longitude	0.0018	0.0015	0.0028	0.0028
		Latitude	0.0011	0.0034	0.0016	0.0016

prediction effect on dataset D2. The accuracy of different datasets is determined by the characteristics of the dataset. When the dataset changes dramatically and the non-stationarity is strong, the prediction accuracy of the BPNN (single data) is low. When the original dataset tends to be stable, the BPNN (single data) has a better prediction effect.

In order to better analyze the prediction performance of the BPNN (single data) for three datasets, Table 5.3 gives some statistical indices of the forecasting results of BPNN (single data) for three kinds of datasets. According to Table 5.3, it can draw the following conclusions:

(a) The BPNN (single data) has the best prediction effect for dataset D3. The MAE, MAPE, RMSE, and SDE of forecasting results for longitude data in D1 are 0.0013°, 0.001%, 0.0021°, and 0.0021°, respectively. The MAE, MAPE, RMSE, and SDE of forecasting results for latitude data in D1 are 0.0029°, 0.0094%, 0.0063°, and 0.0063°, respectively. The MAE, MAPE, RMSE, and

SDE of forecasting results for longitude data in D2 are 0.0041°, 0.0034%, 0.0081°, and 0.0080°, respectively. The MAE, MAPE, RMSE, and SDE of forecasting results for latitude data in D2 are 0.0023°, 0.0075%, 0.0036°, and 0.0036°, respectively. The MAE, MAPE, RMSE, and SDE of forecasting results for longitude data in D3 are 0.0018°, 0.0015%, 0.0028°, and 0.0028°, respectively. The MAE, MAPE, RMSE, and SDE of forecasting results for latitude data in D3 are 0.0011°, 0.0034%, 0.0015°, and 0.0016°, respectively. The statistical indices of forecasting results of dataset D3 by BPNN (single data) are far smaller than that of the other two datasets, which has the best prediction effect. This is consistent with the conclusion got from the figures, which reflects the rigor of the experiment.

(b) From the data in the tables, it can be seen that the BPNN (single data) has a better prediction effect on the test dataset D3 than on the test dataset D1. This is because dataset D3 compared to dataset D1, its data fluctuations are smoother so that the prediction accuracy of the BPNN (single data) is higher.

5.3.2 *Trajectory Prediction of Urban Vehicles Based on Multiple Data*

5.3.2.1 Framework of Model

The model based on multi-data input takes into account the input features of more dimensions. In this part, besides trajectory data, velocity data and angle data are also input data of the model. For example, the input data are composed of the trajectory data, velocity data, and angle data of three historical points. The output data are also trajectory data at future time points. In this way, the influence of multiple factors on vehicle track can be fully considered. Figure 5.13 shows a block diagram of the method.

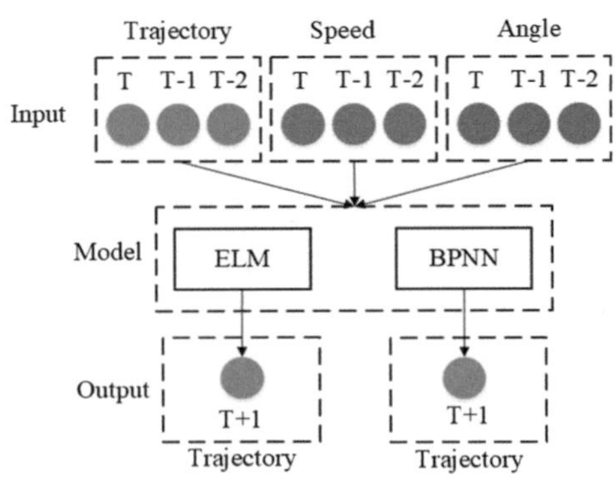

Fig. 5.13 The general framework of this section

5.3.2.2 Steps of Modeling

The steps for using multiple data to predict trajectory data are as follows:

(a) The training sample set and test sample set are divided according to the method in Sect. 5.2.2. In this step, the training set and test set of the three trajectory datasets will be obtained, respectively;
(b) Setting up the necessary training parameters of the ELM model and BPNN model.
(c) Conducting model training of predictors. Ten historical track sample points are used as the input of the model. Take the trajectory sample points at a future time as output. The rolling prediction training is completed by using the training sample set obtained in the previous step;
(d) Testing model. The model test after training is carried out by using the test sample set. The rolling prediction strategy is also used for testing. Finally, track data for the last 4 h are obtained and compared.

5.3.2.3 Forecasting Results

Forecasting Results of the ELM Based on Multiple Data

In the part, the longitude, latitude, speed, and angle of each dataset are used as input data of the training model, and the corresponding longitude and latitude are output data. Figures 5.14, 5.15, and 5.16 show the prediction result curves of the ELM

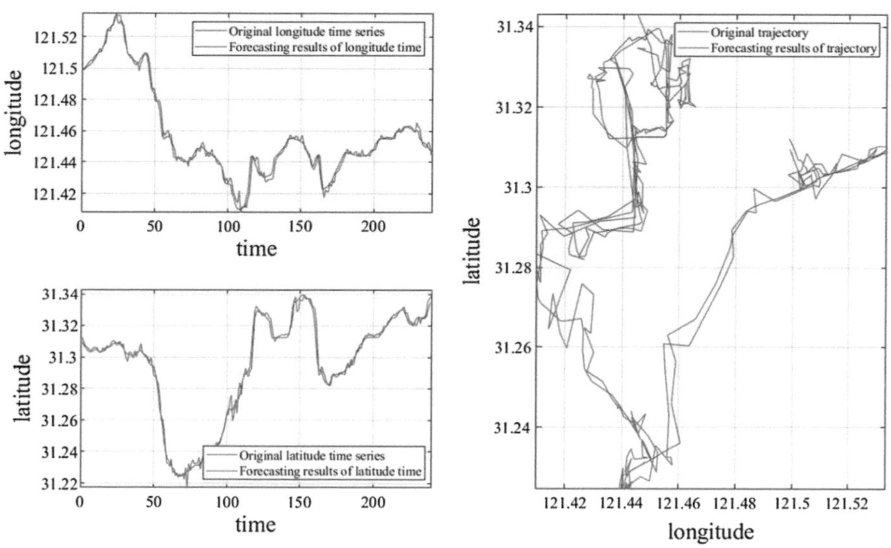

Fig. 5.14 Trajectory prediction results of the ELM model (multiple data) for D1

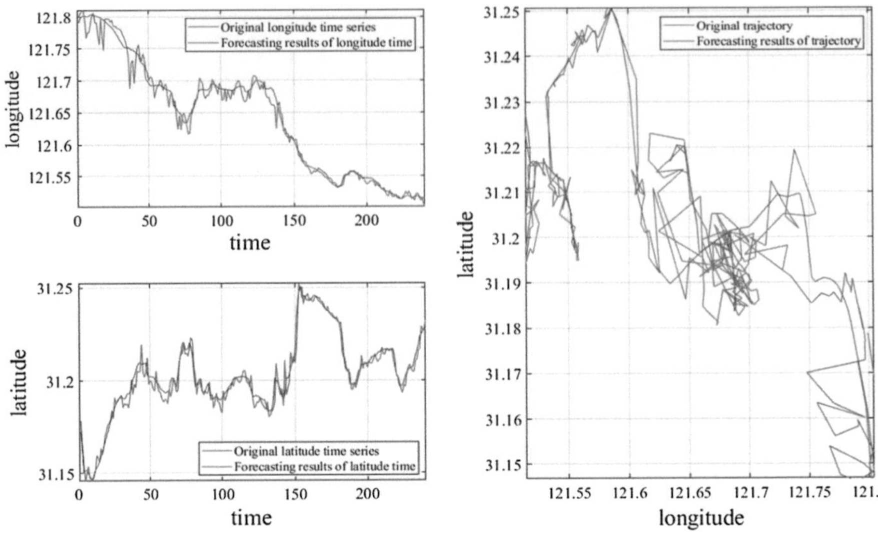

Fig. 5.15 Trajectory prediction results of the ELM model (multiple data) for D2

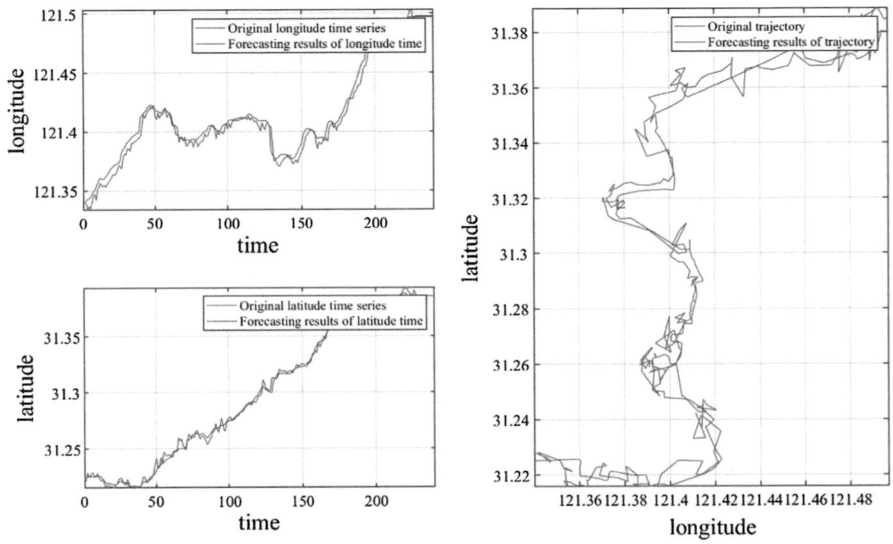

Fig. 5.16 Trajectory prediction results of the ELM model (multiple data) for D3

model (multiple data) for three test sample sets. From the figures it can draw the following conclusions:

(a) The ELM model (multiple data) has an average prediction effect. The forecast curve only reflects the original vehicle trajectory curve from the overall trend. Taking dataset D1 as an example, the prediction curve fluctuates continuously around the original longitude data and the original latitude data. It can also be seen that the final composite prediction trajectory curve has a large prediction error. The error distribution of datasets D2 and D3 is similar to D1, so the analysis will not be repeated here.

(b) The ELM model (multiple data) has the worst prediction effect on dataset D2. This may be due to the severe data fluctuation of dataset D2. In comparison, the ELM model (multiple data) is much better at predicting datasets D1 and D3 than D2. Among them, dataset D3 has the highest prediction accuracy

In order to better analyze the prediction performance of the ELM model (multiple data) for three kinds of noise datasets, Table 5.4 gives some statistical indices of the forecasting results of ELM model (multiple data) for three kinds of datasets. According to Table 5.4, it can draw the following conclusions:

ELM model (multiple data) has the best prediction effect for the trajectory dataset. The MAE, MAPE, RMSE, and SDE of forecasting results for longitude data in D1 are 0.0029°, 0.0024%, 0.0038°, and 0.0038°, respectively. The MAE, MAPE, RMSE, and SDE of forecasting results for latitude data in D1 are 0.0021°, 0.0066%, 0.0030°, and 0.0029°, respectively. The MAE, MAPE, RMSE, and SDE of forecasting results for longitude data in D2 are 0.0133°, 0.0015%, 0.0167°, and 0.0125°, respectively. The MAE, MAPE, RMSE, and SDE of forecasting results for latitude data in D2 are 0.0043°, 0.0137%, 0.0056°, and 0.0054°, respectively. The MAE, MAPE, RMSE, and SDE of forecasting results for longitude data in D3 are 0.0032°, 0.0026%, 0.0042°, and 0.0041°, respectively. The MAE, MAPE, RMSE, and SDE of forecasting results for latitude data in D3 are 0.0026°, 0.0083%, 0.0036°, and 0.0036°, respectively. The statistical indices of forecasting results of dataset D1 by ELM model (multiple data) are far smaller than that of other two datasets, which has the best prediction effect. This is consistent with the conclusion got from the figures, which reflects the rigor of the experiment.

Forecasting Results of BPNN Based on Multiple Data

The test sample set is input into the trained BPNN (multiple data) to get the predicted results. Figures 5.17, 5.18, and 5.19 show the prediction result curves of the BPNN (multiple data) model for three test sample sets. From the figures it can draw the following conclusions:

Table 5.4 The forecasting performance indices of the ELM (multiple data) model

Model	Datasets	Data types	MAE (°)	MAPE (%)	RMSE (°)	SDE (°)
ELM (multiple data)	D1	Longitude	0.0029	0.0024	0.0038	0.0038
		Latitude	0.0021	0.0066	0.0030	0.0029
	D2	Longitude	0.0133	0.015	0.0167	0.0125
		Latitude	0.0043	0.0137	0.0056	0.0054
	D3	Longitude	0.0032	0.0026	0.0042	0.0041
		Latitude	0.0026	0.0083	0.0036	0.0036

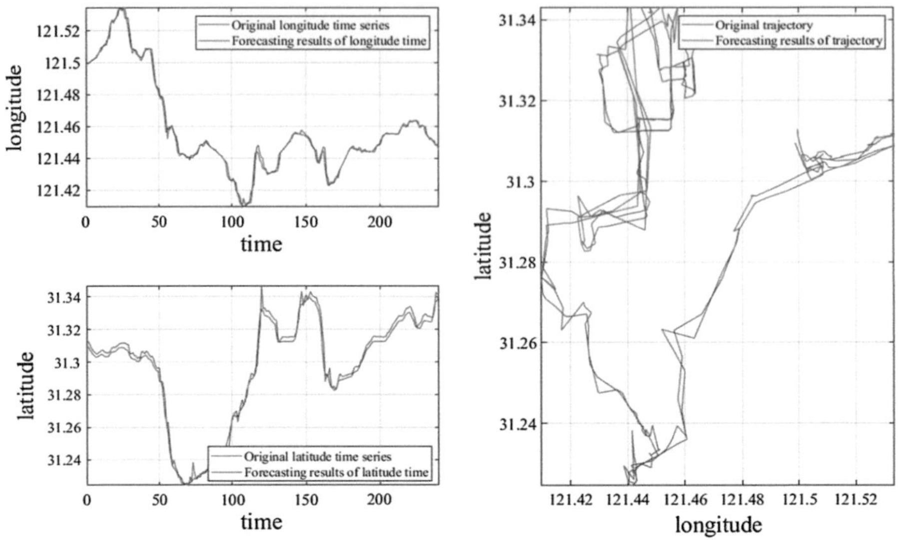

Fig. 5.17 Trajectory prediction results of the BPNN (multiple data) for D1

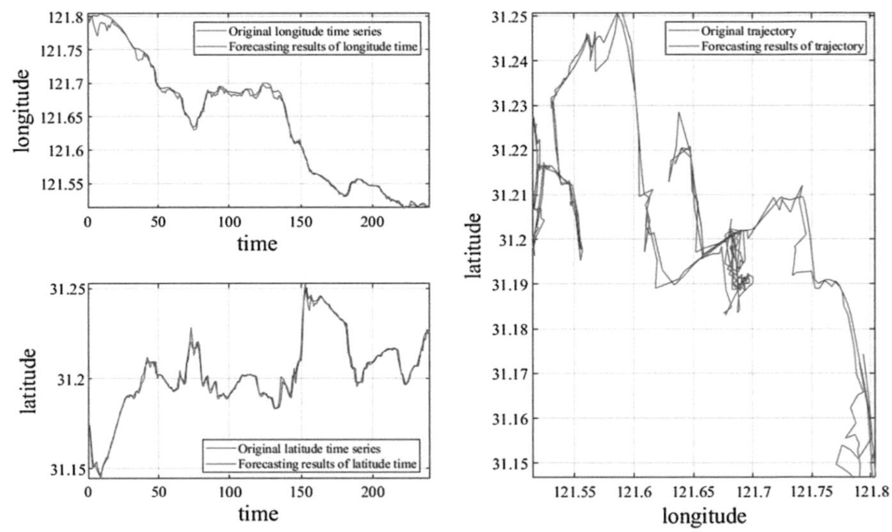

Fig. 5.18 Trajectory prediction results of the BPNN (multiple data) for D2

(a) The BPNN (multiple data) model has better prediction accuracy for dataset D1 and D3. However, the prediction accuracy of the BPNN (multiple data) on dataset D2 is low. As can be seen from the figures, although the accuracy of longitude prediction and latitude prediction of dataset D2 has small errors. When the two prediction curves are combined into a trajectory curve, the errors increase

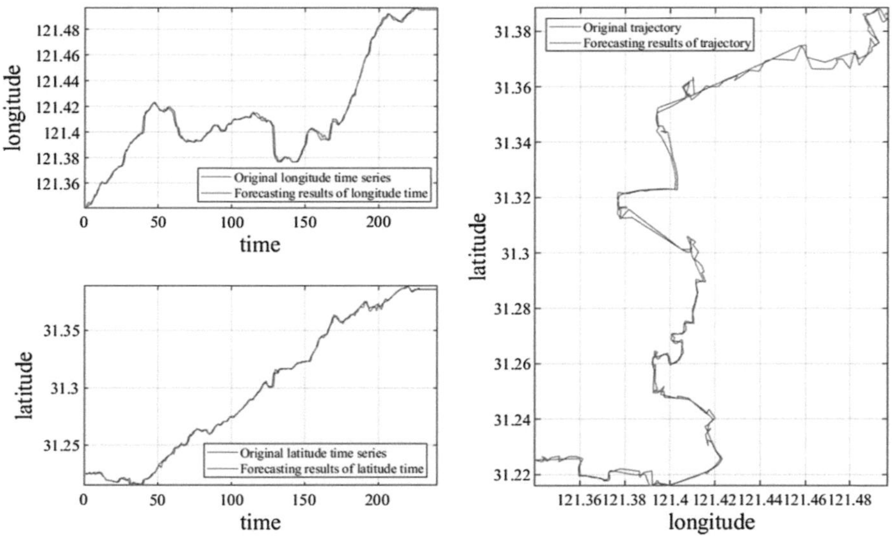

Fig. 5.19 Trajectory prediction results of the BPNN (multiple data) for D3

geometrically. It can be seen that data-driven trajectory prediction puts forward higher requirements for prediction models.

(b) Comparing the prediction accuracy of dataset D1 and D3, it can be seen that the prediction accuracy of dataset D3 is slightly higher than that of dataset D1. It can be seen from the figures that the trajectory curve of dataset D3 is smoother than that of dataset D1. This may account for the higher accuracy of the D3 prediction.

(c) From the figure, it can be concluded that the BPNN (multiple data) does not have a high prediction accuracy for the situation where the data changes dramatically before and after. This is also the reason why the BPNN (multiple data) has the highest prediction accuracy to test set D1 and the lowest to test set D2 in the experiment. The waveform of test set D1 changes gently and the curve is smooth. The fluctuation of dataset D2 is severe, and the data jump before and after is large.

In order to better analyze the prediction performance of BPNN (multiple data) for three kinds of trajectory datasets, Table 5.5 gives some statistical indices of the forecasting results of BPNN (multiple data) for three kinds of datasets. According to Table 5.5, it can draw the following conclusions:

BPNN (multiple data) has the best prediction effect for the trajectory dataset. The MAE, MAPE, RMSE, and SDE of forecasting results for longitude data in D1 are 0.0016°, 0.0013%, 0.0024°, and 0.0024°, respectively. The MAE, MAPE, RMSE, and SDE of forecasting results for latitude data in D1 are 0.0026°, 0.0084%, 0.0032°, and 0.0026°, respectively. The MAE, MAPE, RMSE, and SDE of forecasting results for longitude data in D2 are 0.0061°, 0.0050%, 0.0560°, and 0.0510°,

Table 5.5 The forecasting performance indices of the BPNN (multiple data) model

Model	Dataset	Data types	MAE (°)	MAPE (%)	RMSE (°)	SDE (°)
BP	D1	Longitude	0.0016	0.0013	0.0024	0.0024
		Latitude	0.0026	0.0084	0.0032	0.0026
	D2	Longitude	0.0061	0.0050	0.0560	0.0510
		Latitude	0.0017	0.0053	0.0024	0.0024
	D3	Longitude	0.0014	0.0011	0.0023	0.0023
		Latitude	0.0008	0.0028	0.0014	0.0014

respectively. The MAE, MAPE, RMSE, and SDE of forecasting results for latitude data in D2 are 0.0017°, 0.0053%, 0.0024°, and 0.0024°, respectively. The MAE, MAPE, RMSE, and SDE of forecasting results for longitude data in D3 are 0.0014°, 0.0011%, 0.0023°, and 0.0023°, respectively. The MAE, MAPE, RMSE, and SDE of forecasting results for latitude data in D3 are 0.0008°, 0.0028%, 0.0014°, and 0.0014°, respectively. The statistical indices of forecasting results of dataset D3 by BPNN (multiple data) are far smaller than that of other two datasets, which has the best prediction effect. This is consistent with the conclusion got from the figures, which reflects the rigor of the experiment.

5.3.3 *Trajectory Prediction of Urban Vehicles Under EWT Decomposition Framework*

5.3.3.1 The Theoretical Basis of the EWT

In this paper, the EWT algorithm is used for data pre-processing. The EWT proposed by Gilles (2013) is a novel signal processing technique, which builds the wavelets adaptively. The EWT is based on the theoretical framework of wavelet transform but overcomes the shortage of EMD theory and the problem of signal aliasing. The EWT adaptively divides the Fourier spectrum and selects the appropriate wavelet filter banks. The empirical scaling function and the empirical wavelets can be expressed as Eqs. (5.5) and (5.6):

$$\hat{\phi}_n(\omega) = \begin{cases} 1 & \text{if} |\omega| \le (1-\gamma)\omega_n \\ \cos\left[\frac{\pi}{2}\beta\left(\frac{1}{2\gamma\omega_n}(|\omega| - (1-\gamma)\omega_n)\right)\right] & \text{if} (1-\gamma)\omega_n \le |\omega| \le (1+\gamma)\omega_n \\ 0 & \text{otherwise} \end{cases} \tag{5.5}$$

$$\hat{\psi}_n(\omega)=\begin{cases}1 & \text{if}(1+\gamma)\omega_n \le |\omega| \le (1-\gamma)\omega_{n+1} \\ \cos\left[\frac{\pi}{2}\beta\left(\frac{1}{2\gamma\omega_{n+1}}(|\omega|-(1-\gamma)\omega_{n+1})\right)\right] & \text{if}(1-\gamma)\omega_{n+1} \le |\omega| \le (1+\gamma)\omega_{n+1} \\ \sin\left[\frac{\pi}{2}\beta\left(\frac{1}{2\gamma\omega_n}(|\omega|-(1-\gamma)\omega_n)\right)\right] & \text{if}(1-\gamma)\omega_n \le |\omega| \le (1+\gamma)\omega_n \\ 0 & \text{otherwise}\end{cases} \tag{5.6}$$

where the approximation coefficients and the detail coefficients are computed according to Eqs. (5.7) and (5.8).

$$\begin{aligned} w_f^{\varepsilon}(0,t) &= \langle f,\phi_1\rangle = \int f(\tau)\overline{\phi_1(\tau-t)}d\tau \\ &= \left(\hat{f}(\omega)\overline{\hat{\phi}_1(\omega)}\right)^{\vee} \end{aligned} \tag{5.7}$$

$$\begin{aligned} w_f^{\varepsilon}(n,t) &= \langle f,\psi_n\rangle = \int f(\tau)\overline{\psi_n(\tau-t)}d\tau \\ &= \left(\hat{f}(\omega)\overline{\hat{\psi}_n(\omega)}\right)^{\vee} \end{aligned} \tag{5.8}$$

in addition, there are some constraints to keep the effectiveness of proposed algorithm. For example, the ratio γ in Eqs. (5.5) and (5.6) is restricted to a small value as $\gamma < \min_n(\omega_{n+1} - \omega_n)/(\omega_{n+1} + \omega_n)$ to ensure the empirical scaling function and the empirical wavelets is a tight frame of $L^2(R)$. The $\beta(x)$ is usually defined as $\beta(x) = x^4(35 - 84x + 70x^2 - 20x^3)$.

5.3.3.2 Framework of Model

The input and output of the decomposition framework prediction method are the same as that of the single-input model prediction method. The difference is how the predictors process the data. The prediction model under the decomposition framework firstly decomposes the original data into more stable sub-layers. Then, predictors are established for each sub-layer. Finally, the final prediction result is obtained by adding the prediction results of each predictor. Figure 5.20 shows a block diagram of the method.

5.3.3.3 Steps of Modeling

The steps of using EWT decomposition framework to predict trajectory data are as follows:

(a) The training sample set and test sample set are divided according to the method in Sect. 5.2.2. In this step, the training set and test set of the three trajectory datasets will be obtained, respectively;

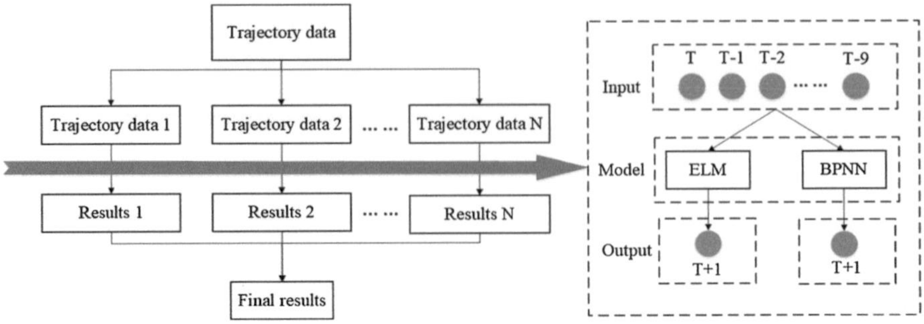

Fig. 5.20 The general framework of this chapter

(b) Setting up the necessary training parameters of the EWT algorithm, ELM model and BPNN model;
(c) Data decomposition. EWT algorithm is used to divide the original trajectory data into several more stable sub-sequence data. This process can also be used as a feature extraction process;
(d) Conducting model training of predictors. A prediction model is trained for each sub-layer. Ten historical track sample points are used as the input of the model. Take the trajectory sample points at a future time as output. The rolling prediction training is completed by using the training sample set obtained in the previous step;
(e) Model combination. The final trajectory prediction data are obtained by adding the prediction results of each sub-predictor.
(f) Testing model. The model test after training is carried out by using the test sample set. The rolling prediction strategy is also used for testing. Finally, track data of the last 4 h are obtained and compared.

5.3.3.4 Forecasting Results

Forecasting Results of EWT-ELM Based on Single Data

The test sample set is input into the trained EWT-ELM model to get the predicted results. Figures 5.21, 5.22, and 5.23 show the prediction result curves of EWT-ELM model for three test sample sets. From the figures it can draw the following conclusions:

(a) EWT-ELM model has a good prediction effect on three testing sets. The fitting degree of both the main trend component and the peak and trough of the wave is close to the real value of the test set. It can be concluded that EWT-ELM model has good effect on trajectory prediction.
(b) From the figures, it can be concluded that the EWT-ELM model does not have a high prediction accuracy for the situation where the data changes dramatically

before and after. The waveform of test set D2 changes gently and the curve is smooth. The fluctuation of dataset D1 is severe, and the data jump before and after is large.

In order to better analyze the prediction performance of the EWT-ELM model for three kinds of noise datasets, Table 5.6 gives some statistical indices of the forecasting results of EWT-ELM model for three kinds of datasets. According to Table 5.6, it can draw the following conclusions:

(a) EWT-ELM model has the best prediction effect for the trajectory dataset. The MAE, MAPE, RMSE, and SDE of forecasting results for longitude data in D1 are 0.0017°, 0.0014%, 0.0027°, and 0.0027°, respectively. The MAE, MAPE, RMSE, and SDE of forecasting results for latitude data in D1 are 0.0017°, 0.0054%, 0.0029°, and 0.0029°, respectively. The MAE, MAPE, RMSE, and SDE of forecasting results for longitude data in D2 are 0.0076°, 0.0063%, 0.0114°, and 0.0092°, respectively. The MAE, MAPE, RMSE, and SDE of forecasting results for latitude data in D2 are 0.0020°, 0.0065%, 0.0032°, and 0.0032°, respectively. The MAE, MAPE, RMSE, and SDE of forecasting results for longitude data in D3 are 0.0015°, 0.0013%, 0.0025°, and 0.0025°, respectively. The MAE, MAPE, RMSE, and SDE of forecasting results for latitude data in D3 are 0.0050°, 0.0033%, 0.0015°, and 0.0014°, respectively. The statistical indices of forecasting results of dataset D3 by the EWT-ELM model are far smaller than that of other two datasets, which has the best prediction effect. This is consistent with the conclusion got from the figures, which reflects the rigor of the experiment.

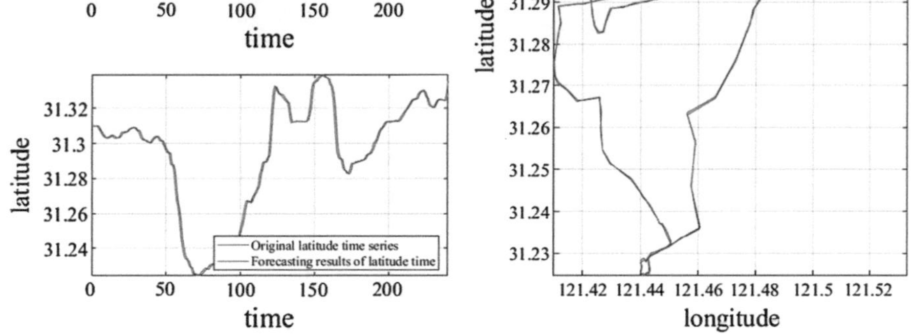

Fig. 5.21 Trajectory prediction results of the EWT-ELM model for D1

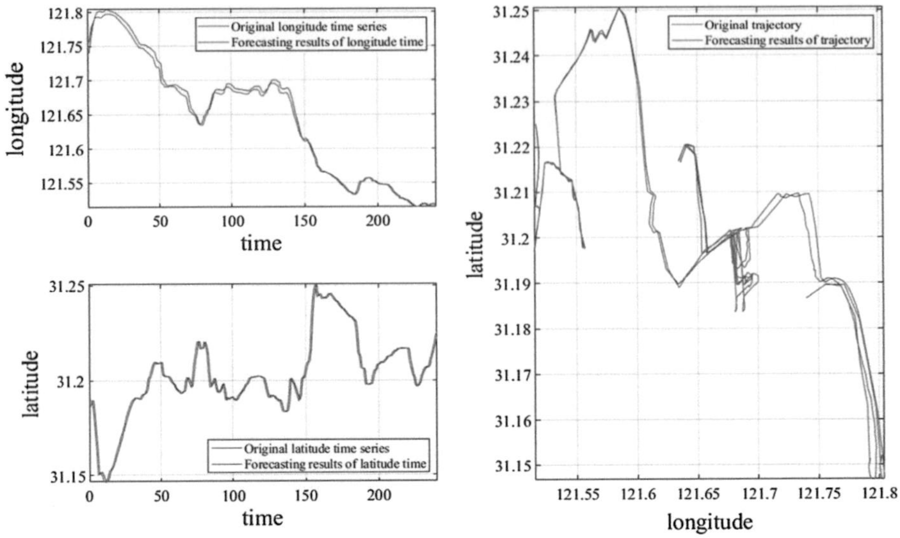

Fig. 5.22 Trajectory prediction results of the EWT-ELM model for D2

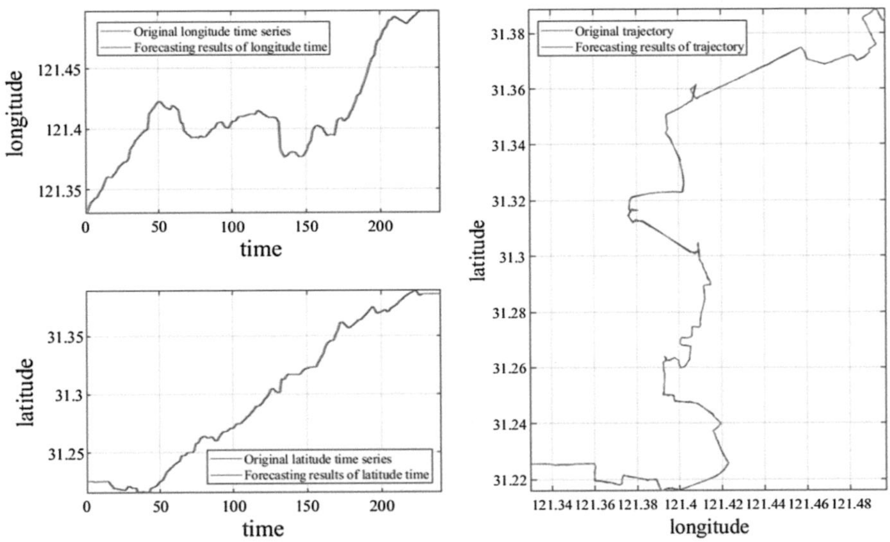

Fig. 5.23 Trajectory prediction results of the EWT-ELM model for D3

(b) From the data in the tables, it can be seen that the EWT-ELM model (single data) has a better prediction effect on the test dataset D3 than on the test dataset D1.This is because dataset D3 compared to dataset D1, its data fluctuations are smoother so that the prediction accuracy of the EWT-ELM (single data) is higher.

Table 5.6 The forecasting performance indices of EWT-ELM model

Model	Dataset	Data types	MAE (°)	MAPE (%)	RMSE (°)	SDE (°)
EWT-ELM	D1	Longitude	0.0017	0.0014	0.0027	0.0027
		Latitude	0.0017	0.0054	0.0029	0.0029
	D2	Longitude	0.0076	0.0063	0.0114	0.0092
		Latitude	0.0020	0.0065	0.0032	0.0032
	D3	Longitude	0.0015	0.0013	0.0025	0.0025
		Latitude	0.0050	0.0033	0.0015	0.0014

Forecasting Results of the EWT-BPNN Based on Single Data

The test sample set is input into the trained EWT-BPNN to get the predicted results. Figures 5.24, 5.25, and 5.26 show the prediction result curves of the EWT-BPNN for three test sample sets. From the figures it can draw the following conclusions:

(a) The EWT-BPNN has a good prediction effect on three test sets. The fitting degree of both the main trend component and the peak and trough of the wave is close to the real value of the test set. It can be concluded that the EWT-BPNN has good effect on trajectory prediction.
(b) From the figures, it can be concluded that the EWT-BPNN does not have a high prediction accuracy for the situation where the data changes dramatically before and after. The waveform of testing set D2 changes gently and the curve is smooth. The fluctuation of dataset D1 is severe, and the data jump before and after is large.

In order to better analyze the prediction performance of the EWT-BPNN for three kinds of noise datasets, Table 5.7 gives some statistical indices of the forecasting results of EWT-BPNN for three kinds of datasets. According to Table 5.7, it can draw the following conclusions:

(a) The EWT-BPNN has the best prediction effect for the trajectory dataset. The MAE, MAPE, RMSE, and SDE of forecasting results for longitude data in D1 are 0.0017°, 0.0014%, 0.0027°, and 0.0027°, respectively. The MAE, MAPE, RMSE, and SDE of forecasting results for latitude data in D1 are 0.0017°, 0.0054%, 0.0029°, and 0.0029°, respectively. The MAE, MAPE, RMSE, and SDE of forecasting results for longitude data in D2 are 0.0076°, 0.0063%, 0.0114°, and 0.0092°, respectively. The MAE, MAPE, RMSE, and SDE of forecasting results for latitude data in D2 are 0.0020°, 0.0065%, 0.0032°, and 0.0032°, respectively. The MAE, MAPE, RMSE, and SDE of forecasting results for longitude data in D3 are 0.0015°, 0.0013%, 0.0025°, and 0.0025°, respectively. The MAE, MAPE, RMSE, and SDE of forecasting results for latitude data in D3 are 0.0050°, 0.0033%, 0.0015°, and 0.0014°, respectively. The statistical indices of forecasting results of dataset D2 by EWT-BPNN are far smaller than that of other two datasets, which has the best prediction effect. This is consistent with the conclusion got from the figures, which reflects the rigor of the experiment.

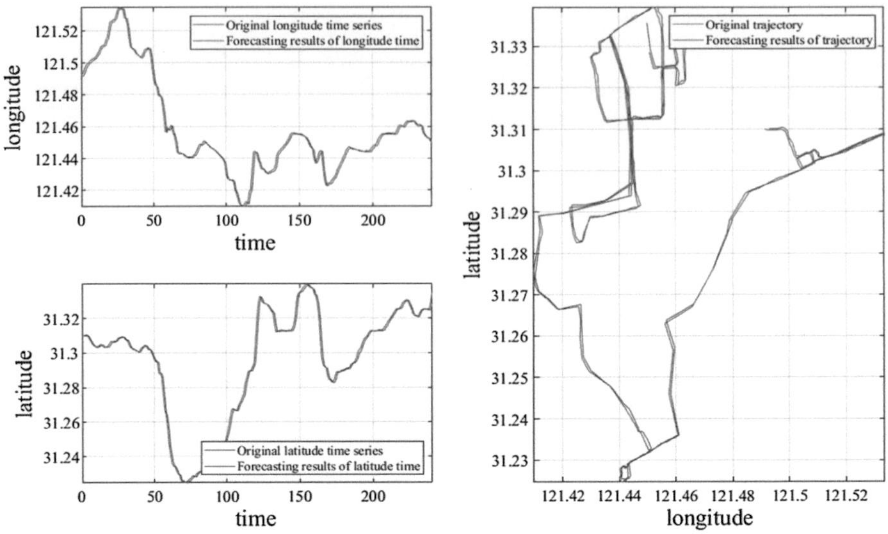

Fig. 5.24 Trajectory prediction results of the EWT-BPNN for D1

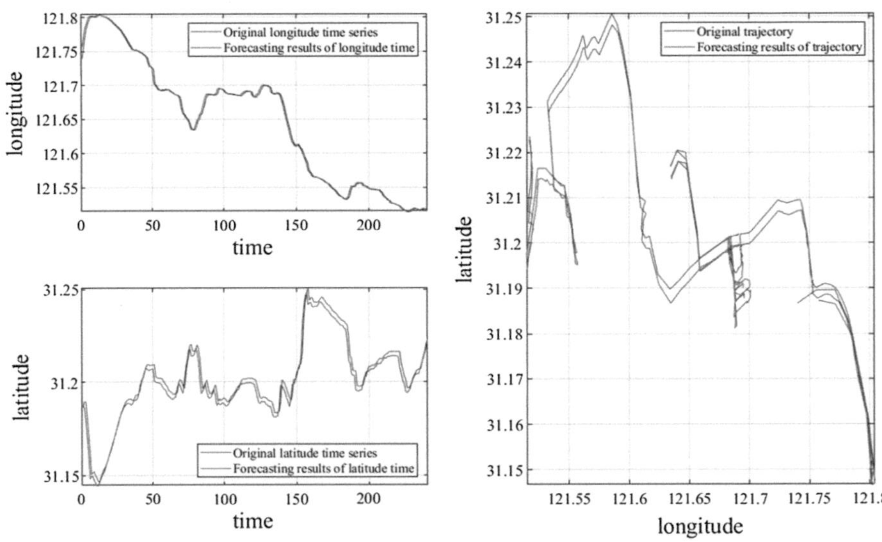

Fig. 5.25 Trajectory prediction results of the EWT-BPNN for D2

(b) From the data in the tables, it can be seen that the EWT-BPNN has a better prediction effect on the test dataset D3 than that on the test dataset D1. This is because dataset D3 compared to dataset D1, its data fluctuations are smoother so that the prediction accuracy of the EWT-BPNN is higher.

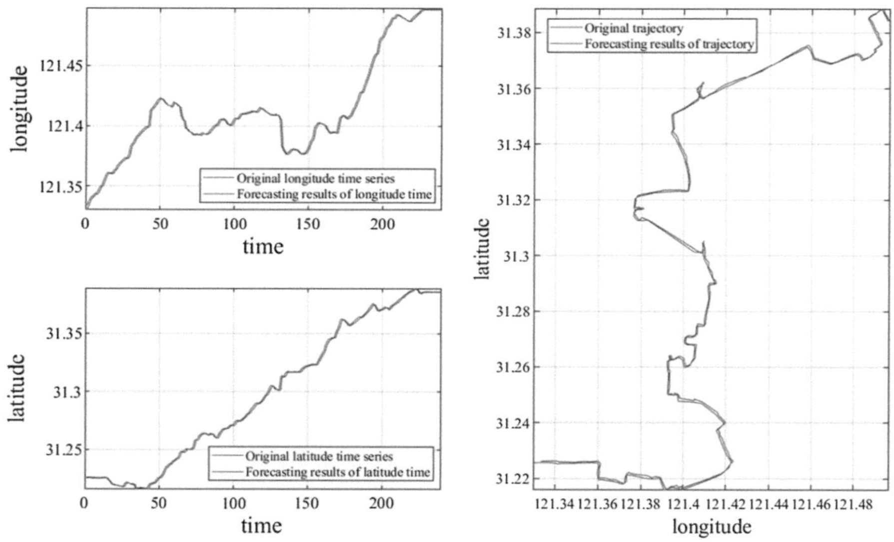

Fig. 5.26 Trajectory prediction results of the EWT-BPNN for D3

Table 5.7 The forecasting performance indices of EWT-BPNN model

Model	Dataset	Data types	MAE (°)	MAPE (%)	RMSE (°)	SDE (°)
EWT-BP	D1	Longitude	0.0017	0.0014	0.0029	0.0029
		Latitude	0.0018	0.0057	0.0027	0.0027
	D2	Longitude	0.0024	0.0020	0.0039	0.0039
		Latitude	0.0031	0.050	0.0040	0.0032
	D3	Longitude	0.0019	0.0015	0.0028	0.0025
		Latitude	0.0014	0.0046	0.0019	0.0015

5.3.4 Comparative Analysis of Forecasting Performance

The following conclusions can be drawn from Figs. 5.7, 5.8, 5.9, 5.14, 5.15, 5.16, 5.21, 5.22, 5.23 and Table 5.8.

(a) In general, the ELM model has a good prediction accuracy for vehicle trajectory data. Take the prediction result of ELM (multiple data) on dataset D2, which has the lowest prediction accuracy, for example, its prediction accuracy has reached a good level. The MAE, MAPE, RMSE, and SDE of forecasting results for longitude data in D2 are 0.0133°, 0.0015%, 0.0167°, and 0.0125°, respectively. The MAE, MAPE, RMSE, and SDE of forecasting results for latitude data in D2 are 0.0043°, 0.0137%, 0.0056°, and 0.0054°, respectively. Therefore, it can be concluded that the ELM has the potential for vehicle trajectory prediction.

(b) The prediction accuracy of the ELM model (multiple data) is lower than that of ELM model (single data). From the prediction curves in Figs. 5.14, 5.15, and 5.16, it can be seen that the main reason for the decrease in accuracy is the severe overfitting of ELM model (multiple data). The reason for this phenomenon may be that the latitude of multi-feature input data is too complex, which increases the difficulty of model training. Feature selection and feature extraction algorithms can effectively reduce the problem of overfitting caused by multiple input features.
(c) The EWT-ELM model has excellent prediction accuracy. It can be seen from Table 5.9 that the prediction accuracy of the ELM model is improved after data pre-processing with the EWT decomposition algorithm. This conclusion can be more intuitively reflected in Figs. 5.7, 5.8, 5.9, 5.21, 5.22, and 5.23. The EWT-ELM model fixes the prediction deviation of the ELM model on longitude data of dataset D2. Overall, the trajectory curve fitting effect of the EWT-ELM model is much better than that of ELM model (single data).

The following conclusions can be drawn from Figs. 5.10, 5.11, 5.12, 5.17, 5.18, 5.19, 5.24, 5.25, 5.26 and Table 5.9.

(a) In general, the BPNN model has a good prediction accuracy for vehicle trajectory data. Take the prediction result of BPNN (multiple data) on dataset D2, which has the lowest prediction accuracy, for example, its prediction accuracy has reached a good level. The MAE, MAPE, RMSE, and SDE of forecasting results for longitude data in D2 are 0.0133°, 0.0015%, 0.0167°, and 0.0125°,

Table 5.8 The comprehensive forecasting performance indices of the ELM models

Model	Datasets	Data types	MAE (°)	MAPE (%)	RMSE (°)	SDE (°)
ELM (single data)	D1	Longitude	0.0015	0.0012	0.0023	0.0023
		Latitude	0.0013	0.0041	0.0021	0.0021
	D2	Longitude	0.0143	0.0118	0.0188	0.0130
		Latitude	0.0017	0.0056	0.0027	0.0027
	D3	Longitude	0.0014	0.0011	0.0023	0.0023
		Latitude	0.0009	0.0030	0.0015	0.0015
ELM (multiple data)	D1	Longitude	0.0029	0.0024	0.0038	0.0038
		Latitude	0.0021	0.0066	0.0030	0.0029
	D2	Longitude	0.0133	0.015	0.0167	0.0125
		Latitude	0.0043	0.0137	0.0056	0.0054
	D3	Longitude	0.0032	0.0026	0.0042	0.0041
		Latitude	0.0026	0.0083	0.0036	0.0036
EWT-ELM	D1	Longitude	0.0017	0.0014	0.0027	0.0027
		Latitude	0.0017	0.0054	0.0029	0.0029
	D2	Longitude	0.0076	0.0063	0.0114	0.0092
		Latitude	0.0020	0.0065	0.0032	0.0032
	D3	Longitude	0.0015	0.0013	0.0025	0.0025
		Latitude	0.0050	0.0033	0.0015	0.0014

respectively. The MAE, MAPE, RMSE, and SDE of forecasting results for latitude data in D2 are 0.0043°, 0.0137%, 0.0056°, and 0.0054°, respectively. Therefore, it can be concluded that BPNN has the potential for vehicle trajectory prediction.

(b) The prediction accuracy of the BPNN model (multiple data) is lower than that of BPNN model (single data). From the prediction curves in Figs. 5.14, 5.15, and 5.16, it can be seen that the main reason for the decrease in accuracy is the severe overfitting of the BPNN model (multiple data). The reason for this phenomenon may be that the latitude of multi-feature input data is too complex, which increases the difficulty of model training. Feature selection and feature extraction algorithms can effectively reduce the problem of overfitting caused by multiple input features.

(c) EWT-BPNN model has excellent prediction accuracy. It can be seen from Table 5.9 that the prediction accuracy of the BPNN model is improved after data pre-processing with the EWT decomposition algorithm. This conclusion can be more intuitively reflected in Figs. 5.7, 5.8, 5.9, 5.21, 5.22, and 5.23. EWT-BPNN model fixes the prediction deviation of the BPNN model on longitude data of dataset D2. Overall, the trajectory curve fitting effect of the EWT-BPNN model is much better than that of BPNN model (single data).

The following conclusions can be drawn from Tables 5.8 to 5.9.

(a) The predicted longitude of the BPNN model is higher than that of ELM model. Leaving aside the local data, which may be an error in forecasting volatility.

Table 5.9 The comprehensive forecasting performance indices of the BP models

Model	Datasets	Data types	MAE (°)	MAPE (%)	RMSE (°)	SDE (°)
BP (single data)	D1	Longitude	0.0013	0.0011	0.0021	0.0021
		Latitude	0.0029	0.0094	0.0063	0.0063
	D2	Longitude	0.0041	0.0034	0.0081	0.0080
		Latitude	0.0023	0.0075	0.0036	0.0036
	D3	Longitude	0.0018	0.0015	0.0028	0.0028
		Latitude	0.0011	0.0034	0.0016	0.0016
BP (multiple data)	D1	Longitude	0.0016	0.0013	0.0024	0.0024
		Latitude	0.0026	0.0084	0.0032	0.0026
	D2	Longitude	0.0061	0.0050	0.056	0.051
		Latitude	0.0017	0.0053	0.0024	0.0024
	D3	Longitude	0.0014	0.0011	0.0023	0.0023
		Latitude	0.0008	0.0028	0.0014	0.0014
EWT-BP	D1	Longitude	0.0017	0.0014	0.0029	0.0029
		Latitude	0.0018	0.0057	0.0027	0.0027
	D2	Longitude	0.0024	0.0020	0.0039	0.0039
		Latitude	0.0031	0.050	0.0040	0.0032
	D3	Longitude	0.0019	0.0015	0.0028	0.0025
		Latitude	0.0014	0.0046	0.0019	0.0015

The overall longitude prediction of the two models is consistent with this conclusion. It can be seen that the fitting effect of neural network for nonlinear systems is better than that of traditional machine learning algorithms. In this experiment, the parameters of BPNN model are selected based on empirical values, leading to the phenomenon of overfitting in the prediction results. BPNN model will have better prediction effect when its parameters are optimized.

(b) In both the ELM model and BPNN model, the prediction accuracy of correlation model adopting multi-feature input strategy is lower than that of the model adopting a single-feature input strategy. It can be seen that although multiple data contain more track movement information, higher sample dimensions also increase the difficulty of model training. In order to run as much sample information as possible without increasing data latitude, it can carry out feature engineering by feature selection algorithm or feature extraction algorithm.

(c) EWT decomposition algorithm can optimize both the ELM model and BPNN model. From the data in Tables 5.8 and 5.9, it can be seen that the prediction accuracy of the two models presents an overall increasing trend after the EWT decomposition algorithm is added. The optimization effect of the EWT decomposition algorithm may be reflected in the optimization of data distribution. The known experimental models from the previous analysis have higher accuracy in predicting time series with smoother distribution. By decomposing the original time series into several more stable sub-layers, the EWT decomposition algorithm improves the prediction accuracy of the original model.

5.4 Conclusion

In the chapter, the ELM model and BPNN model are used to discuss their possibilities in vehicle trajectory prediction. In this chapter, the trajectory data of three different cars in one day are selected for prediction experiments. The experiment is divided into three parts, which are single-model prediction based on single data, multiple-model prediction based on multiple data, and model prediction based on decomposition algorithm. The following conclusions can be drawn from the prediction experiments in this chapter:

(a) The prediction of vehicle trajectory data is more difficult than that of the conventional time series. Vehicle track data has stronger irregularity, and due to the use of latitude and longitude data alone prediction method for track prediction, the error product effect is serious. All of these require the prediction model owning higher prediction accuracy.

(b) The ELM model and BPNN model both have good prediction accuracy in the dataset selected in this chapter. It can be considered that both models can predict the trajectory of cars in smart cities in real life.

(c) Simply adding the feature dimension of the input data does not increase the prediction accuracy of the model. Because it increases the training difficulty of the model while increasing the feature dimension. The characteristic engineering method may well solve this contradiction. This remains for further study in the future.
(d) Both the ELM model and BPNN model have higher prediction accuracy for datasets with smoother changes.
(e) Based on the conclusion (d), the EWT decomposition algorithm is used to decompose the original trajectory series into smoother sub-sequences. It can be seen from the experimental results that the model based on decomposition algorithm has higher prediction accuracy than the single model.

References

Dong H, Li M, Zhang S, Han L, Li J, Su X (2017) Short-term power load forecasting based on kernel principal component analysis and extreme learning machine. Journal of Electronic Measurement and Instrument 32:188–193

Gilles J (2013) Empirical wavelet transform. IEEE Transactions on Signal Processing 61(16): 3999–4010

Granda F, Azpilicueta L, Vargasrosales C, Celayaecharri M, Lopeziturri P, Aguirre E et al (2018) Deterministic propagation modeling for intelligent vehicle communication in smart cities. Sensors 18(7):4898

Hecht-Nielsen R (1992) Theory of the backpropagation neural network. *Neural networks for perception*. Academic Press, 65–93

Huang G-B, Chen L (2007) Convex incremental extreme learning machine. Neurocomputing 70(16–18):3056–3062

Huang GB, Zhou H, Ding X, Rui Z (2012) Extreme learning machine for regression and multiclass classification. IEEE Transactions on Systems Man Cybernetics Part B 42(2):513–529

Huang G-B, Zhu Q-Y, Siew C-K (2004) Extreme learning machine: A new learning scheme of feedforward neural networks. Neural Networks 2:985–990

Huang G-B, Zhu Q-Y, Siew C-K (2006) Extreme learning machine: Theory and applications. Neurocomputing 70(1–3):489–501

Hui L, Wu H, Li Y (2018) Smart wind speed forecasting using EWT decomposition, GWO evolutionary optimization, RELM learning and IEWT reconstruction. Energy Conversion Management 161:266–283

Kai Z, Jia-Jun X, Fan LI, Ting-Ting FU (2018) Bayesian trajectory prediction for a hypersonic gliding reentry vehicle based on intent inference. Journal of Astronautics 39(11):1258–1265

Wang H, Qiang L, Feng Y, Zhi L, Sun L (2016) Influential spatial facility prediction over large scale cyber-physical vehicles in smart city. EURASIP Journal on Wireless Communications Networking 2016(1):103

Wang LL, Chen ZG, Wu J (2019) Vehicle trajectory prediction algorithm in vehicular network. Wireless Networks 25(4):2143–2156

Woo H, Yonghoon JI, Kono H, Tamura Y, Asama H (2017) Lane-change detection based on vehicle-trajectory prediction. IEEE Robotics 2(2):1109–1116

Xia Z, Hu Z, Luo J (2017) UPTP vehicle trajectory prediction based on user preference under complexity environment. Wireless Personal Communications 97(12):1–15

Xie G, Gao H, Qian L, Huang B, Li K, Wang J (2018) Vehicle trajectory prediction by integrating physics- and maneuver-based approaches using interactive multiple models. IEEE Transactions on Industrial Electronics 65(7):5999–6008

Zhang W, Zhang S, Shuai Z, Yu D, Huang NN (2018) A novel method based on FTS with both GA-FCM and multifactor BPNN for stock forecasting. Soft Computing 3(3):1–16

Zhu X, Li P, Havyarimana V, Georges HM, Dong W, Li K (2018) GOI: A novel design for vehicle positioning and trajectory prediction under urban environments. IEEE Sensors Journal PP(99):1-1

Chapter 6
Prediction Model of Traffic Flow Driven Based on Single Data in Smart Traffic Systems

Abstract The traffic flow series is a typical time series with some time-varying rules. Because the living and working hours of the house have certain rules, the traffic flow series has a strong periodic variation law. Deterministic predictions can more accurately find the change values of traffic flows, while interval predictions can find the range of changes in traffic flows. Both prediction methods play an important role in improving intelligent transportation systems. In this chapter, traffic flow prediction models are built using historical data from the traffic flow series. The models are divided into the deterministic prediction model and interval prediction model, which are composed of BP prediction model, WD-BP prediction model, BP-GARCH interval prediction model, and WD-BP-GARCH interval prediction model. It can be seen from the experimental results that the predictive performance of the WD-BP predictive model is higher than the BP predictive model in the deterministic forecast of traffic flow. In the interval prediction of traffic flow, BP neural network is used to establish a deterministic prediction model, and the GARCH model is used to calculate the uncertainty of forecasting traffic flow.

6.1 Introduction

The accurate prediction of traffic flow is an important step to realize the intelligent traffic system. There are many factors that affect traffic flow, such as weather, time, traffic accidents and traffic flow in other segments, etc. In this chapter, the single data-driven traffic flow prediction model is established. The single data refers to the use of historical data of traffic flow series for modeling without considering other external factors. The single data-driven traffic flow prediction model can better establish the relationship between the historical traffic flow data of the road segment and the future data.

Traffic flow prediction is mainly divided into deterministic prediction and interval prediction. Deterministic traffic flow prediction is the value of traffic flow at a certain time in the future, and the predicted result is a certain traffic flow value. The

H. Liu, *Smart Cities: Big Data Prediction Methods and Applications*,
https://doi.org/10.1007/978-981-15-2837-8_6

prediction of traffic flow interval is the range of traffic flow variation at a certain time in the future, and the prediction result is an interval range.

Most of the researches on traffic flow forecasting at this stage are vehicular-time series deterministic predictions. *Williams et al.* (1998) established a seasonal autoregressive comprehensive moving average and winter index smoothing model and tested it. Tan et al. (2009) proposed a traffic flow prediction method based on moving average (MA), exponential smoothing (ES), autoregressive integrated moving average (ARIMA) and neural network. Wu et al. (2018) established a mixed model of traffic flow prediction using convolutional neural networks and cyclic neural networks. Mackenzie et al. (2018) studied and evaluated the application of stratified time memory in short-term forecasting of main road traffic in the Adelaide metropolitan area of South Australia.

Deterministic prediction accuracy has strong volatility over time and is generally lower than average prediction accuracy (El-Fouly et al., 2006). For the relevant decision-making bodies, the prediction bias caused by the uncertainty of the point prediction result may affect the final decision. In some cases, the prediction interval with high confidence has higher practical significance than the simple point prediction value. With the rapid development of time series prediction, interval prediction as a method for predicting the uncertainty of series point prediction volatility has gradually gained more attention. At present, interval prediction theory is widely used in wind speed time series prediction (Khosravi et al., 2013), financial index time series prediction (Bernardo et al. 2012), and power load forecasting (Guan et al. 2013).

Common time series interval prediction methods mainly include heuristic algorithms and statistical methods. The heuristic algorithm uses a neural network to directly predict the upper and lower bounds of the time series confidence interval. In related research, Qin et al. (2015) used the CSO-BPNN hybrid model for confidence interval estimation after wavelet decomposition of wind speed time series. The essence of the heuristic algorithm is still the traditional point prediction algorithm, ignoring the statistical relationship of the point prediction model and prediction error, and cannot give the accurate confidence of the prediction result.

From the statistical point of view, statistical methods are used to perform interval fitting on the original series or regression analysis on the point prediction residual series under certain confidence conditions. Common statistical methods include fuzzy time series identification, hidden Markov model regression, conditional heteroscedastic model regression (Liu et al. 2011), quantile regression method, Bootstrap method, etc.. Fuzzy time series (FTS) is a nonparametric model introduced by Song and Chissom (1994) based on the fuzzy set theory. These models are easy to implement and flexible to handle both numerical and non-numeric data. Zhang and Wan (2007) trained fuzzy interval neural networks based on statistical interval datasets with weekly mean, and estimated maximum error and standard deviation to achieve currency exchange fluctuation range prediction. Jiang et al. (2013) used Bayesian structural fracture model to predict the wind direction, and all of them had good prediction results.

In this chapter, the prediction model of traffic flow is established driven by single data. The main contributions are organized as follows. Section 6.2 introduces the original traffic flow series for prediction. Section 6.3 shows the traffic flow deterministic prediction model driven by single data based on BP neural network. In Sect. 6.4, the traffic flow interval prediction model driven by single data based on the BP neural network and the ARCH model is established. Section 6.5 shows the traffic flow interval prediction model under the decomposition framework. Section 6.6 introduces the big data prediction architecture of traffic flow. Section 6.7 analyzes and compares the predictive performance of the models involved. In Sect. 6.8, conclusions are obtained.

6.2 Original Traffic Flow Series for Prediction

The data used in this chapter is the traffic flow series of a segment in Shenzhen, China from south to north. For convenience, the traffic flow series is named as $\{V_{1t}\}$. Figure 6.1 shows the distribution of traffic flow series $\{V_{1t}\}$, which is collected continuously from November 1st, 2018 to November 8th, 2018, with a sampling interval of 20 min and a total of 576 samples.

Some regular features of the traffic flow series $\{V_{1t}\}$ can be observed through Fig. 6.1. First of all, the traffic flow series $\{V_{1t}\}$ has a strong periodicity, the sampling time is 20 min, and there are 72 samples per day. From Fig. 6.1, the daily traffic flow during the 8 days (576 samples) can be seen. The cycle change law is presented, that is, the traffic flow is low at night, while it is high during the daytime. Then, in the daily traffic flow (every 72 samples), there is a tendency to rise first, then fall and then rise again. This is because the traffic flow has the characteristics of morning and evening peaks, 8:00–9:00 and 17:00–18:00 is the peak period for commuting. Finally, only the fourth day (samples between 216 and 288) in Fig. 6.1

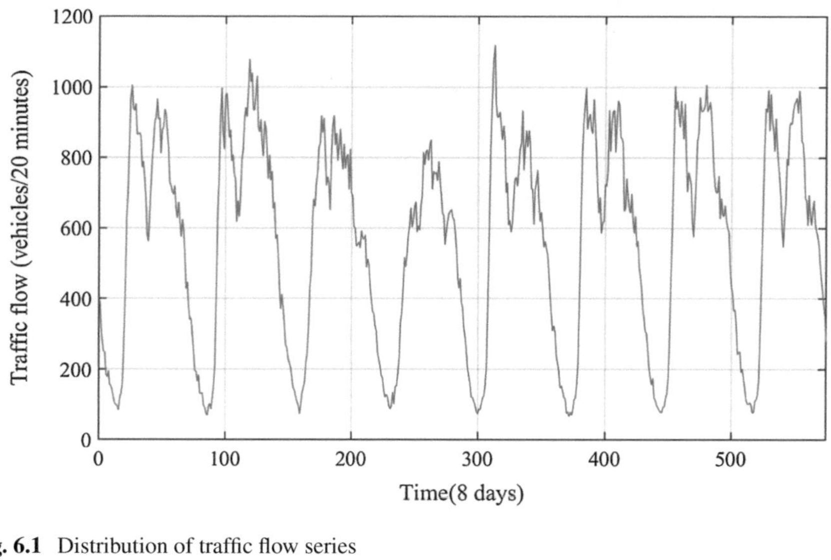

Fig. 6.1 Distribution of traffic flow series

does not have the morning and evening peak characteristics mentioned above, because this day happens to be Sunday (November 4, 2018), people do not have to go to work, so the peak characteristics in the morning and evening are not obvious.

Table 6.1 shows the mathematical statistics of the traffic flow series $\{V_{1t}\}$, including minimum, maximum, mean, standard deviation, skewness, kurtosis, and length. This information can help to understand the statistics of the series better. It can be seen from Table 6.1 that the traffic flow series $\{V_{1t}\}$ has a large fluctuation range, the minimum value is 66, the maximum value is 1119, and the standard deviation is 294.42. The skewness is less than 0 for the left skew state, and the kurtosis is greater than 0, indicating that the distribution of the series $\{V_{1t}\}$ is steeper than the Normal distribution.

6.3 Traffic Flow Deterministic Prediction Driven by Single Data

6.3.1 Modeling Process

The BP neural network is used to establish the traffic flow prediction model based on single data, which includes four steps (Li et al. 2012): (step 1) setting the input/output data structure, (step 2) dividing the training/testing set, (step 3) training the BP prediction model, and (step 4) testing the prediction performance of the model.

(Step 1) Setting the Structure of Input/Output Data

As for predicting one specific segment traffic flow driven by single data, the chapter only uses the historical data of its traffic flow series without any other features. It is independent from other types of data like weather and independent from traffic flows of other segments.

Figure 6.2 shows the prediction structure that uses the first four historical data to predict the *m*th data in the future, this is called the *m*-step prediction. This chapter uses the previous four historical data as input data to make predictions. The advance of step is set 1-step to 5-step, respectively. Figure 6.2 shows the structure of input/output data.

(Step 2) Dividing the Training/Test Set

As can be seen in Sect. 6.2, the traffic flow series contains a total of 8 days with a total of 576 samples. And this traffic flow series has obvious periodic characteristics. As shown in Fig. 6.3, samples from the first 4 days (samples between 1 and 288) are used as training data in this section to train the BP neural network. The samples from the last 4 days (samples between 289 and 576) are used as test data to test the predictive performance of the model. Since the traffic flow series $\{V_{1t}\}$

Table 6.1 The mathematical statistical information of original series

Series	Minimum	Maximum	Mean	Standard deviation	Skewness	Kurtosis	Length
$\{V_{1t}\}$	66	1119	571.77	294.42	−0.34	1.78	576

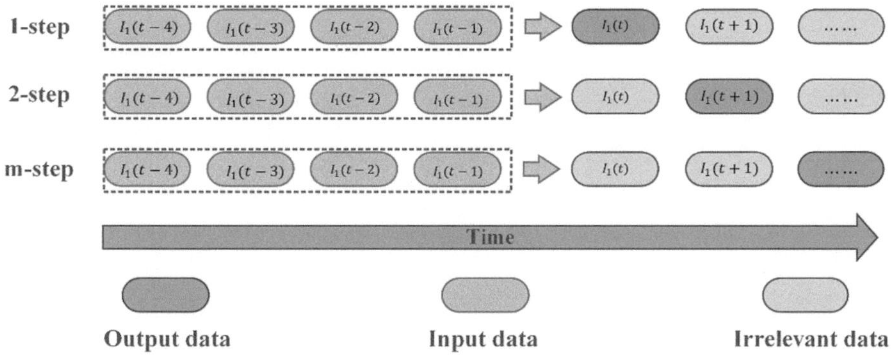

Fig. 6.2 The input/output data structure of single data

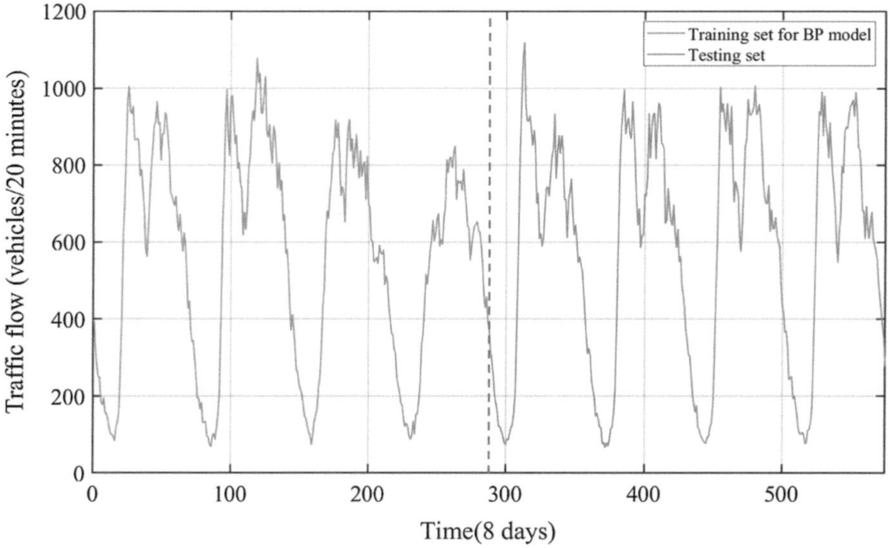

Fig. 6.3 The training set and testing set

are predicted, respectively, from 1-step to the 5-step, five training/test sets are established.

(Step 3) Training the BP Prediction Model

The BP neural network is trained using the training data in step 2. The number of input layer nodes is set to 4, the number of output layer nodes is set to 1, the number of hidden layer nodes is set to 300, the number of iterations is set to 3000, the error threshold is set to 0.00001, and the activation function uses the Sigmoid function. Since the traffic flow series $\{V_{1t}\}$ is predicted from 1-step to the 5-step, five BP prediction models are obtained.

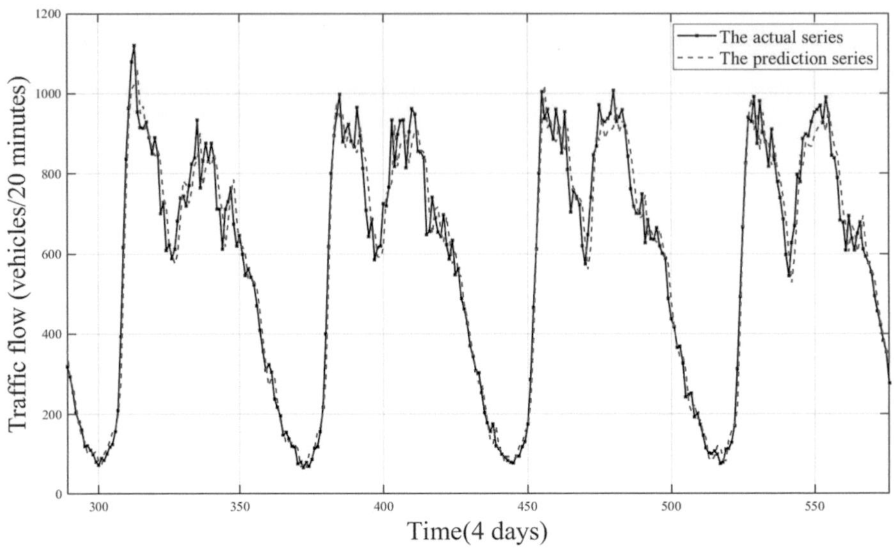

Fig. 6.4 The results of the 1-step prediction for the traffic flow series

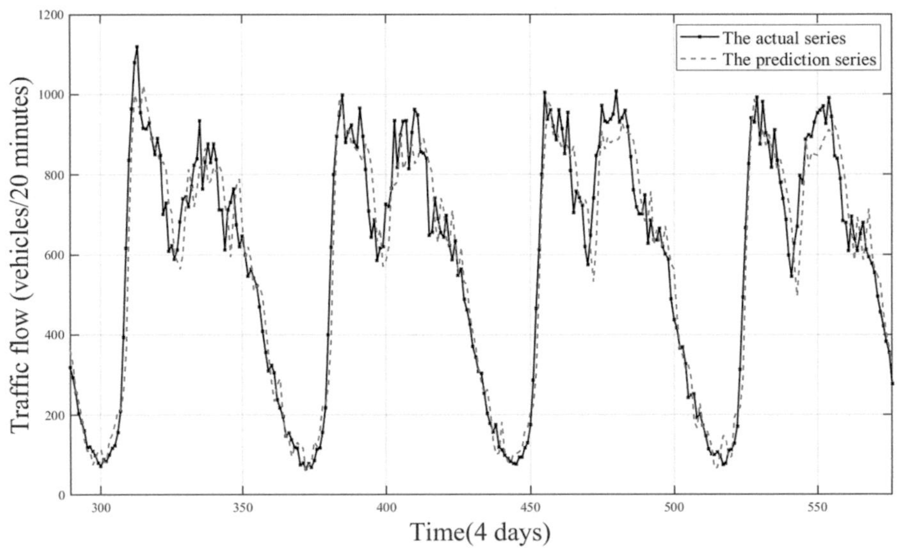

Fig. 6.5 The results of the 2-step prediction for the traffic flow series

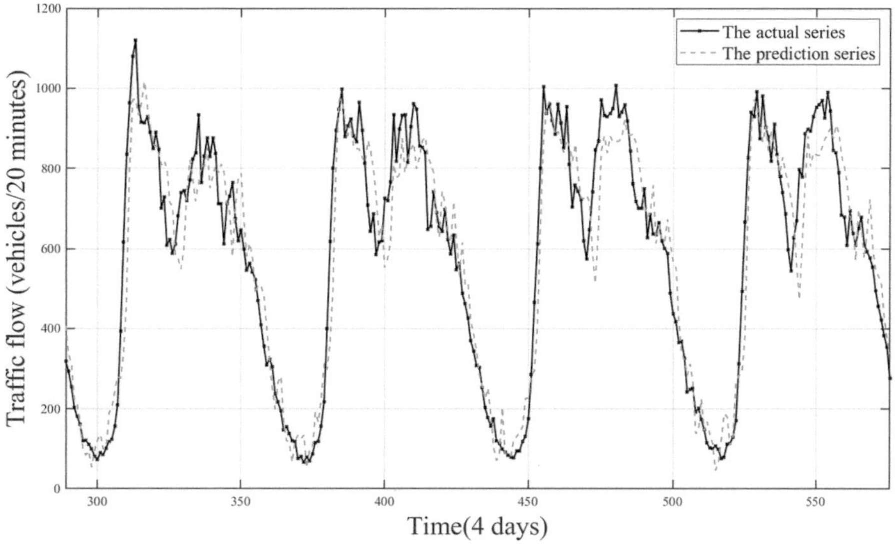

Fig. 6.6 The results of the 3-step prediction for the traffic flow series

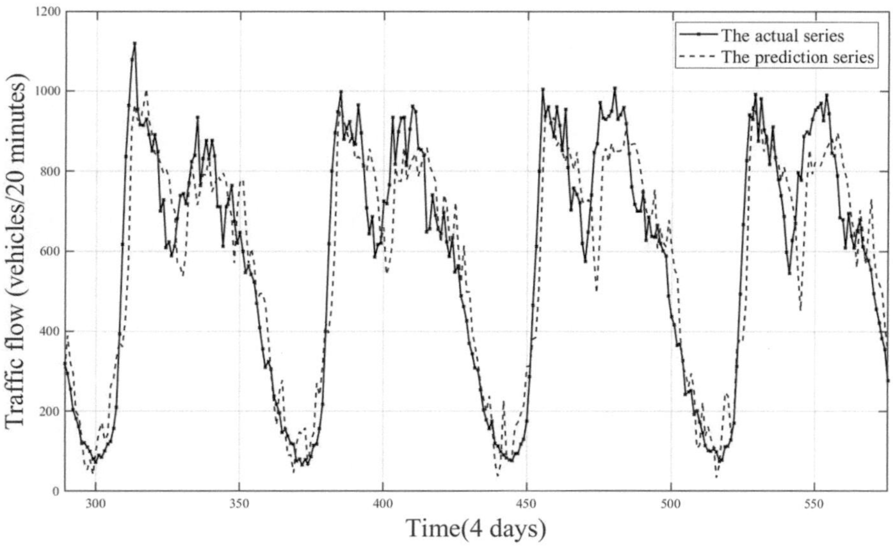

Fig. 6.7 The results of the 4-step prediction for the traffic flow series

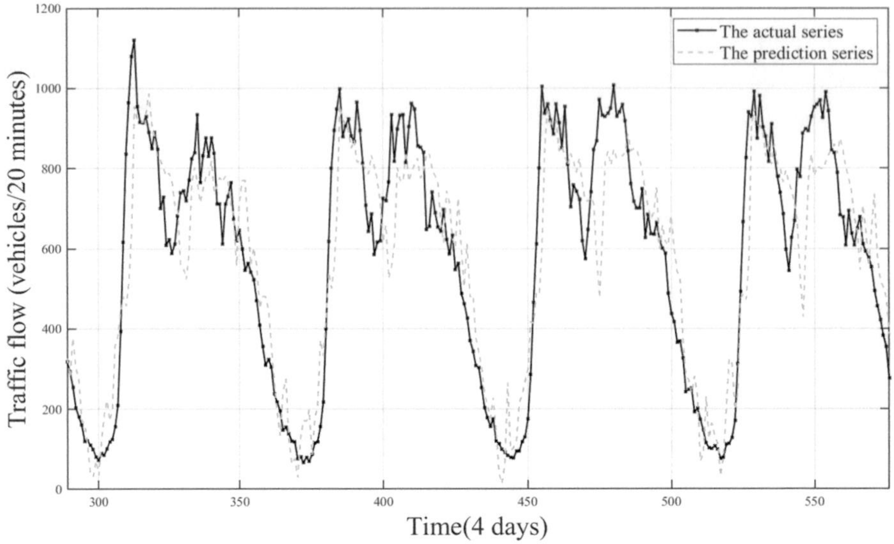

Fig. 6.8 The results of the 5-step prediction for the traffic flow series

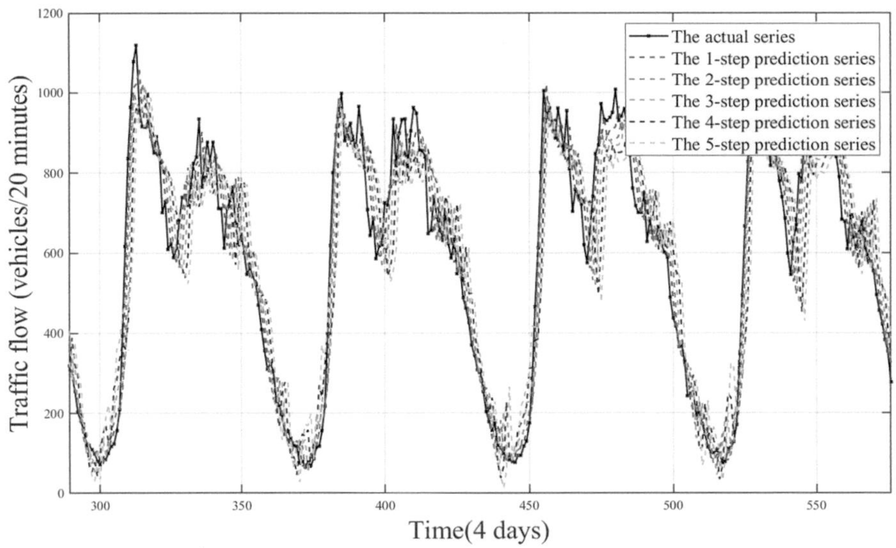

Fig. 6.9 Comparison of prediction results based on different prediction steps

(Step 4) Testing the Prediction Performance of the Model

Five testing sets obtained in step 2 are, respectively, input into the corresponding BP prediction model, and the predicted results of the traffic flow series from 1-step to 5-step are, respectively, obtained. And the five error evaluation index values of MAE, MAPE, RMSE, SDE, and R are calculated to evaluate the performance of the BP prediction model.

6.3.2 *The Prediction Results*

Figures 6.4, 6.5, 6.6, 6.7, 6.8, and 6.9 show the predicted results from 1-step to 5-step of the vehicle volume obtained using the BP prediction model, respectively. Table 6.2 shows the predictive performance of the BP prediction models based on different advanced prediction steps. As can be seen from these figures and table:

(a) As can be seen from Figs. 6.4, 6.5, 6.6, 6.7 and 6.8, the BP prediction model driven based on single data has a good effect on traffic flow prediction, and the result of 1-step prediction is the best. The traffic flow is unstable during the morning and evening peaks, and the BP prediction model does not predict the jump points of these unstable regions as well as other sample regions.
(b) From Fig. 6.9, comparing the influence of different prediction steps on the prediction results, it can be found that the prediction effect will decrease as the number of advanced prediction steps increases, and more steps advanced, the delay phenomenon is more serious, that is, the prediction curve is like the right shift of the real curve. Combined with the performance estimates in Table 6.2, this phenomenon can be found more intuitively. Taking 1-step prediction and 5-step prediction as an example, the MAE, MAPE, RMSE, SDE, and R of 1-step prediction are 42.4613, 9.2225%, 54.5249, 54.5151, and 0.9838, respectively. The MAE, MAPE, RMSE, SDE, and R of 5-step prediction are 110.6633, 32.5640%, 136.8818, 136.6744, and 0.8930, respectively.

6.4 Traffic Flow Interval Prediction Model Driven by Single Data

6.4.1 *The Framework of the Interval Prediction Model*

In Sect. 6.3, the BP prediction model is used to forecast traffic flow, which is a specific predictive value of future traffic flows. In this section, a generalized autoregressive conditional heteroscedasticity (GARCH) is introduced to fit the predicted residuals of BP prediction model established in Sect. 6.3, so as to construct an interval prediction model based on GARCH.

Table 6.2 The performance estimating results of the BP prediction model

Model	Step	MAE	MAPE	RMSE	SDE	R
BP	1	42.4613	9.2225	54.5249	54.5151	0.9838
	2	59.3250	13.8013	75.2769	75.2596	0.9689
	3	76.2445	19.4015	96.6068	96.5943	0.9480
	4	92.7374	25.3732	118.2233	118.1690	0.9211
	5	110.6633	32.5640	136.8817	136.6744	0.8930

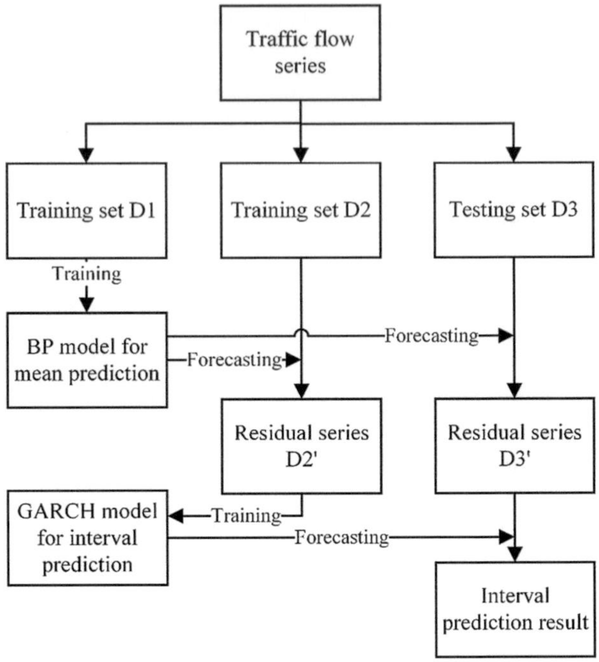

Fig. 6.10 Interval prediction model flow

The GARCH model is an improved model based on the ARCH model. In order to achieve an ideal fitting effect, the ARCH model often needs to select a larger order q, which greatly increases the calculation amount of the model. For heteroscedasticity series with long-term autoregressive characteristics, its practicability is poor, which affects the prediction accuracy of the overall model (Katsiampa 2017). Compared with the ARCH model, the GARCH (Bollerslev 1986) model increases the autoregressive term of heteroscedasticity under historical conditions, reduces the requirement of the model for order q, and thus reduces the calculation amount of the model on the premise of guaranteeing the accuracy of the model. The formula of GARCH is expressed as follows:

$$\begin{cases} x_t = x_{\text{mean}}^{t} + \mu_t \\ \mu_t = \sigma_t \rho_t \\ \sigma_t^2 = \theta + \sum_{i=1}^{p} \alpha_j \sigma_{t-i}^2 + \sum_{j=1}^{q} \beta_j \varepsilon_{t-j}^2 \end{cases} \tag{6.1}$$

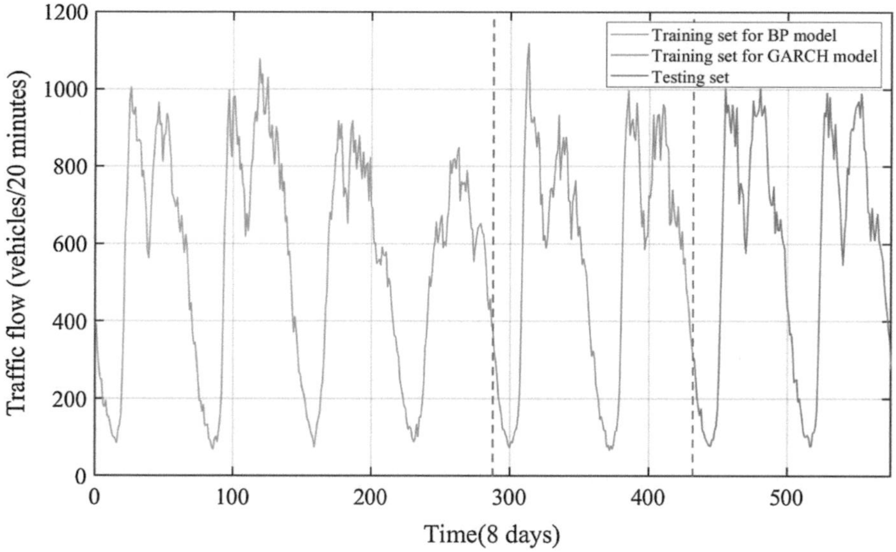

Fig. 6.11 The division of datasets

Restrictive conditions of related parameters of the GARCH model are expressed as follows:

$$\begin{cases} \theta > 0 \\ \alpha_j \geq 0, \beta_j \geq 0 \\ \sum_{i=1}^{p} \alpha_j + \sum_{j=1}^{q} \beta_j < 1 \end{cases} \tag{6.2}$$

Figure 6.10 shows the forecasting process for interval prediction in this section. First, the traffic flow data $\{V_{1t}\}$ are divided into three parts, namely, training set D1, training set D2, and test set D3. In the second step, the BP neural network is trained by the training set D1 to establish a prediction model, and the training set D2 and the test set D3 are predicted, then the residual series of D2 and D3 are obtained, respectively. In the third step, the boundary prediction model is built by using the predicted series, and residual series and the actual series of D2 are used to train the GARCH model. Finally, the upper and lower boundaries of the testing set D3 are predicted by the GARCH model, and the interval prediction results are obtained.

6.4.2 *Modeling Process*

The above outlines the process of establishing interval prediction. The interval prediction model based on the BP neural network and GACRH model can be divided into the following steps: (step 1) Dividing datasets, (step 2) establishing the BP prediction model, (step 3) establishing the GARCH model, and (step 4) testing the prediction performance of the model.

(Step 1) Dividing Datasets

Before building the model, divide the traffic flow series $\{V_{1t}\}$ into three parts, which are the training set of the BP prediction model, the training set of the ARCH model, and the testing set of the BP-ARCH interval model, respectively. Figure 6.11 shows the dataset partitioning of the traffic flow series $\{V_{1t}\}$. Data from days 1 to 4 (samples between 1 and 288, 288 samples) are used as a training set for the BP prediction model, data on days 5 and 6 (samples between 289 and 432, 144 samples) are used as the training set of the GARCH model, and data from the last 2 days (samples between 433 and 576, 144 samples) are used as the testing set to test the performance of the BP-GARCH interval model.

(Step 2) Establishing the BP Prediction Model

Since the training set of the BP prediction model divided in (step 1) is the same as that of Sect. 6.3, the BP prediction model established in this section is the same as the BP prediction model established in Sect. 6.3. The steps to establish a BP prediction model are given in Sect. 6.3.

(Step 3) Establishing the GARCH Model

The BP prediction model established in (step 2) is used to predict the training set of the GARCH model, and the predicted values of the samples between 289 and 432 are obtained. The actual value of the sample between 289 and 432 is subtracted from the predicted value to obtain the residual value. The significance levels are set to 0.1, 0.05, and 0.01, respectively, and three GARCH models are trained using the residual value at different significance levels.

(Step 4) Testing the Prediction Performance of the Model

The testing set is predicted using the BP prediction model established in (step 2), and the predicted values of the sample between 433 and 576 are obtained. The actual values of the sample between 433 and 576 are subtracted from the predicted values to obtain the residual values. The GARCH model is used to predict the prediction interval of the testing set at the significance level of 0.1, 0.05, and 0.01, respectively.

The performance evaluation indices of interval prediction are mainly to measure the effectiveness of the range of variation. There are three main indicators, namely the Prediction Interval Coverage Probability (PICP), the Prediction Interval Normalized Average Width (PINAW), and the Coverage Width-based Criterion (CWC). The three indicators are used to evaluate the performance of the interval prediction model in this section.

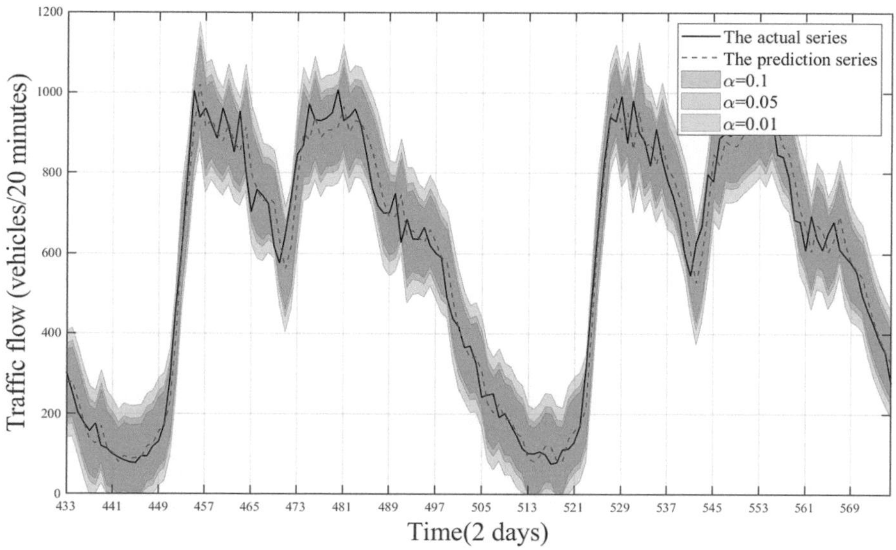

Fig. 6.12 The results of the 1-step interval prediction for the traffic flow series

The PICP calculates the proportion of the true value in the prediction interval and most directly measures the validity of the interval prediction model. The calculation formula is as follows:

$$\text{PICP} = \frac{1}{n}\sum_{t=1}^{n} c_t \tag{6.3}$$

where if $t(x) \in [L_t, U_t]$, then $c_t = 1$, otherwise $c_t = 0$. If the PICP index is less than the corresponding confidence level $1 - \alpha$, the model is invalid.

The PINAW indicator calculates the width of the prediction interval and normalizes it in consideration of the range of variation of the predicted value in the calculation process. The calculation formula is as follows:

$$\text{PINAW} = \frac{1}{nR}\sum_{t=1}^{n} \left(U_t - L_t\right) \tag{6.4}$$

where R is the difference between the maximum and minimum values of the predicted values, and the degree of change in the predicted values is measured.

The PINAW and PICP indices are highly correlated. If the PINAW indicator is high, the PICP indicator will be high, but this prediction cannot provide enough information for the actual situation. The CWC indicator can comprehensively

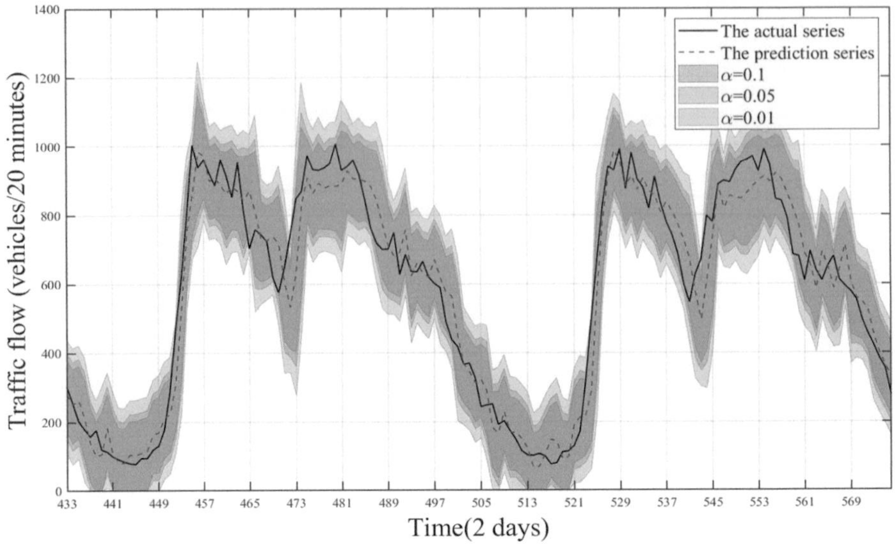

Fig. 6.13 The results of the 2-step interval prediction for the traffic flow series

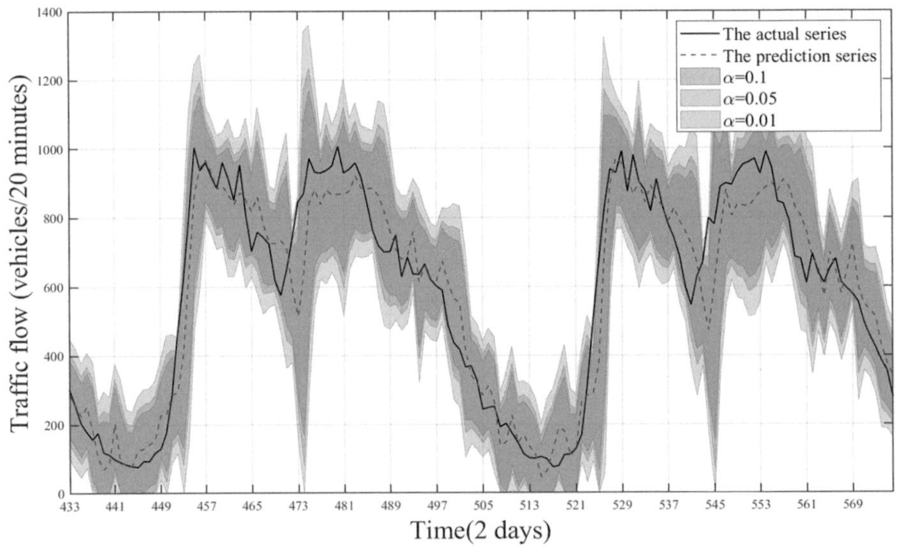

Fig. 6.14 The results of the 3-step interval prediction for the traffic flow series

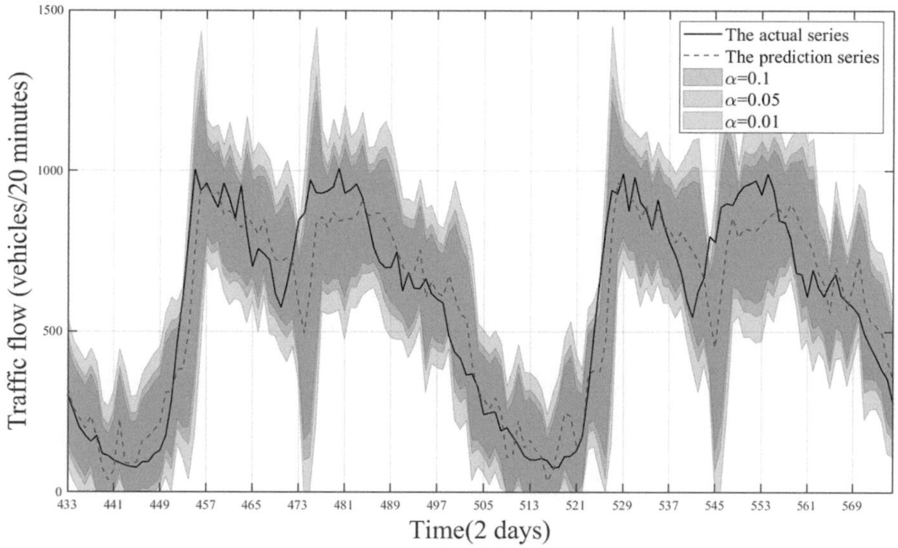

Fig. 6.15 The results of the 4-step interval prediction for the traffic flow series

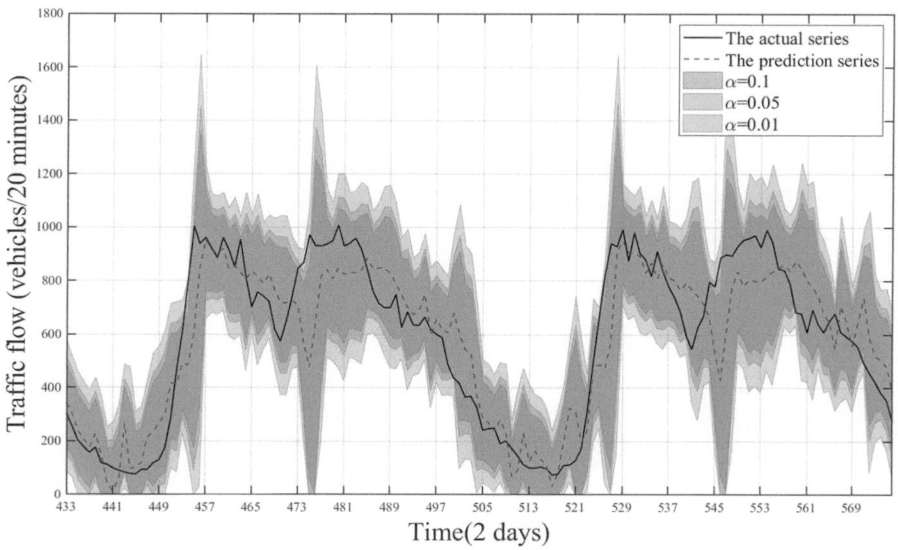

Fig. 6.16 The results of the 5-step interval prediction for the traffic flow series

Table 6.3 The performance estimating results of the interval prediction model

Model	Step	Significant level	PICP	PINAW	CWC
BP-ARCH	1	$\alpha = 0.1$	0.9028	0.2034	0.2034
		$\alpha = 0.05$	0.9653	0.2411	0.2411
		$\alpha = 0.01$	1.0000	0.3135	0.3135
	2	$\alpha = 0.1$	0.9167	0.2766	0.2766
		$\alpha = 0.05$	0.9583	0.3275	0.3275
		$\alpha = 0.01$	0.9861	0.4249	0.9410
	3	$\alpha = 0.1$	0.9097	0.3541	0.3541
		$\alpha = 0.05$	0.9583	0.4182	0.4182
		$\alpha = 0.01$	0.9861	0.5386	1.1929
	4	$\alpha = 0.1$	0.9236	0.4298	0.4298
		$\alpha = 0.05$	0.9583	0.5068	0.5068
		$\alpha = 0.01$	0.9861	0.6496	1.4387
	5	$\alpha = 0.1$	0.8889	0.4935	1.3537
		$\alpha = 0.05$	0.9444	0.5789	1.3432
		$\alpha = 0.01$	0.9931	0.7382	0.7382

consider the PICP and the PINAW to provide an objective and comprehensive evaluation. The calculation formula of CWC is as follows:

$$\mathrm{CWC} = \mathrm{PINAW}\left(1 + \xi\left(\mathrm{PICP}\right)e^{-\eta(\mathrm{PICP}-\xi)}\right) \tag{6.5}$$

where if PICP $\geq \xi$, then ξ(PICP) = 0, otherwise ξ(PICP) = 1. There are two important parameters to be set in the calculation process of the CWC indicator, namely ξ and η. Generally set to the confidence level $1 - \alpha$ corresponding to the prediction interval ξ. η measures the degree of punishment of the PICP below the significance level, generally taking the number between 50 and 100. In this section, η is set to 50.

6.4.3 The Prediction Results

Figures 6.12, 6.13, 6.14, 6.15, and 6.16 show the predicted results from 1-step to 5-step of the vehicle volume obtained using the BP-GARCH interval prediction models, respectively. Table 6.3 shows the predictive performance of the BP-GARCH interval prediction models based on different advanced prediction steps at different significance levels. As can be seen from these figures and table:

(a) It can be seen from these figures that during the prediction of traffic flow series $\{V_{1t}\}$, as the number of prediction steps increases, so does the interval width, which means that the accuracy of model prediction decreases.

(b) A more detailed model prediction performance can be seen in Table 6.3. The prediction effect is worse with the increase in the predicted number of steps. Taking the significance level as 0.1 as an example, the models with different advance prediction steps are compared. The PICP, PINAW, and CWC indicators of the 1-step prediction are 0.9028, 0.2034, and 0.2034, respectively. The PICP, PINAW, and CWC indicators of the 2-step prediction are 0.9167, 0.2766, and 0.2766, respectively. The PICP, PINAW, and CWC indicators of the 3-step prediction are 0.9097, 0.3541, and 0.3541, respectively. The PICP, PINAW, and CWC indicators of the 4-step prediction are 0.9236, 0.4298, and 0.4298, respectively. The PICP, PINAW, and CWC indicators of the 5-step prediction are 0.8889, 0.4935, and 1.3537, respectively. This shows that the gradual increase of the mean prediction residual caused by the increase of the predicted step number definitely affects the robustness and accuracy of the model.

6.5 Traffic Flow Interval Prediction Under Decomposition Framework

6.5.1 The Framework of the WD-BP-GARCH Prediction Model

Wavelet transform is developed on the basis of the short-time Fourier transform (Daubechies 1990). A change window that can be translated in the time axis is obtained to realize the observation of the time-frequency characteristics of the time sequence.

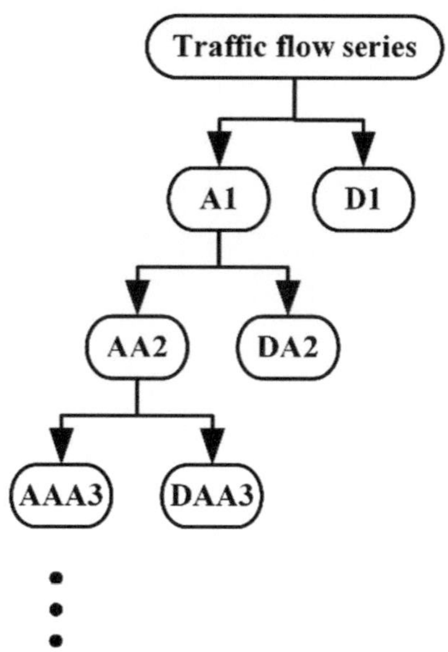

Fig. 6.17 The structure of the WD

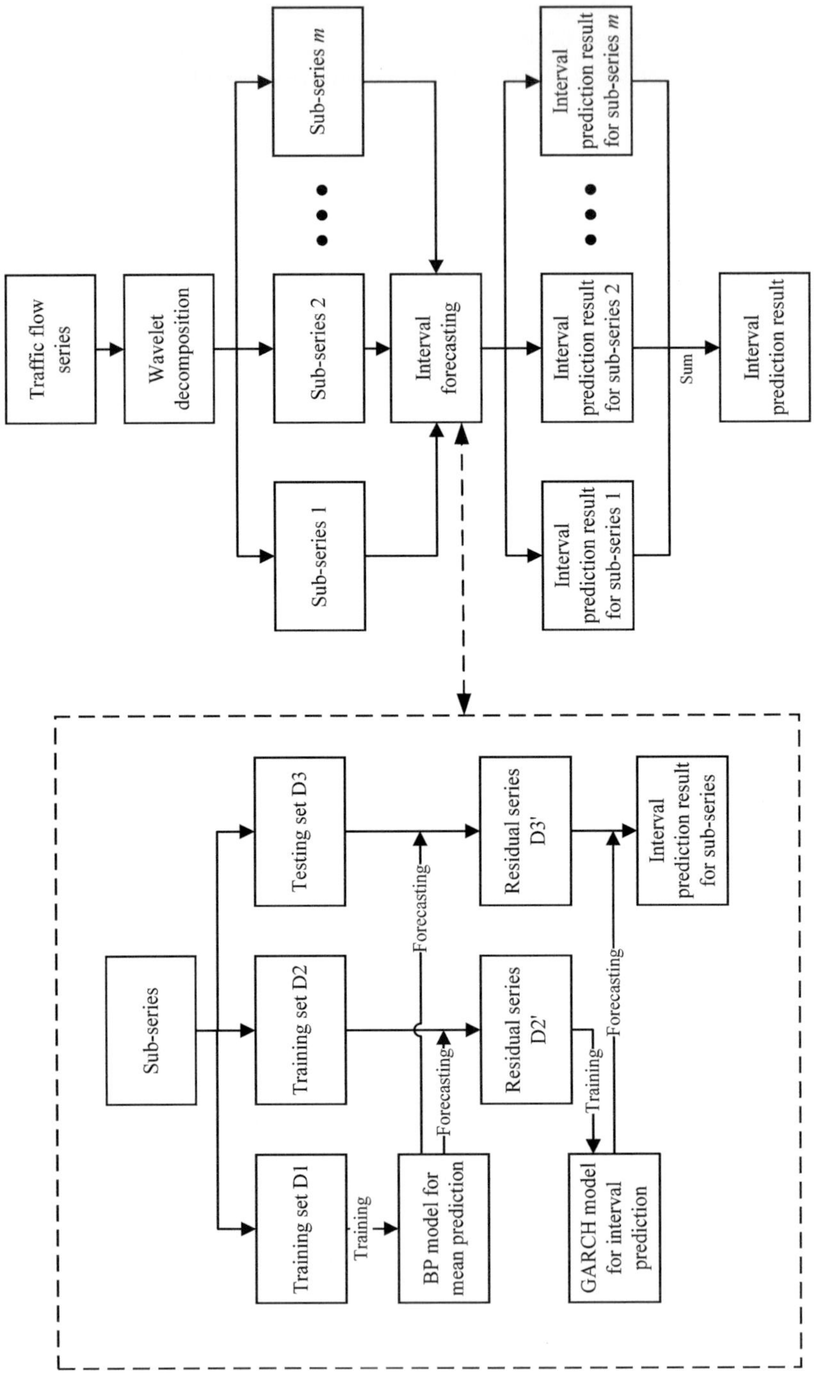

Fig. 6.18 The prediction process of the WD-BP-GARCH model for traffic flow series

The change window varies with the frequency of the proposed frequency, and the sampling step size used in the time domain for different frequency components is variable. A small sampling step size for high-frequency components can be used to extract high-frequency information more finely. A larger sampling step size for low-frequency components can obtain more frequent low-frequency fluctuation characteristics.

Wavelet decomposition includes continuous wavelet transform (CWT) and discrete wavelet transform (DWT). Since the time sequence in the actual engineering application is usually a discrete point series, the discrete wavelet transform is more effective. The discrete wavelet transform is divided into two forms. The first one is that the signal to be decomposed is discrete (such as the traffic flow series used in this chapter) and $f(qt_0)$ is used to characterize the discrete series, where $q = 1, 2, \ldots, N$, N is the total number of sample points and t_0 is the sampling interval. The transformation formula of discrete wavelet is as follows:

$$D(\alpha,\beta)=|\alpha|^{-\frac{1}{2}}t\sum_{q=1}^{N}f(qt_0)\psi\left(\frac{t-\beta}{\alpha}\right) \tag{6.6}$$

where α and β represent the scale factor and the translation factor, respectively. The scale factor reflects the scaling performance of the wavelet basis function, and the translation factor reflects the translation performance of the wavelet basis function along the time axis.

The second form of expression is to discretize α and β to obtain a discrete wavelet transform. Take $\alpha=\alpha_0^p, \beta=q\beta_0^p; p,q\in Z$, it can discretize the wavelet transform as follows:

$$D(p,q)=|\alpha_0|^{-p/2}\int_{-\infty}^{+\infty}f(t)\,\psi\left(\alpha_0^{-p}t-q\beta_0^p\right)dt \tag{6.7}$$

If $\alpha_0 = 2$, $\beta_0 = 1$, a binary discretization wavelet transform of the original signal is obtained, which is expressed as follows:

$$D(p,q)=2^{-p/2}\int_{-\infty}^{+\infty}f(t)\,\psi\left(2^{-p}t-q\right)dt \tag{6.8}$$

Figure 6.17 shows the structure of the wavelet decomposition. Figure 6.18 shows the prediction process of the WD-BP-GARCH interval prediction model based on wavelet decomposition in this section. In this section, the traffic flow series is decomposed into a series of sub-series, and then each sub-series is subjected to the BP-GARCH interval prediction. The structure of the BP-GARCH interval prediction model is the same as that of Sect. 6.4. Finally, the predicted results of each sub-series are summed to get the final predicted results.

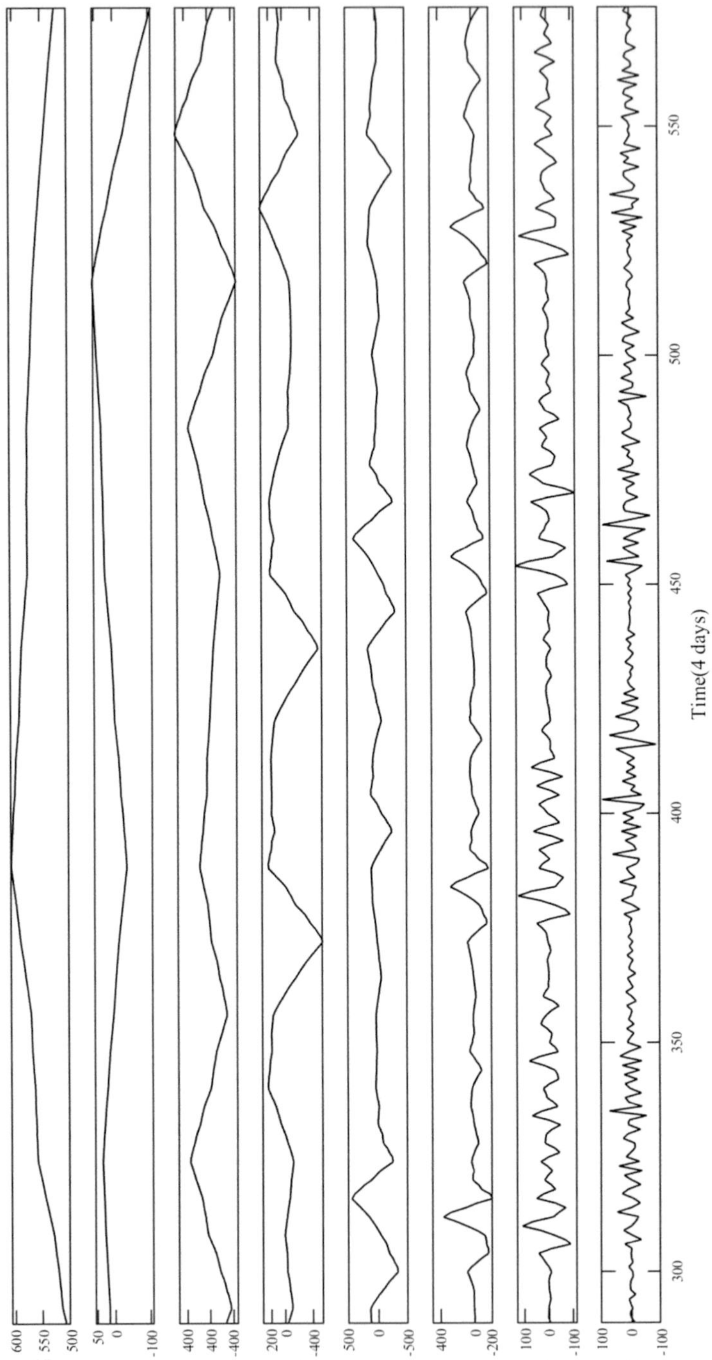

Fig. 6.19 The eight sub-series of the traffic flow series based on the WD

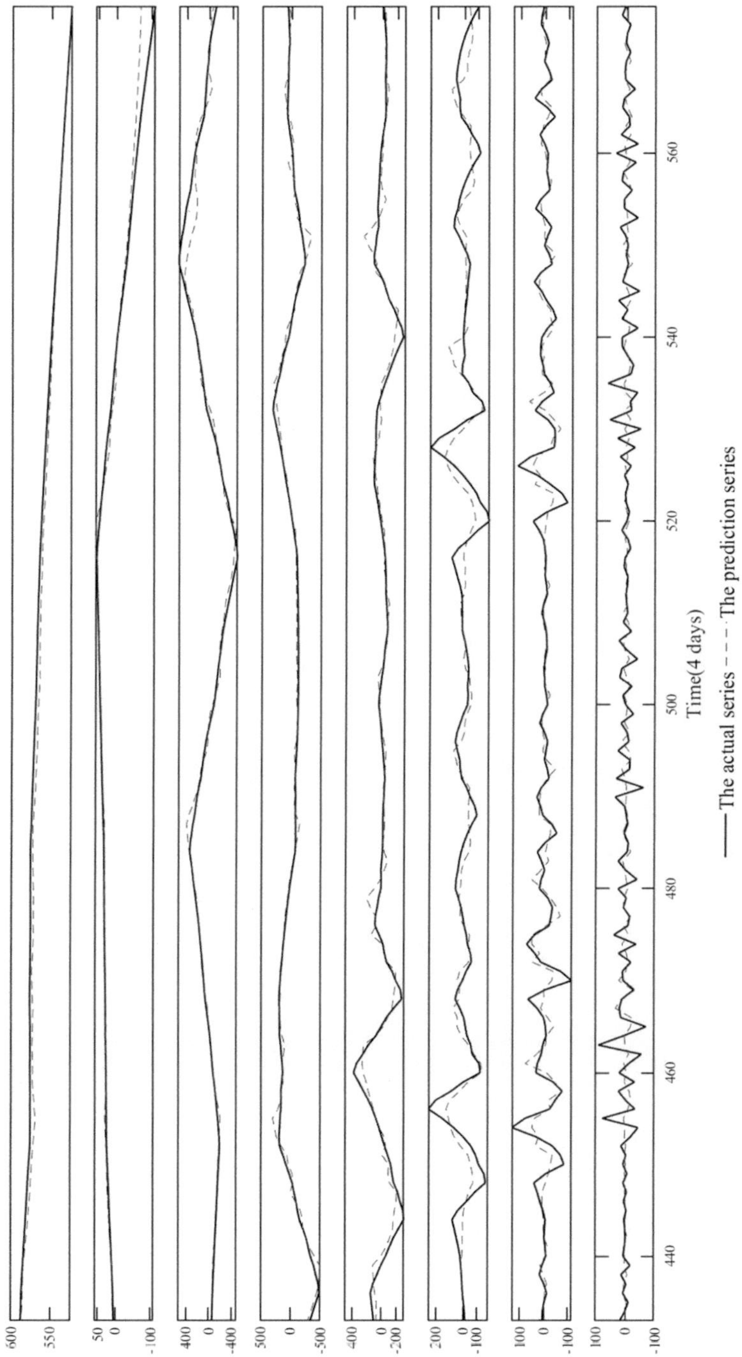

Fig. 6.20 The prediction results of 3-step for each sub-series based on the BP prediction model

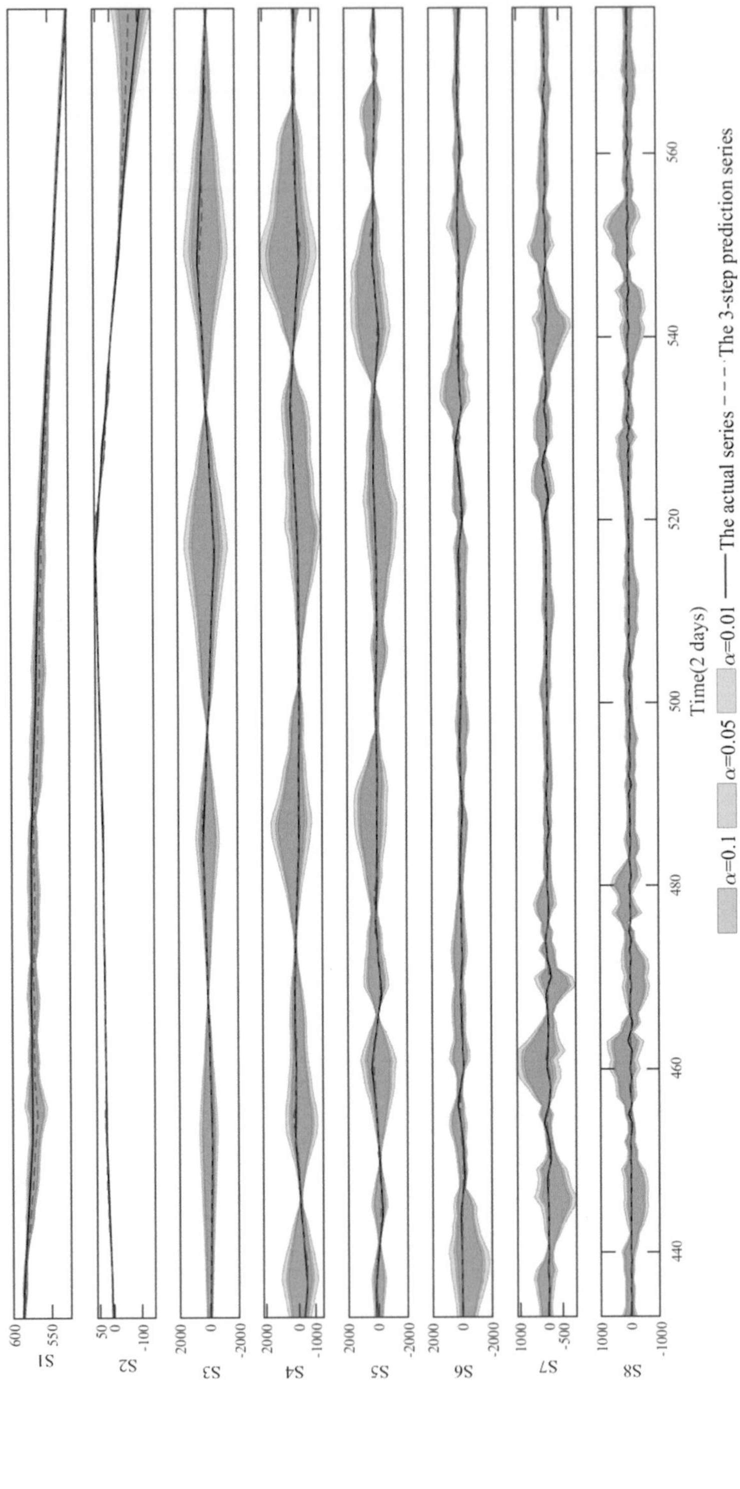

Fig. 6.21 The prediction results of 3-step for each sub-series based on the BP-GARCH interval model

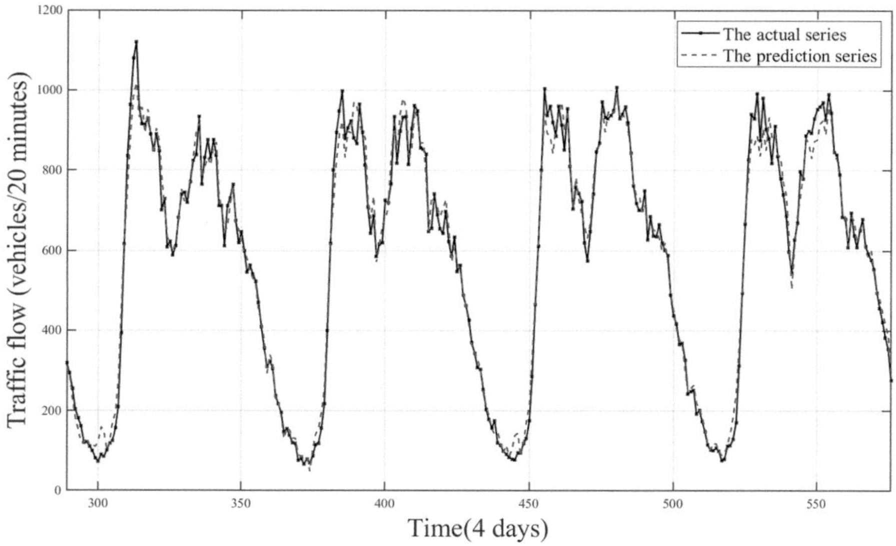

Fig. 6.22 The results of the 1-step prediction for the traffic flow series

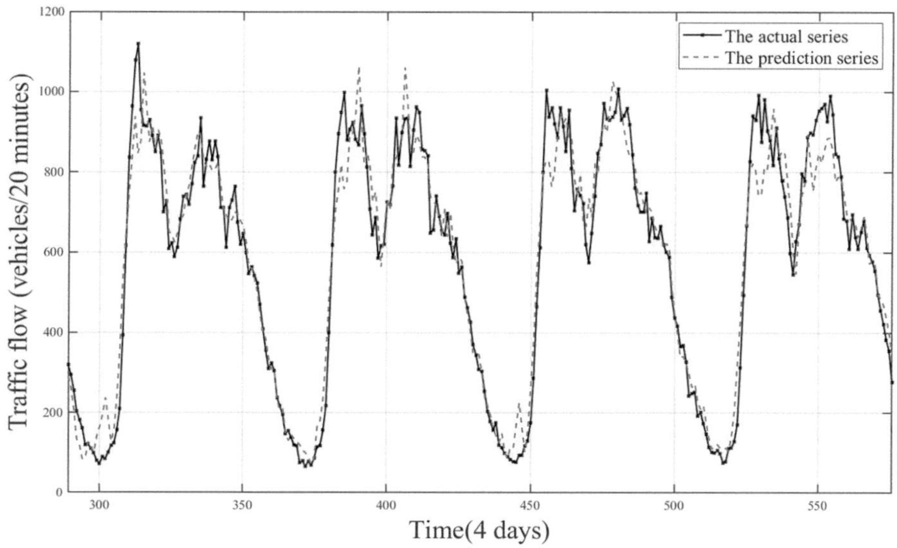

Fig. 6.23 The results of the 2-step prediction for the traffic flow series

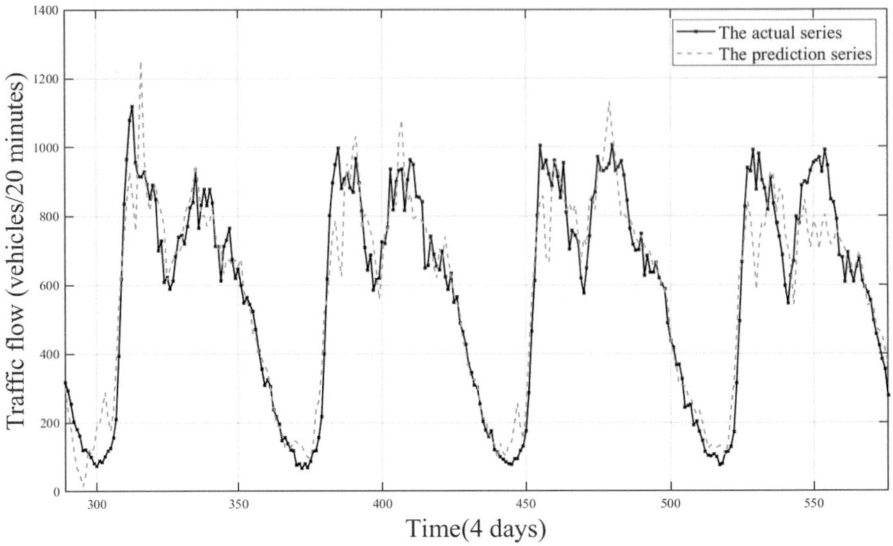

Fig. 6.24 The results of the 3-step prediction for the traffic flow series

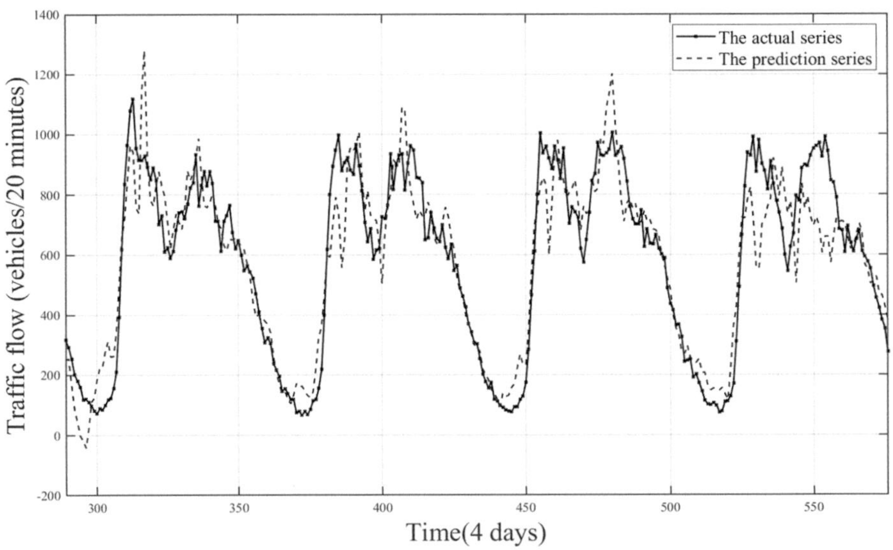

Fig. 6.25 The results of the 4-step prediction for the traffic flow series

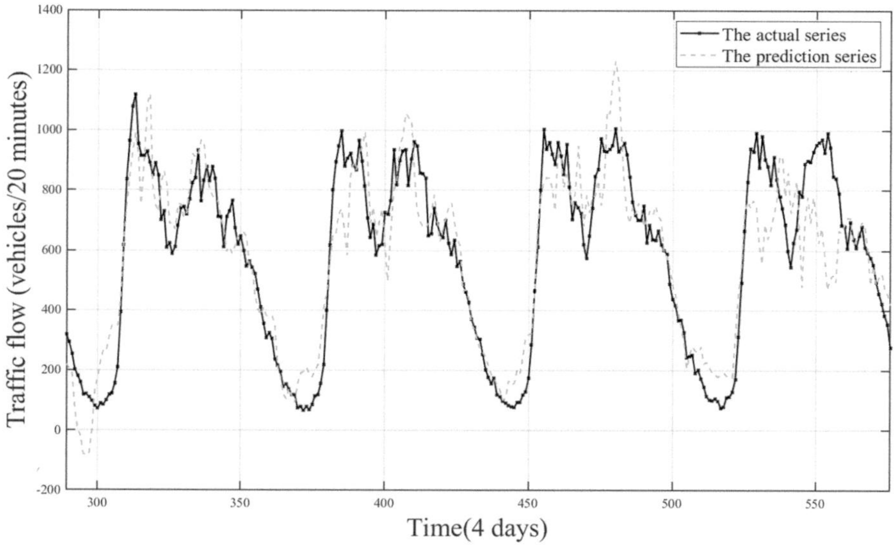

Fig. 6.26 The results of the 5-step prediction for the traffic flow series

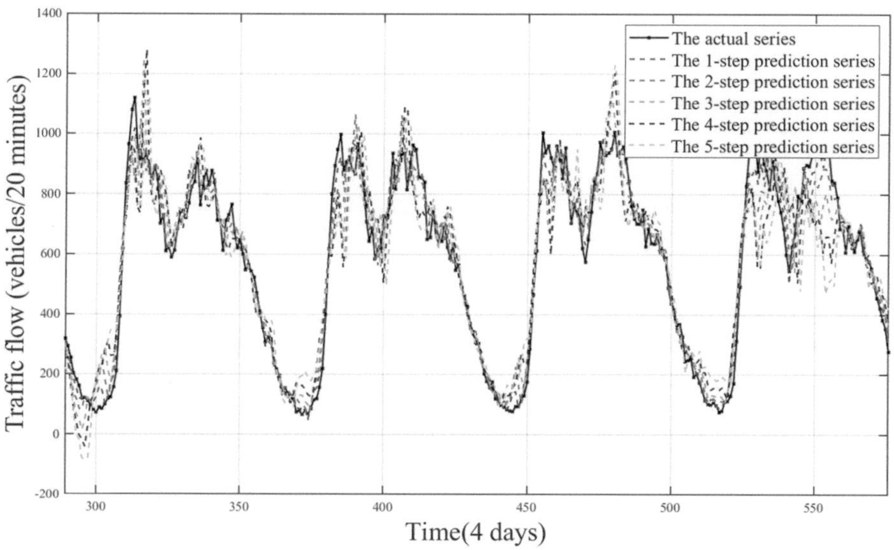

Fig. 6.27 Comparison of prediction results based on different prediction steps

Table 6.4 The performance estimating results of the WD-BP prediction model

Model	Step	MAE	MAPE	RMSE	SDE	R
WD-BP	1	24.3219	7.7160	31.4853	31.1830	0.9951
	2	45.5881	13.7582	61.0405	60.6710	0.9807
	3	61.5208	18.0669	78.3052	78.1725	0.9676
	4	77.2389	22.0487	98.9294	98.8916	0.9465
	5	90.0611	25.5334	113.6245	113.4140	0.9280

6.5.2 *Modeling Process*

The wavelet decomposition algorithm, the BP neural network, and GARCH model are used to establish the WD-BP-GARCH interval prediction model driven by single data, which includes five steps: (step 1) decomposing the traffic flow series into eight sub-series by the WD, (step 2) establishing the BP-GARCH interval prediction model for each series, (step 3) interval prediction for each sub-series, (step 4) summing the prediction results of each sub-series, and (step 5) testing the prediction performance.

(Step 1) Decomposing the Traffic Flow Series Into Eight Sub-Series by the WD
The WD is used to decompose the traffic flow series into eight sub-series. Figure 6.19 shows the decomposition series of the traffic flow series $\{V_{1t}\}$ in samples between 288 and 576. The eight decomposition sub-series are represented by S1–S8, respectively.

(Step 2) Establishing the BP-GARCH Interval Prediction Model for Each Series
Each sub-series are treated as a research series, and eight BP-GARCH interval prediction models of the eight sub-series are established. The BP-GARCH interval prediction model still uses the BP-ARCH interval prediction model described in Sect. 6.4.

(Step 3) Interval Prediction for Each Sub-Series
Figure 6.20 shows the prediction results for each sub-series based on the BP prediction model. Figure 6.21 shows the prediction results for each sub-series based on the BP-GARCH interval model. It can be seen from the figures that after adding wavelet decomposition, whether it is based on the BP neural network deterministic prediction or interval prediction model based on interval prediction, the prediction effect of each sub-series is good.

(Step 4) Summing the Prediction Results of Each Sub-Series
The prediction results of the eight sub-series are summed to obtain the final prediction result.

(Step 5) Testing the Prediction Performance
Section 6.4 introduces three indices that can well measure the prediction performance of the interval prediction model, which are PICP, PINAW, and CWC. The three indicators are also used to evaluate the performance of the WD-BP-GARCH interval prediction model.

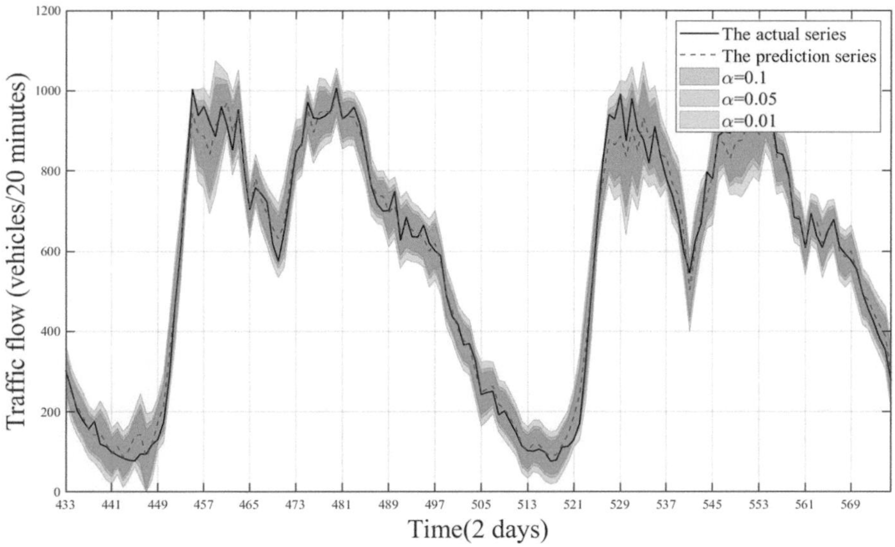

Fig. 6.28 The results of the 1-step interval prediction for the traffic flow series

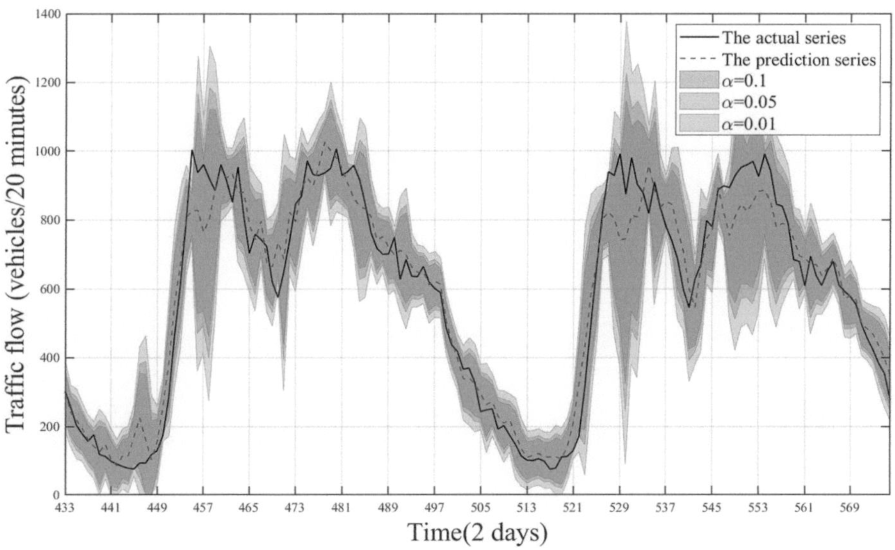

Fig. 6.29 The results of the 2-step interval prediction for the traffic flow series

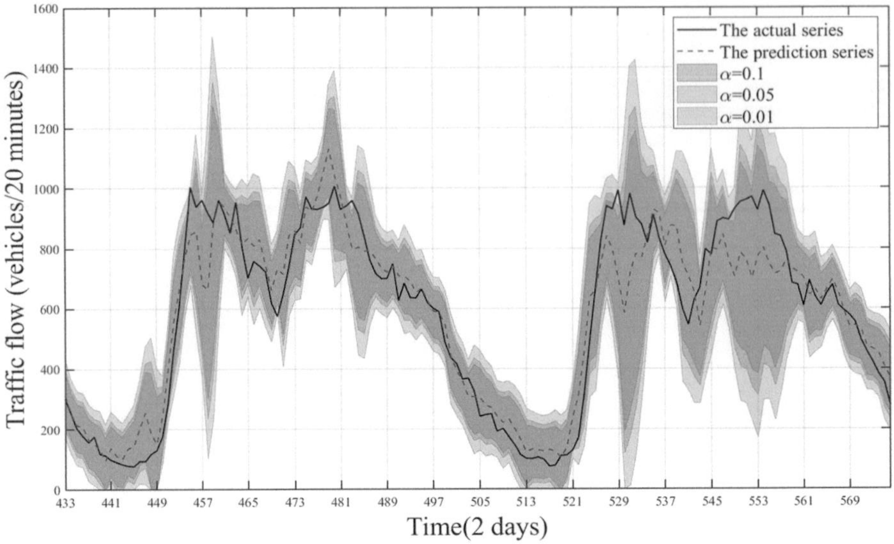

Fig. 6.30 The results of the 3-step interval prediction for the traffic flow series

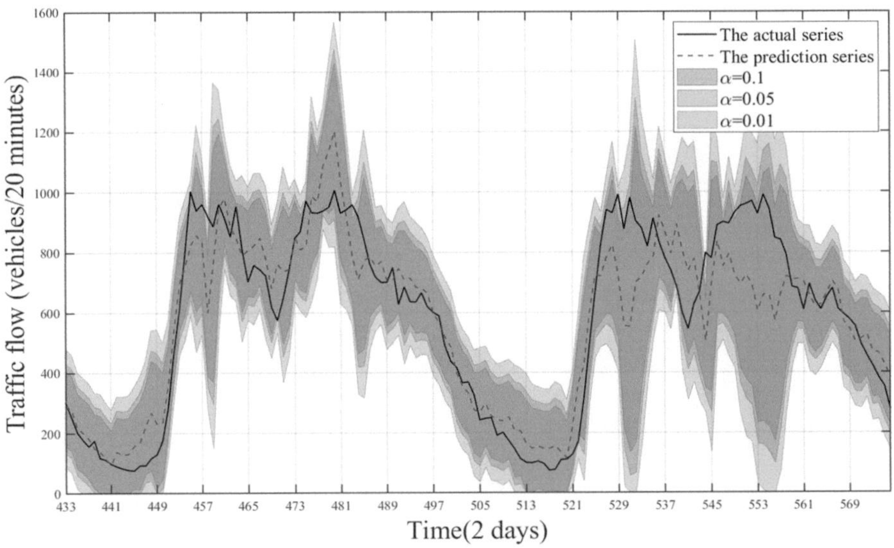

Fig. 6.31 The results of the 4-step interval prediction for the traffic flow series

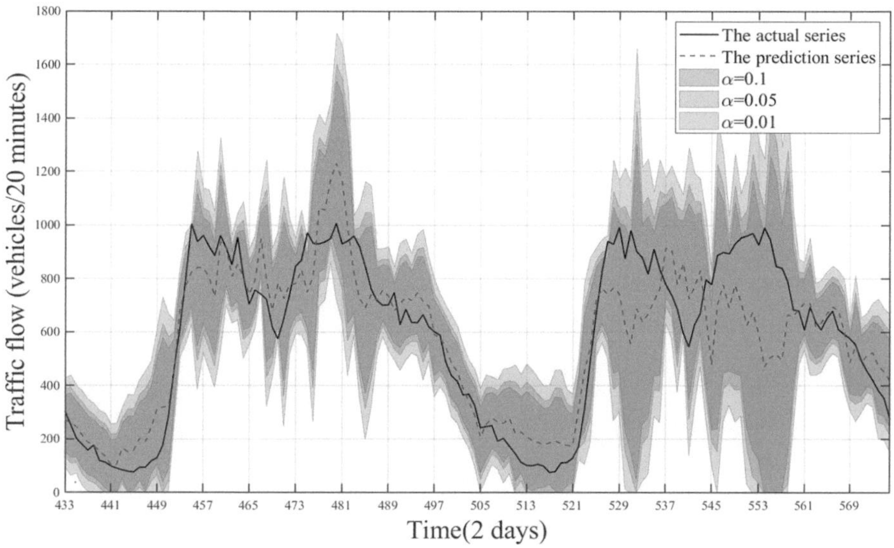

Fig. 6.32 The results of the 5-step interval prediction for the traffic flow series

6.5.3 The Prediction Results

Figures 6.22, 6.23, 6.24, 6.25, 6.26, and 6.27 show the predicted results from 1-step to 5-step of the vehicle volume obtained using the WD-BP prediction model, respectively. Table 6.4 shows the predictive performance of the WD-BP prediction models based on different advanced prediction steps. As can be seen from these figures and table:

(a) As can be seen from Figs. 6.22, 6.23, 6.24, 6.25, 6.26, and 6.27, the WD-BP prediction model based on single data has a good effect on traffic flow prediction, and the result of the advance of one-step prediction is the best.

    The traffic flow is unstable in the morning and evening rush hours, and the WD-BP prediction model cannot predict the jump point of these unstable regions like other sample regions.

(b) From Fig. 6.27, comparing the influence of different prediction steps on the prediction results, it can be found that the prediction effect will decrease as the number of advanced prediction steps increases, and more steps advanced, the delay phenomenon is more serious, that is, the prediction curve is like the right shift of the real curve. Combined with the performance estimates in Table 6.4, this phenomenon can be found more intuitively. Taking 1-step prediction and 5-step prediction as an example, the MAE, MAPE, RMSE, SDE, and R of 1-step prediction are 24.3219, 7.7160%, 31.4853, 31.1830, and 0.9951, respectively. The MAE, MAPE, RMSE, SDE, and R of 5-step prediction are 90.0611, 25.5334%, 113.6245, 113.4140, and 0.9280, respectively.

Table 6.5 The performance estimating results of the WD-BP-GARCH prediction model

Model	Step	Significant level	PICP	PINAW	CWC
WD-BP-GARCH	1	$\alpha = 0.1$	0.9028	0.1213	0.1213
		$\alpha = 0.05$	0.9514	0.1446	0.1446
		$\alpha = 0.01$	0.9861	0.1899	0.4206
	2	$\alpha = 0.1$	0.9097	0.2332	0.2332
		$\alpha = 0.05$	0.9444	0.2776	0.6440
		$\alpha = 0.01$	0.9861	0.3626	0.8030
	3	$\alpha = 0.1$	0.8889	0.2914	0.7994
		$\alpha = 0.05$	0.9306	0.3468	1.2635
		$\alpha = 0.01$	1.0000	0.4528	0.4528
	4	$\alpha = 0.1$	0.8958	0.3690	0.8234
		$\alpha = 0.05$	0.9306	0.4365	1.5905
		$\alpha = 0.01$	0.9792	0.5621	1.5282
	5	$\alpha = 0.1$	0.8889	0.4116	1.1289
		$\alpha = 0.05$	0.9444	0.4848	1.1249
		$\alpha = 0.01$	0.9792	0.6172	1.6780

Figures 6.28, 6.29, 6.30, 6.31, and 6.32 show the traffic flow prediction results from 1-step to 5-step obtained by the WD-BP-GARCH interval prediction model, respectively. Table 6.5 shows the predictive performance of the WD-BP-GARCH interval prediction models based on different advanced prediction steps at different significance levels. As can be seen from these figures and tables:

(1) The higher the significance level is, the lower the confidence level will be, leading to the width of prediction interval smaller. As the number of prediction steps increases, the predicted interval width will increase. When the significance level is smaller, the width of the predicted interval will increase more, and the jump will be more intense and unstable. Therefore, the significance level can be set according to the actual traffic situation to better predict the range of traffic flow.
(2) The prediction effect is worse with the increase in the predicted number of steps. Taking the significance level as 0.1 as an example, the models with different prediction steps are compared. The PICP, PINAW, and CWC indices of the 1-step prediction are 0.9028, 0.1213, and 0.1213, respectively. The PICP, PINAW, and CWC indicators of 2-step prediction are 0.9097, 0.2332, and 0.2332, respectively. The PICP, PINAW, and CWC indices of 3-step prediction are 0.8889, 0.2914, and 0.7994, respectively. The PICP, PINAW, and CWC indices of 4-step prediction are 0.8958, 0.3690, and 0.8234, respectively. The PICP, PINAW, and CWC indices of 5-step prediction are 0.8889, 0.4116, and 1.1289, respectively.

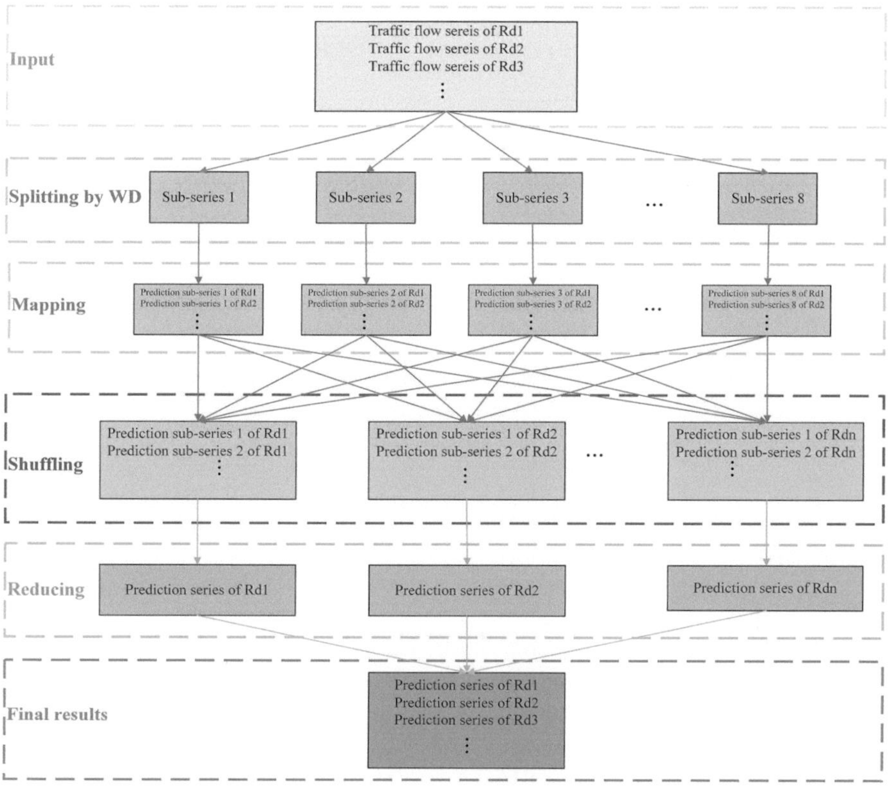

Fig. 6.33 The MapReduce prediction architecture of traffic flow

Table 6.6 The performance estimating results of the deterministic prediction models

Model	Step	MAE	MAPE	RMSE	SDE	R
BP	1	42.4613	9.2225	54.5249	54.5151	0.9838
	2	59.3250	13.8013	75.2769	75.2596	0.9689
	3	76.2445	19.4015	96.6068	96.5943	0.9480
	4	92.7374	25.3732	118.2233	118.1690	0.9211
	5	110.6633	32.5640	136.8817	136.6744	0.8930
WD-BP	1	24.3219	7.7160	31.4853	31.1830	0.9951
	2	45.5881	13.7582	61.0405	60.6710	0.9807
	3	61.5208	18.0669	78.3052	78.1725	0.9676
	4	77.2389	22.0487	98.9294	98.8916	0.9465
	5	90.0611	25.5334	113.6245	113.4140	0.9280

Table 6.7 The performance estimating results of the interval prediction model

Model	Step	Significant level	PICP	PINAW	CWC
BP-GARCH	1	$\alpha = 0.1$	0.9028	0.2034	0.2034
		$\alpha = 0.05$	0.9653	0.2411	0.2411
		$\alpha = 0.01$	1.0000	0.3135	0.3135
	2	$\alpha = 0.1$	0.9167	0.2766	0.2766
		$\alpha = 0.05$	0.9583	0.3275	0.3275
		$\alpha = 0.01$	0.9861	0.4249	0.9410
	3	$\alpha = 0.1$	0.9097	0.3541	0.3541
		$\alpha = 0.05$	0.9583	0.4182	0.4182
		$\alpha = 0.01$	0.9861	0.5386	1.1929
	4	$\alpha = 0.1$	0.9236	0.4298	0.4298
		$\alpha = 0.05$	0.9583	0.5068	0.5068
		$\alpha = 0.01$	0.9861	0.6496	1.4387
	5	$\alpha = 0.1$	0.8889	0.4935	1.3537
		$\alpha = 0.05$	0.9444	0.5789	1.3432
		$\alpha = 0.01$	0.9931	0.7382	0.7382
WD-BP-ARCH	1	$\alpha = 0.1$	0.9028	0.1213	0.1213
		$\alpha = 0.05$	0.9514	0.1446	0.1446
		$\alpha = 0.01$	0.9861	0.1899	0.4206
	2	$\alpha = 0.1$	0.9097	0.2332	0.2332
		$\alpha = 0.05$	0.9444	0.2776	0.6440
		$\alpha = 0.01$	0.9861	0.3626	0.8030
	3	$\alpha = 0.1$	0.8889	0.2914	0.7994
		$\alpha = 0.05$	0.9306	0.3468	1.2635
		$\alpha = 0.01$	1.0000	0.4528	0.4528
	4	$\alpha = 0.1$	0.8958	0.3690	0.8234
		$\alpha = 0.05$	0.9306	0.4365	1.5905
		$\alpha = 0.01$	0.9792	0.5621	1.5282
	5	$\alpha = 0.1$	0.8889	0.4116	1.1289
		$\alpha = 0.05$	0.9444	0.4848	1.1249
		$\alpha = 0.01$	0.9792	0.6172	1.6780

6.6 Big Data Prediction Architecture of Traffic Flow

It is very easy and convenient to forecast the traffic flow of a certain segment. However, in order to make real-time traffic flow prediction for each segment of a region, it is necessary to obtain daily traffic flow data and establish traffic flow prediction model at the same time. There are thousands of roads in an area, and the amount of data that needs to be processed is very large. Therefore, it is necessary to establish a big data prediction architecture of traffic flow to realize intelligent transportation system.

The Hadoop distributed file system is an Apache Software Foundation that supports a reliable, scalable, and open-source distributed computing framework

(Shvachko et al. 2010). Hadoop's parallel distributed computing structure can efficiently process big data of traffic flow and realize real-time prediction of traffic flow of each road segment in the whole road network.

MapReduce is a programming model in the Hadoop for parallel computation of big datasets (Dean and Ghemawat 2008). The big data prediction architecture of traffic flow is mainly established under the MapReduce model. Figure 6.33 shows the Map/Reduce prediction architecture of traffic flow. Firstly, the wavelet decomposition algorithm is used to decompose the traffic flow series of all segments into eight sub-series. Then, eight Maps are established, and each Map is used to process traffic flow sub-series of a frequency range. In each map, predictive data corresponding to traffic flow series of all road segments can be obtained. Finally, the predicted traffic flow value of each road segment is obtained by combining the predicted sub-series of each road segment with the Reduce module.

6.7 Comparative Analysis of Forecasting Performance

In Sect. 6.3, the BP prediction model used for deterministic prediction of traffic flow is established. In Sect. 6.4, the BP-GARCH interval prediction model is established based on the BP prediction model for deterministic prediction and combined with the GARCH model. In Sect. 6.5, combined with wavelet decomposition algorithm, the WD-BP prediction model is established for deterministic prediction and the GARCH model is used for interval prediction, which is called WD-BP-GARCH interval prediction model.

The performance comparison of the above models is divided into two aspects. On the one hand, the performance of the mean prediction model is compared, that is, the performance of the deterministic prediction model. Deterministic prediction models include the BP prediction model and WD-BP prediction model. Table 6.6 compares the prediction performance of these two models. On the other hand, interval prediction performance of interval prediction models is compared, including the BP-GARCH interval prediction model and the WD-BP-GARCH interval prediction model. Table 6.7 shows the interval prediction performance of these two interval prediction models.

From these tables, the following conclusions can be drawn:

(a) The BP prediction model and the WD-BP prediction model are used to predict the mean value of traffic flow. It can be seen from Table 6.6 that, regardless of the value of prediction steps, the prediction performance of the WD-BP prediction model is always higher than that of the BP prediction model. Taking 1-step prediction as an example, the MAE, MAPE, RMSE, SDE, and R of the BP prediction model are 42.4613, 9.2225%, 54.5249, 54.5151, and 0.9838, respectively. The MAE, MAPE, RMSE, SDE, and R of the WD-BP prediction model are 24.3219, 7.7160 %, 31.4853, 31.1830, and 0.9951, respectively.

(b) In deterministic prediction, whether the prediction model is, the prediction performance decreases with the increase of prediction steps. Detailed performance index values can be seen in Table 6.6. Taking the BP prediction model as an example, the MAE, MAPE, RMSE, SDE, and R of 1-step prediction are 42.4613, 9.2225%, 54.5249, 54.5151, and 0.9838, respectively. The MAE, MAPE, RMSE, SDE, and R of 2-step prediction are 59.3250, 13.8013%, 75.2769, 75.2596, and 0.9689, respectively. The MAE, MAPE, RMSE, SDE, and R of 3-step prediction are 76.2445, 19.4015%, 96.6068, 96.5943, and 0.9480, respectively. The MAE, MAPE, RMSE, SDE, and R of 4-step prediction are 92.7374, 25.3732%, 118.2233, 118.1690, and 0.9211, respectively. The MAE, MAPE, RMSE, SDE, and R of 5-step prediction are 110.6633, 32.5640%, 136.8817, 136.6744, and 0.8930, respectively.

(c) The BP-GARCH interval prediction model and the WD-BP-GARCH interval prediction model are used to predict the range of the interval of traffic flow. It can be seen from Table 6.7 that, regardless of the value of prediction steps, the prediction performance of the WD-BP-GARCH interval prediction model is always higher than that of the BP-GARCH interval prediction model. Taking 1-step prediction at the significance level of 0.1 as an example, the PICP, PINAW, and CWC of the BP-GARCH interval prediction model are 0.9028, 0.2034, and 0.2034, respectively. The PICP, PINAW, and CWC of the WD-BP-GARCH interval prediction model are 0.9028, 0.1213, and 0.1213, respectively.

(d) In interval prediction, whether it is the BP-GARCH interval prediction model or the WD-BP-GARCH interval prediction model, the prediction performance decreases with the increase of prediction steps. Detailed performance index values can be seen in Table 6.7. Taking the BP-GARCH interval prediction model at the significance level of 0.1 as an example, the PICP, PINAW, and CWC of 1-step prediction are 0.9028, 0.2034, and 0.2034, respectively. The PICP, PINAW, and CWC of 2-step prediction are 0.9167, 0.2766, and 0.2766, respectively. The PICP, PINAW, and CWC of 3-step prediction are 0.9097, 0.3541, and 0.2034, respectively. The PICP, PINAW, and CWC of 4-step prediction are 0.9028, 0.2034, and 0.4298, respectively. The PICP, PINAW, and CWC of 5-step prediction are 0.8889, 0.4935, and 1.3537, respectively.

(e) With the decrease of significance level, the prediction performance of the interval prediction model also decreases. Detailed performance index values can be seen in Table 6.7. Taking the WD-BP-GARCH interval prediction model for 1-step prediction as an example, the PICP, PINAW, and CWC at significant level 0.1 are 0.9028, 0.1213, and 0.1213, respectively. The PICP, PINAW, and CWC at significant level 0.05 are 0.9514, 0.1446, and 0.1446, respectively. The PICP, PINAW, and CWC at significant level 0.01 are 0.9861, 0.1899, and 0.4206, respectively.

6.8 Conclusion

This chapter studies the traffic flow prediction model based on single data drive. Traffic flow series is a typical time series data, which has some variation rules in time. In the running process of traffic vehicles, the historical data of traffic flow have a direct impact on future traffic flow. Because residential lives and work time has a certain rule, traffic flow series has a strong periodic change rule, such as the large traffic flow in the morning and evening rush hours. Deterministic prediction can more accurately find the change value of traffic flow, and interval prediction can find the range of change of traffic flow. Both prediction methods have an important role in the improvement of intelligent transportation systems.

In this chapter, the traffic flow prediction models are established by using the historical data of the traffic flow series. The models are divided into deterministic prediction model and interval prediction model, which are, respectively, composed of the BP prediction model, the WD-BP prediction model, the BP-GARCH interval prediction model, and the WD-BP-GARCH interval prediction model. According to the experimental results, the following conclusions can be drawn:

(a) In the deterministic prediction of traffic flow, the prediction performance of the WD-BP prediction model is higher than that of the BP prediction model, and the WD-BP prediction model can better deal with jump points and has a better prediction effect. This shows that using wavelet decomposition algorithm to decompose traffic flow series according to different frequencies and predicting each sub-series, respectively, can improve the prediction effect.
(b) In the interval prediction of traffic flow, the BP neural network is used to establish the deterministic prediction model, and the GARCH model is used to calculate the uncertainty of the predicted traffic flow. By comparing the prediction performance of the interval prediction models, the performance of the WD-BP-GARCH interval prediction model is better than that of the BP-GARCH interval prediction model. The value of significance level also has an influence on the prediction accuracy of the interval prediction models. The significance level is smaller, and the prediction performance of the interval prediction model is lower.
(c) Whether it is deterministic prediction or interval prediction, the wavelet decomposition algorithm can improve the prediction performance of the models and provide more accurate prediction results. And as the number of predicted steps increases, the predictive performance of the model will decrease.

References

Bernardo, D., Hagras, H., & Tsang, E. (2012). An interval type-2 fuzzy logic system for the modeling and prediction of financial applications. In *International conference on autonomous and intelligent systems* (pp. 95–105). Springer.

Bollerslev T (1986) Generalized autoregressive conditional heteroskedasticity. Journal of Econometrics 31(3):307–327
Daubechies I (1990) The wavelet transform, time-frequency localization and signal analysis. IEEE Transactions on Information Theory 36(5):961–1005
Dean J, Ghemawat S (2008) MapReduce: Simplified data processing on large clusters. Communications of the ACM 51(1):107–113
El-Fouly T, El-Saadany E, Salama M (2006) One day ahead prediction of wind speed using annual trends. In: 2006 IEEE power engineering society general meeting. IEEE
Guan C, Luh PB, Michel LD, Chi Z (2013) Hybrid Kalman filters for very short-term load forecasting and prediction interval estimation. IEEE Transactions on Power Systems 28(4):3806–3817
Jiang Y, Song Z, Kusiak A (2013) Very short-term wind speed forecasting with Bayesian structural break model. Renewable Energy 50:637–647
Katsiampa P (2017) Volatility estimation for Bitcoin: A comparison of GARCH models. Economics Letters 158:3–6
Khosravi A, Nahavandi S, Creighton D (2013) Prediction intervals for short-term wind farm power generation forecasts. IEEE Transactions on Sustainable Energy 4(3):602–610
Li J, Cheng J-H, Shi J-Y, Huang F (2012) Brief introduction of back propagation (BP) neural network algorithm and its improvement. In: Advances in computer science and information engineering. Springer, pp 553–558
Liu H, Erdem E, Shi J (2011) Comprehensive evaluation of ARMA–GARCH (-M) approaches for modeling the mean and volatility of wind speed. Applied Energy 88(3):724–732
Mackenzie J, Roddick JF, Zito R (2018) An evaluation of HTM and LSTM for short-term arterial traffic flow prediction. IEEE Transactions on Intelligent Transportation Systems 20(5):1847–1857
Qin S, Liu F, Wang J, Song Y (2015) Interval forecasts of a novelty hybrid model for wind speeds. Energy Reports 1:8–16
Shvachko K, Kuang H, Radia S, Chansler R (2010) The Hadoop distributed file system. 2010. In: IEEE 26th symposium on mass storage systems and technologies (MSST). IEEE, pp 1–10
Song Q, Chissom BS (1994) Forecasting enrollments with fuzzy time series—Part II. Fuzzy Sets and Systems 62(1):1–8
Tan M-C, Wong SC, Xu J-M, Guan Z-R, Zhang P (2009) An aggregation approach to short-term traffic flow prediction. IEEE Transactions on Intelligent Transportation Systems 10(1):60–69
Williams BM, Durvasula PK, Brown DE (1998) Urban freeway traffic flow prediction: Application of seasonal autoregressive integrated moving average and exponential smoothing models. Transportation Research Record 1644(1):132–141
Wu Y, Tan H, Qin L, Ran B, Jiang Z (2018) A hybrid deep learning based traffic flow prediction method and its understanding. Transportation Research Part C: Emerging Technologies 90:166–180. http://www.sciencedirect.com/science/article/pii/S0968090X18302651
Zhang Y-Q, Wan X (2007) Statistical fuzzy interval neural networks for currency exchange rate time series prediction. Applied Soft Computing 7(4):1149–1156

Chapter 7
Prediction Models of Traffic Flow Driven Based on Multi-Dimensional Data in Smart Traffic Systems

Abstract During the running of the traffic vehicle, the traffic from other road sections to the target road section is uninterrupted, and the traffic of the adjacent road sections will affect the future traffic flow of the target road section. Therefore, this chapter uses the traffic changes of adjacent road sections and the traffic changes of the target road sections to build multi-dimensional datasets. Based on different model frameworks, four traffic flow prediction models are established, namely the Elman model, LSTM model, WPD-Elman model, and WPD-LSTM model. According to the experimental results, the traffic flow prediction performance of the WPD-Elman model using multi-dimensional data is the best. When building a traffic flow prediction model, the wavelet packet algorithm can effectively improve the accuracy of the model.

7.1 Introduction

There are many factors affecting traffic flow. This chapter considers the impact of traffic flow on adjacent road segments and establishes a multi-dimensional data-driven traffic flow prediction model.

The research of traffic problems has lasted more than half a century, and its theories are also continuously developing with the increase in the number of vehicles and the progress of science and technology.

Many scholars use multi-dimensional data to study traffic flow problems. Goudarzi et al. (2018) used the three-layer dynamic Bayesian network technology to extract the traffic flow series of five expressway segments and established a traffic flow prediction model. Luo et al. (2019) used the k-nearest neighbor algorithm to capture the spatial characteristics of traffic flow and established a traffic flow prediction model based on a two-layer LSTM network.

Duan et al. (2018) has established a hybrid model based on convolutional neural network and LSTM network, which can extract the spatial characteristics and

H. Liu, *Smart Cities: Big Data Prediction Methods and Applications*,
https://doi.org/10.1007/978-981-15-2837-8_7

temporal characteristics of traffic flow and realize the prediction of traffic flow. Li et al. (2019b) studied the influence of multi-dimensional data on the performance of predictive models in traffic flow prediction and found that using all available data as features does not necessarily improve the predictive performance of the model. Tang et al. (2018) studied the spatial correlation between different observing stations and the intrinsic relationship between different traffic flow parameters and established a spatiotemporal multi-task collaborative learning model for traffic flow prediction. Qu et al. (2019) combined traffic flow history data and environmental factor data to propose a traffic prediction method based on deep neural network. The results showed that environmental factor data such as day of the week, weather, season, and other data had a certain impact on the prediction accuracy of the model. Zhang and Kabuka (2018) considered the influence of weather factors on traffic flow, and established a traffic flow prediction model by using gated recurrent neural network. The results showed that considering the weather factor could improve the prediction accuracy of the model. Song et al. (2018) studied the correlation between traffic flow and environmental factors, and proposed a traffic flow time-series prediction method that integrated environmental factors. Koesdwiady et al. (2016) studied the relationship between weather parameters and traffic flow, and proposed a traffic flow prediction model based on deep belief network combined with weather parameters.

In the chapter, the prediction model of traffic flow driven based on multi-dimensional data is built. The main contributions are organized as follows. Section 7.2 analyzes the traffic flow and its influencing factors. Section 7.3 shows the Elman prediction model of traffic flow based on multi-dimensional data. Section 7.4 uses the LSTM network to build the traffic flow prediction model. In Sect. 7.5, the traffic flow prediction models under wavelet packet decomposition are established, which include the WPD-Elman model and the WPD-LSTM model. Section 7.6 analyzes and compares the predictive performance of the models involved. In Sect. 7.7, the conclusions are drawn.

7.2 Analysis of Traffic Flow and Its Influencing Factors

There are many factors that affect traffic flow, such as time, weather, geographical location, traffic accidents, etc. Regular commuting time can cause roads to be more congested in the morning and evening. Bad weather, such as rain or snow, will affect the speed of the vehicle, which will affect the traffic flow. The geographical location of the road also has a certain impact on traffic flow. For example, the roads in the urban area will be more congested, and the roads in the suburbs will generally be relatively smooth. Traffic accidents will affect traffic flow irregularly. Many scholars have also studied the impact of these factors on traffic flow. Zhu et al. (2016) studied the spatial relationship of traffic flow, combined with traffic parameters such as traffic speed and traffic density, and established a traffic flow prediction model based on linear conditional Gaussian Bayesian network. Ahn et al. (2016) used the three-dimensional Markov random domain to model the spatiotemporal relationship of road traffic flow and its adjacent road segments, and used Bayesian classifier and support vector regression to predict traffic flow. Based on the geographical factors and environmental factors, *Kim* and *Hong* established a hybrid model based on Gaussian mixture model clustering and artificial neural network to predict traffic flow (Kim and Hong 2015). Zheng et al. (2019) proposed a deep embedded learning method for traffic flow prediction, which takes into account the impact of other factors such as road network structure and weather factors on traffic flow.

Traffic flow is a typical time-series data. The traffic flow change of a road segment is affected not only by the traffic flow in its own historical time, but also by the traffic flow of the surrounding road segments. Figure 7.1 shows the traffic flow on the road segment, and the traffic flow of Road1, Road2, and Road3 may partially enters into Road4, which will affect the future traffic flow of Road4. At the same time, the traffic flow of Road4 will also be limited by the traffic volume in historical time. Therefore, the historical traffic volume of its own road segment and adjacent road segments is the influencing factor of traffic flow. This chapter combines the traffic flow of such adjacent road segments to establish multi-dimensional datasets, and uses this multi-dimensional data to establish a traffic flow prediction model to predict traffic flow.

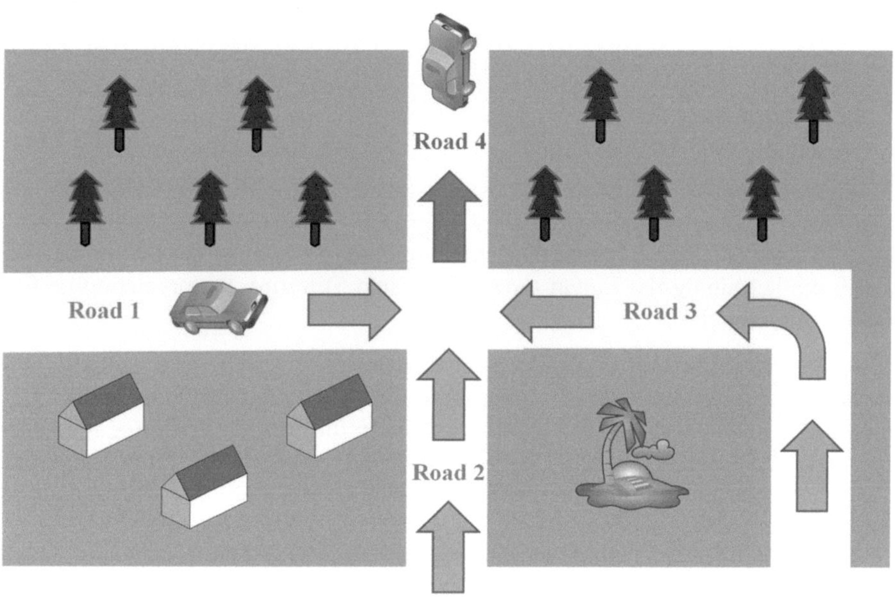

Fig. 7.1 Traffic flow motion diagram

7.3 Elman Prediction Model of Traffic Flow Based on Multiple Data

7.3.1 The Framework of the Elman Prediction Model

In 1990, *Elman* proposed Elman neural network (Chong et al. 2016), which is a typical artificial intelligence algorithm. Elman neural network includes the input layer, hidden layer, feedback layer, and output layer (Li et al. 2019a), and it has short-term memory ability due to the feedback principle (Li et al. 2014). In the chapter, the Elman network is used to predict each sub-layer of error series. The Elman neural network can be expressed as follows:

$$H_n = \sigma_h \left(W_h x_n + U_h h_{n-1} + b_n \right) \tag{7.1}$$

$$y_n = \sigma_y \left(W_h h_n + b_y \right) \tag{7.2}$$

where x_n represents the input series; h_n is the hidden layer vector; y_n denotes the output series; W_h, W_y, and U_h are the weigh matrices; b_y is the bias vector; $\sigma_h(x)$ and $\sigma_y(x)$ are the activation functions, which can be calculated by Eq. (7.3):

$$\sigma(x) = \frac{1}{1+e^{-x}} \tag{7.3}$$

Figure 7.2 shows the principle of the Elman network.

7.3.2 Modeling Process

The Elman network is used to establish the traffic flow prediction model based on multiple data, which includes five steps: (step 1) obtaining multiple traffic flow data, (step 2) setting the input/output data structure, (step 3) dividing the training/testing set, (step 4) training the Elman model, and (step 5) testing the prediction performance of the model.

(Step 1) Obtaining Multiple Data of Traffic Flow

A variety of traffic flow data involves the traffic flow series of multiple road segments at different times. This chapter uses the traffic flow data of three road segments, which are represented by Rd1, Rd2, and Rd3, respectively. These three road segments are taken from Futian District, Shenzhen, China. Their specific meanings are shown in Table 7.1.

The traffic flow data include the three traffic flow series of three segments, which are taken during 8 days from November 1, 2018 to November 8, 2018, and the road situation represented by the data is shown in Fig. 7.3. The sampling time of each

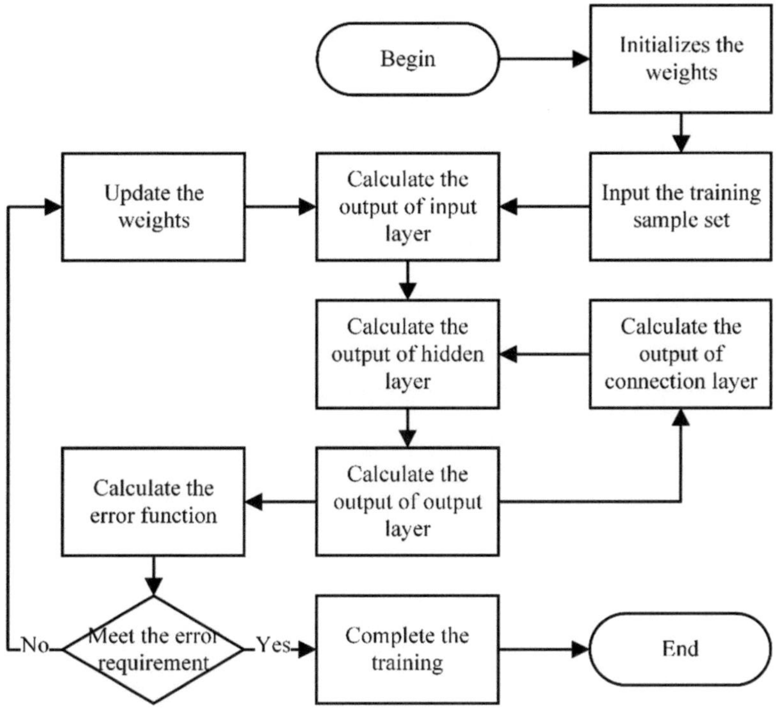

Fig. 7.2 The principle of the Elman network

Table 7.1 The specific meanings of Rd1, Rd2, and Rd3

Symbol	Meaning of these three road segments	Direction of traffic flow
Rd1	The road segment of the west side of the intersection of Fuqiang Road and Jintian Road	West to east
Rd2	The road segment of the west side of the intersection of Fuqiang Road and Jintian Road	East to west
Rd3	The road segment of the north side of the intersection of Fuqiang Road and Jintian Road	South to north

traffic flow series is 5 min with a total of 2304 sample points. The sample distribution of these three traffic flow series is shown in Fig. 7.4.

(Step 2) Setting the Input/Output Data Structure

When building a model, it is important to set the input and output data structure of the model, which determines the structure of the model. Since the traffic volume of the historical time of the predicted segment will affect the traffic volume in the future, the traffic volume of the historical time will be used as part of the input data. In addition, the traffic volume of other adjacent road segments will also affect the

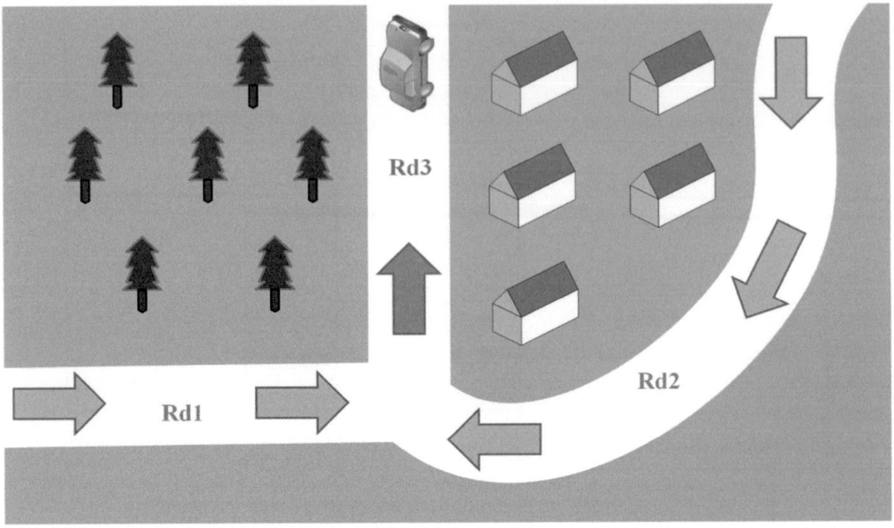

Fig. 7.3 The map of Rd1, Rd2, and Rd3

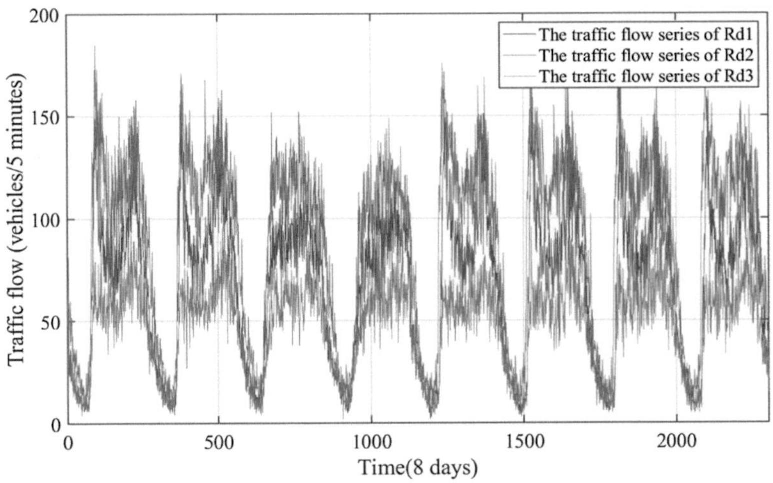

Fig. 7.4 The sample distribution of the three traffic flow series

traffic volume of the predicted segment, so the historical time traffic volume of other adjacent segments should be used as another part of the input data.

In the chapter, data from three road segments, Rd1, Rd2, and Rd3, are used. Among them, the traffic flow of Rd1 and Rd2 flows to Rd3. Therefore, the traffic volume of the first four times of the traffic flow series of Rd1, Rd2, and Rd3 is used as the input data, and the vehicle amount of Rd3 is taken as the output data. The input/output data structure of 1-step prediction is shown in Fig. 7.5.

(Step 3) Dividing the Training/Testing Set
The chapter uses a total of 8 days of traffic flow data for research. It can be clearly seen from Fig. 7.4 that the traffic flow series exhibits a strong periodicity. The daily traffic volume is larger during the day and will be significantly reduced at night. In order to better train the model and test the prediction performance, the first 6 days of data (1728 samples) are used as a training set to train the Elman prediction model, and the last 2 days of data (567 samples) are used as a testing set to test the model performance.

(Step 4) Training the Elman Model
In the chapter, the maximum number of iterations is set to 3000 and the error prediction is 0.00001. The Elman prediction model is trained using the input and output data of the training set.

(Step 5) Testing the Prediction Performance of the Model
The predicted traffic flow can be obtained by inputting the input data of the test set into the Elman prediction model trained by the input value. The predicted traffic flow is compared with the actual traffic flow, and the five error evaluation index values of MAE, MAPE, RMSE, SDE, R and the training time of the Elman prediction model are calculated to evaluate the performance of the prediction model.

7.3.3 *The Prediction Results*

Figures 7.6, 7.7, 7.8, 7.9, and 7.10 show the predicted results from one to five steps of the vehicle volume obtained by using the Elman prediction model, respectively. Table 7.2 shows the predictive performance of the Elman prediction models based on different advanced prediction steps. As can be seen from these figures and table:

The five Elman prediction models based on different advanced prediction steps show that the traffic hopping point cannot be processed well, and the prediction

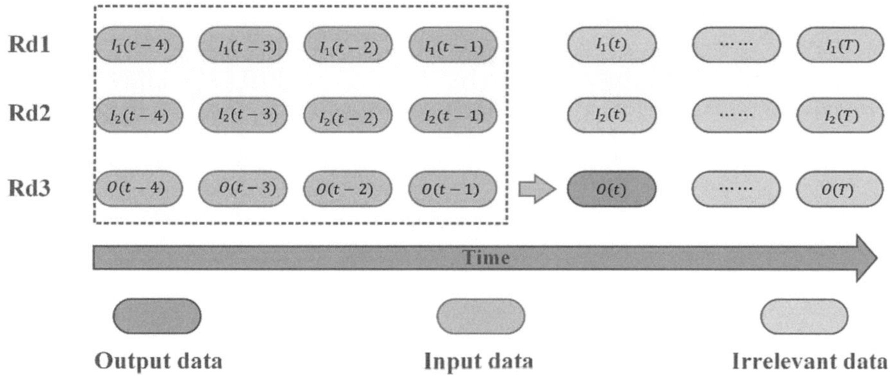

Fig. 7.5 The input/output data structure of one-step prediction

effect of traffic flow needs to be strengthened. Taking the prediction result of the 2-step as an example, the MAE, MAPE, RMSE, SDE, R and the training time of the Elman prediction model are 7.011, 18.17%, 9.3797, 0.9081, 4.355, and 9.3797s, respectively.

The number of advanced steps has little effect on the prediction performance of the Elman prediction model. Taking the prediction result of the 1-step as an example, the MAE, MAPE, RMSE, SDE, R and the training time of the Elman prediction model are 7.0369, 18.04%, 9.3783, 0.9081, 5.593, and 9.3783s, respectively. In comparison, the MAE, MAPE, RMSE, SDE, R and the training time of the 5-step Elman prediction model are 7.2391, 19.21%, 9.5415, 0.9049, 4.279, and 9.5415s, respectively.

7.4 LSTM Prediction Model of Traffic Flow Based on Multiple Data

7.4.1 The Framework of the LSTM Prediction Model

In 1997, *Hochreiter* and *Schmidhuber* proposed the long short-term memory network (LSTM) (Hochreiter and Schmidhuber 1997). The LSTM is derived from the recurrent neural network (RNN), which has great forecasting performance for time series (Graves and Schmidhuber 2005).

As shown in Fig. 7.11, the LSTM includes three components: the input gate, the forget gate, and the output gate. In this chapter, the LSTM network is used to

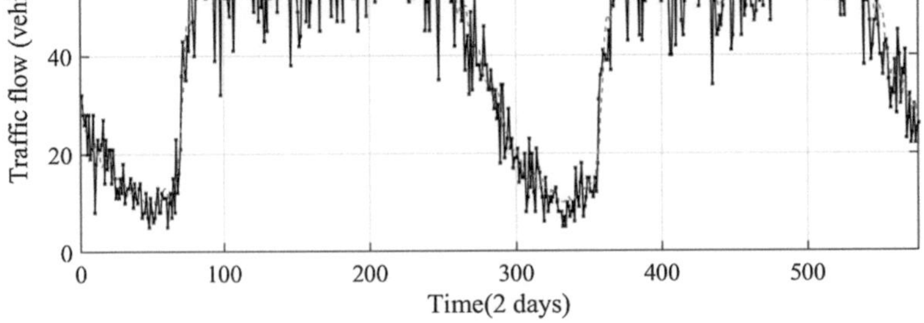

Fig. 7.6 The results of the 1-step prediction for the traffic flow series

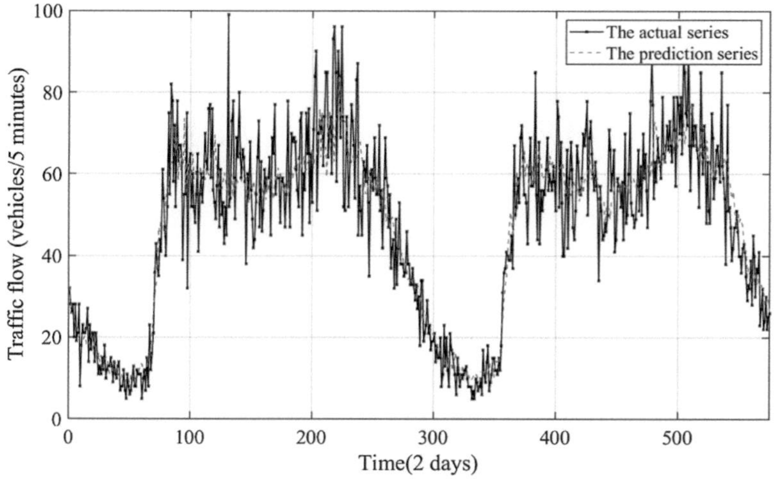

Fig. 7.7 The results of the 2-step prediction for the traffic flow series

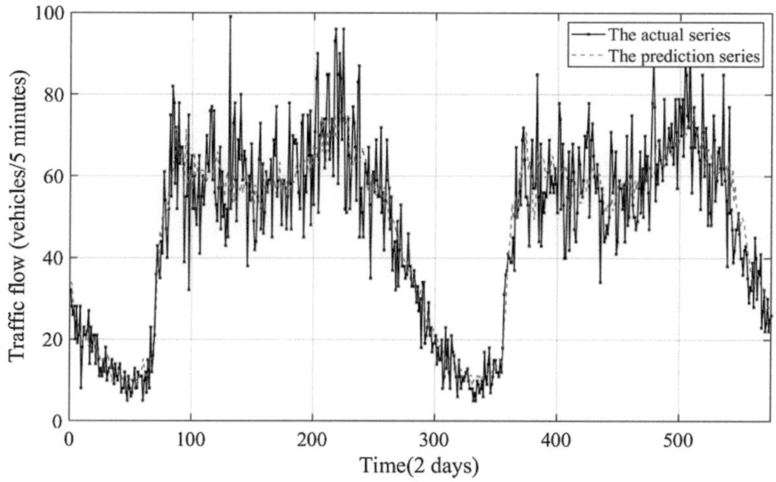

Fig. 7.8 The results of the 3-step prediction for the traffic flow series

establish the prediction model for traffic flow. The detailed computation process of LSTM network can be represented as follows:

$$f_t = \sigma\left(W_{fx}X_t + W_{fy}Y_{t-1} + b_f\right) \tag{7.4}$$

$$i_t = \sigma\left(W_{ix}X_t + W_{iy}Y_{t-1} + b_i\right) \tag{7.5}$$

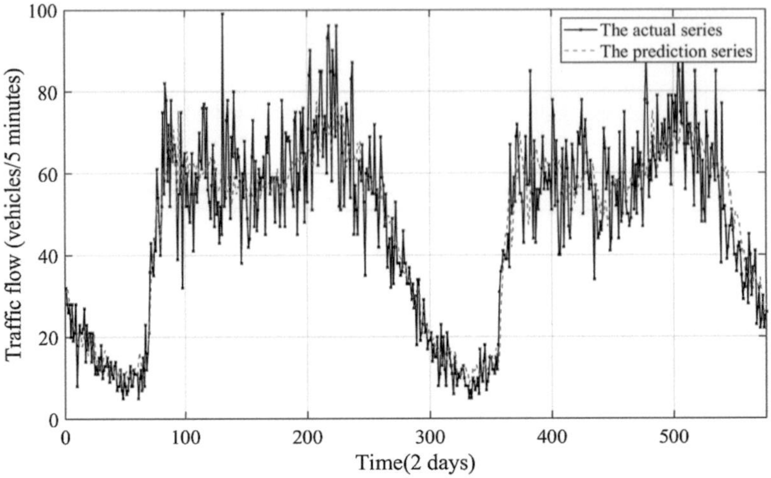

Fig. 7.9 The results of the 4-step prediction for the traffic flow series

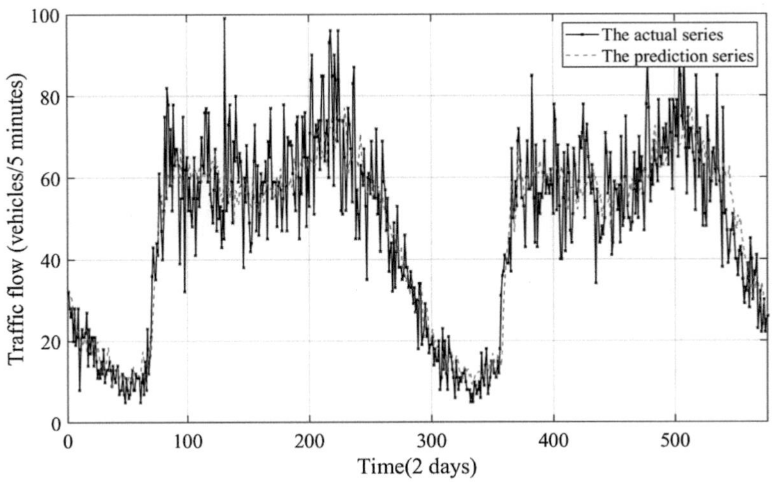

Fig. 7.10 The results of the 5-step prediction for the traffic flow series

$$\tilde{C}_t = \tanh\left(W_{Cx}X_t + W_{Cy}Y_{t-1} + b_C\right) \tag{7.6}$$

$$o_t = \sigma\left(W_{ox}X_t + W_{oy}Y_{t-1} + b_o\right) \tag{7.7}$$

$$C_t = f_t \cdot C_{t-1} + i_t \tilde{C}_t \tag{7.8}$$

Table 7.2 The performance estimating results of the Elman prediction model

Model	Step	MAE	MAPE	RMSE	SDE	R	Training time(s)
Elman	1	7.0369	18.043	9.3783	0.90813	5.593	9.3783
	2	7.011	18.172	9.3797	0.90811	4.355	9.3797
	3	6.9137	18.045	9.263	0.91055	4.127	9.263
	4	7.141	18.582	9.4572	0.90655	4.478	9.4572
	5	7.2391	19.211	9.5415	0.90488	4.279	9.5415

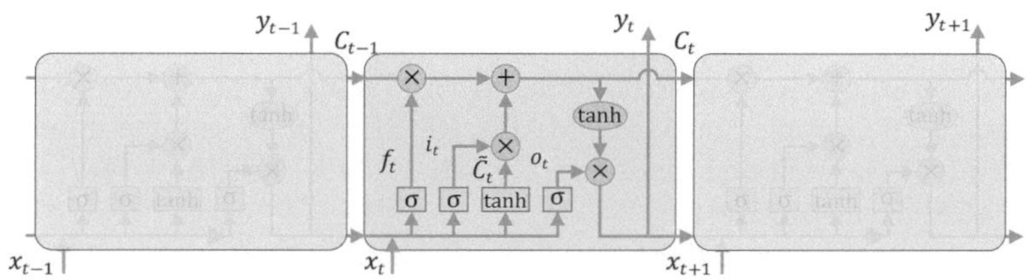

Fig. 7.11 The principle of LSTM network

$$Y_t = o_t \cdot \tanh\left(C_t\right) \tag{7.9}$$

where f_t, i_t, o_t denote the activations of the forget gate, the input gate, and the output gate, respectively; X_t is the input series; Y_t is the output series; C_t is the cell state at time t; W is the weight matrices; b is and bias vectors; "·" represents the dot product; $\sigma(x)$, $\tanh(x)$ are the activation functions, which can be depicted as follows:

$$\sigma\left(x\right) = \frac{1}{1+e^{-x}} \tag{7.10}$$

$$\tanh\left(x\right) = \frac{e^{x}-e^{-x}}{e^{x}+e^{-x}} \tag{7.11}$$

7.4.2 Modeling Process

The LSTM network is used to establish the traffic flow prediction model based on multiple data. The five-step process of the LSTM model modeling is similar to the modeling process of Elman model, which are: (step 1) obtaining multiple traffic flow data, (step 2) setting the input/output data structure, (step 3) dividing the training/testing set, (step 4) training the LSTM model, and (step 5) testing the prediction performance of the model. The traffic flow data still uses the data from Sect. 7.3.

7.4.3 The Prediction Results

Figures 7.12, 7.13, 7.14, 7.15, and 7.16 show the predicted results from 1 to 5 steps of the vehicle volume obtained by using the LSTM prediction model, respectively. Table 7.3 shows the predictive performance of the LSTM prediction models based on different advanced prediction steps. As can be seen from these figures and table:

(a) The five LSTM prediction models based on different advanced prediction steps show that the traffic hopping point cannot be processed well, and the prediction effect of traffic flow needs to be strengthened. Taking the prediction result of the 2-step as an example, the MAE, MAPE, RMSE, SDE, R and the training time of the LSTM prediction model are 7.1341, 17.24%, 9.5016, 9.4917, 0.90786, and 46.895s, respectively.

(b) The number of advanced steps has little effect on the prediction performance of the LSTM prediction model. Take the prediction result of the 1-step as an example, the MAE, MAPE, RMSE, SDE, R and the training time of the LSTM prediction model are 7.1817, 17.22%, 9.5345, 9.5283, 0.9069, and 49.609s, respectively. In comparison, the MAE, MAPE, RMSE, SDE, R and the training time of the 5-step LSTM prediction model are 7.2984, 17.78%, 9.6791, 9.6707, 0.90321, and 50.094s, respectively.

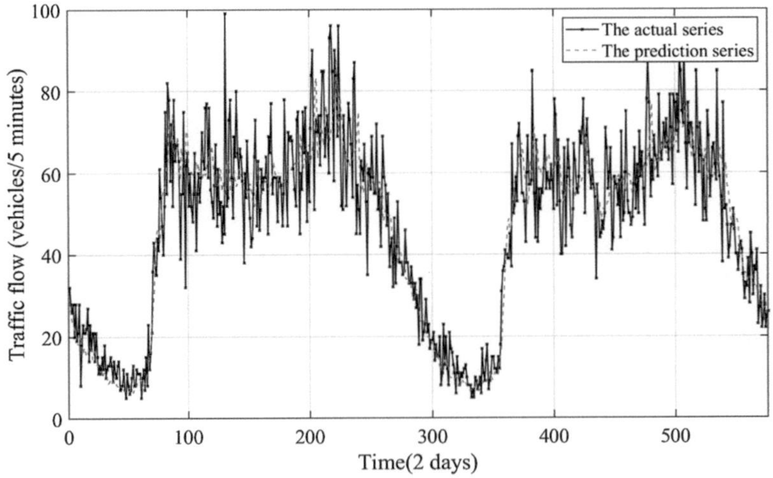

Fig. 7.12 The results of the 1-step prediction for the traffic flow series

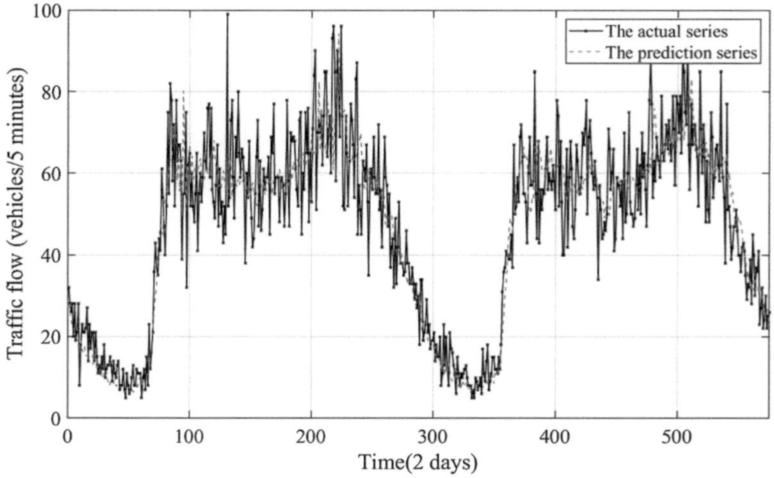

Fig. 7.13 The results of the 2-step prediction for the traffic flow series

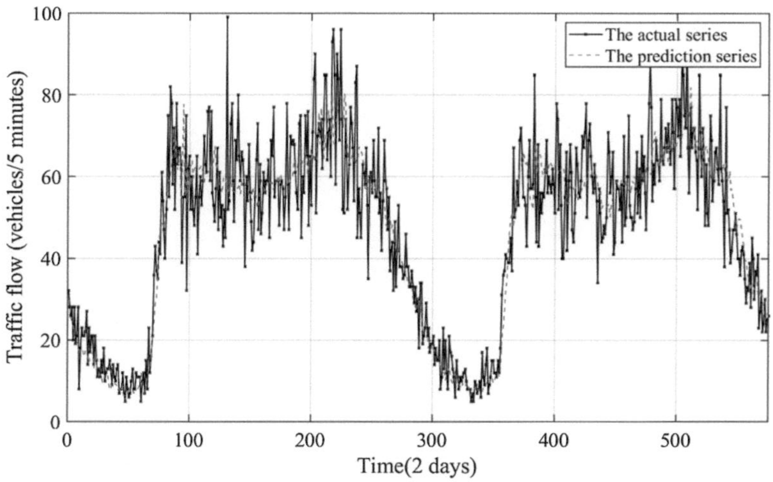

Fig. 7.14 The results of the 3-step prediction for the traffic flow series

7.5 Traffic Flow Prediction Under Wavelet Packet Decomposition

7.5.1 The Framework of the WPD-Prediction Model

Wavelet packet decomposition (WPD) is a well-known signal analysis tool (Sable and Talbar 2016). Its advantage is that it provides a more accurate decomposition in the high-frequency part of the signal, without redundancy or omission (Zhang et al.

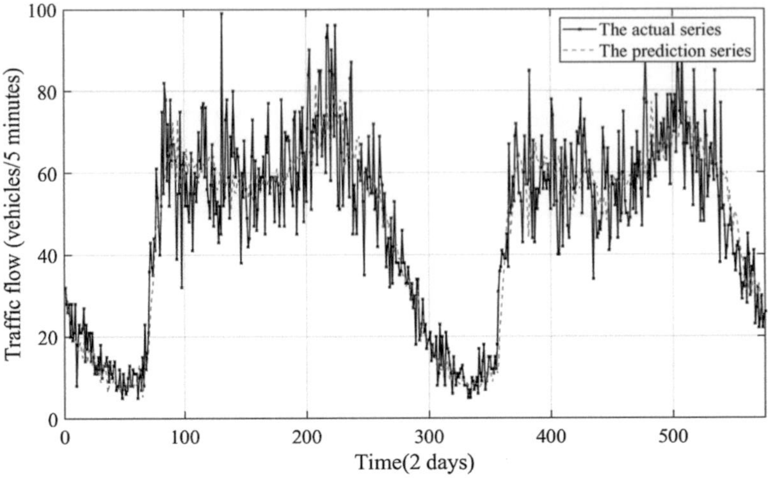

Fig. 7.15 The results of the 4-step prediction for the traffic flow series

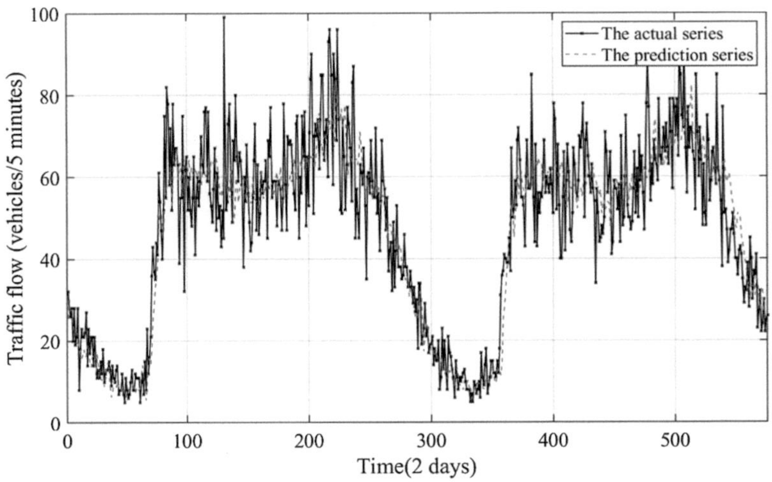

Fig. 7.16 The results of the 5-step prediction for the traffic flow series

2018). Wavelet packet decomposition includes the continuous wavelet transform and the discrete wavelet transforms (Liu et al. 2018). In this chapter, the WPD is used to decompose the error series into a series of sub-layers and the error series is discrete series. The structure of the WPD is shown in Fig. 7.17.

The mathematical expression of the traffic flow series by wavelet packet decomposition can be described as follows:

$$CTW(\alpha,\beta) = X(t), \psi^*_{\alpha,\beta}(t) = \int_{-\infty}^{+\infty} X(t)\psi^*_{\alpha,\beta}(t)dt \tag{7.12}$$

Table 7.3 The performance estimating results of the LSTM prediction model

Model	Step	MAE	MAPE	RMSE	SDE	R	Training time(s)
LSTM	1	7.1817	17.219	9.5345	9.5283	0.9069	49.609
	2	7.1341	17.244	9.5016	9.4917	0.90786	46.895
	3	6.9526	16.714	9.3624	9.3589	0.90994	49.76
	4	7.1379	17.119	9.4919	9.4869	0.90727	50.486
	5	7.2984	17.779	9.6791	9.6707	0.90321	50.094

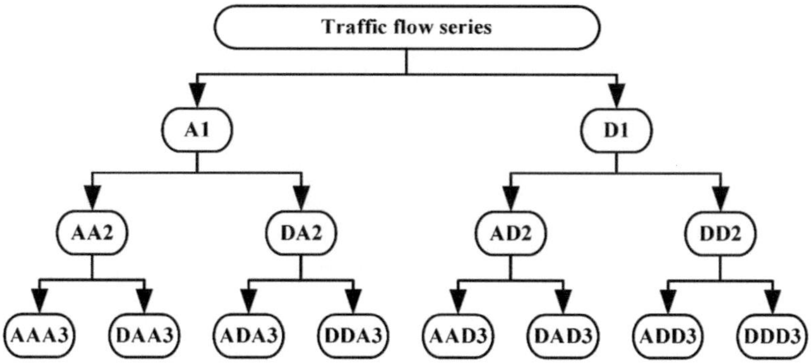

Fig. 7.17 The structure of the WPD

$$\psi^*_{\alpha,\beta}(t) = \psi^*((t-\beta)/\alpha)/\sqrt{\alpha} \tag{7.13}$$

$$\begin{cases} \alpha = 2^j \\ \beta = k2^j \end{cases} \tag{7.14}$$

where $X(t)$ is the traffic flow series; ψ^* denotes the mother function of wavelet; "*" denotes the complex conjugate; α is the scale factor; β is the translation parameter; j is the scale coefficient; and k is the translation coefficient.

Figure 7.18 shows the principle of the hybrid model using the wavelet packet decomposition algorithm and predictor (WPD-prediction model). Firstly, the wavelet packet decomposition algorithm is used to decompose the traffic flow series of three road segments. Each traffic flow series is decomposed into eight sub-series. As shown in Fig. 7.17, each series represents a different frequency type. Then, traffic flow series of each frequency type is used to establish traffic flow prediction model, which is based on the Elman neural network or LSTM network. Finally, the outputs of each traffic flow prediction model are summed to get the final traffic flow prediction result.

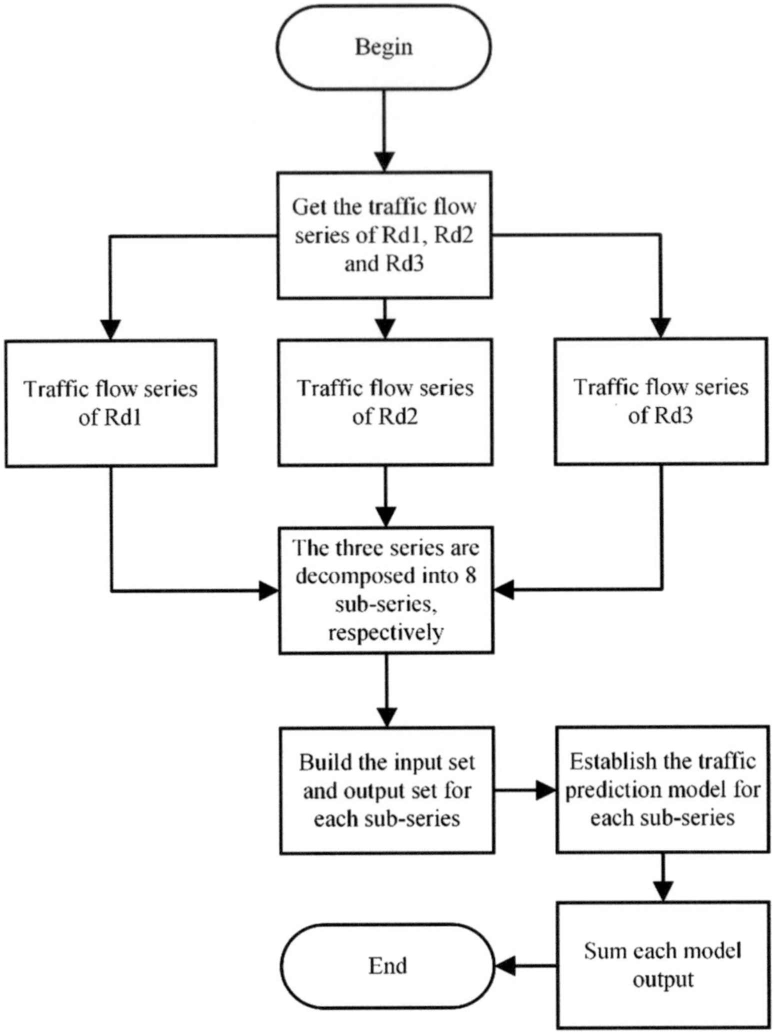

Fig. 7.18 The principle of WPD-prediction model

7.5.2 Modeling Process

The wavelet packet decomposition algorithm and the Elman neural network (or LSTM network) are used to establish the traffic flow prediction model based on multiple data, which includes six steps: (step 1) obtaining multiple traffic flow data, (step 2) decomposing each series into eight sub-series by the WPD, (step 3) setting the input/output data structure for each series, (step 4) dividing the training/testing set, (step 5) training eight prediction model based on the Elman neural network (or LSTM network), and (step 6) test the prediction performance.

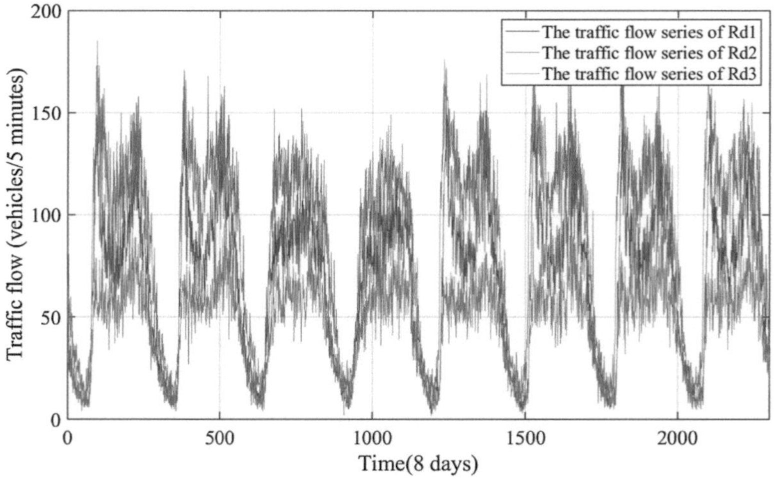

Fig. 7.19 The sample distribution of the three traffic flow series

(Step 1) Obtaining Multiple Traffic Flow Data

The traffic flow data is the same as that of Sect. 7.3, including three traffic flow series of three road segments, which is taken during 8 days from November 1, 2018 to November 8, 2018. The sampling time of each traffic flow series is 5 min, with a total of 2304 sample points. The sample distribution of these three traffic flow series is shown in Fig. 7.19.

(Step 2) Decomposing Each Series Into Eight Sub-Series by WPD

The WPD is used to decompose these three traffic flow series into eight sub-series, respectively. The decomposition level and wavelet function of the WPD is set to 3 and "dmey," respectively.

Figures 7.20, 7.21, and 7.22 show the decomposition series of traffic flow series of Rd1, Rd2, and Rd3, respectively. The eight decomposition sub-series are represented by S1–S8, respectively.

(Step 3) Setting the Input/Output Data Structure for Each Series

In (step 2) each series is decomposed into eight sub-series. Each prediction model will be built for each sub-series, so eight input/output datasets need to be set for the eight prediction models. Figure 7.23 shows the input/output data structure set by taking the S1 sub-series of each traffic flow series as an example.

In each input/output data structure, as in Sect. 7.3, the traffic flow data of the first four historical times of these three S1 sub-series of the Rd1, Rd2, and Rd3 are taken as input. The traffic flow data of the future time of the S1 sub-series of the Rd1 are taken as output.

When the data of the later time are taken as the output data, it is called 1-step prediction. When the data of the second time in the future are taken as the output data, it is called the 2-step prediction, and so forth. Figure 7.23 shows the input/

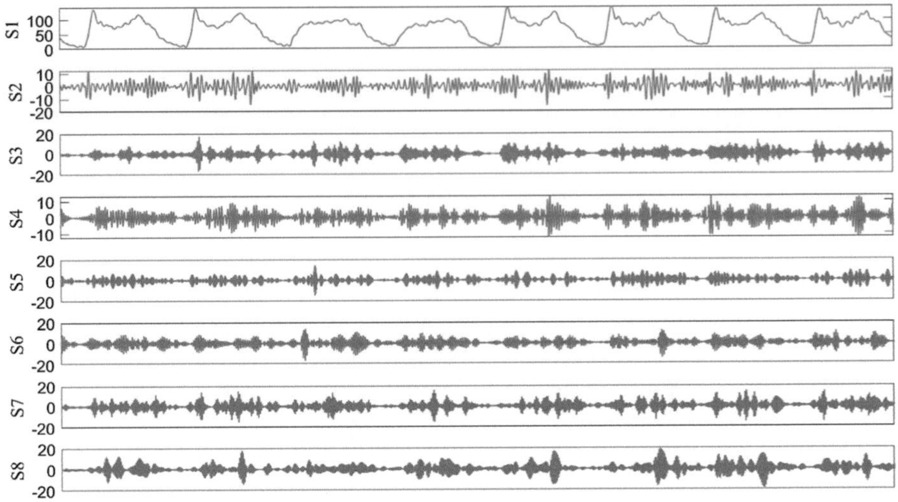

Fig. 7.20 The eight sub-series of Rd1 traffic flow series based on the WPD

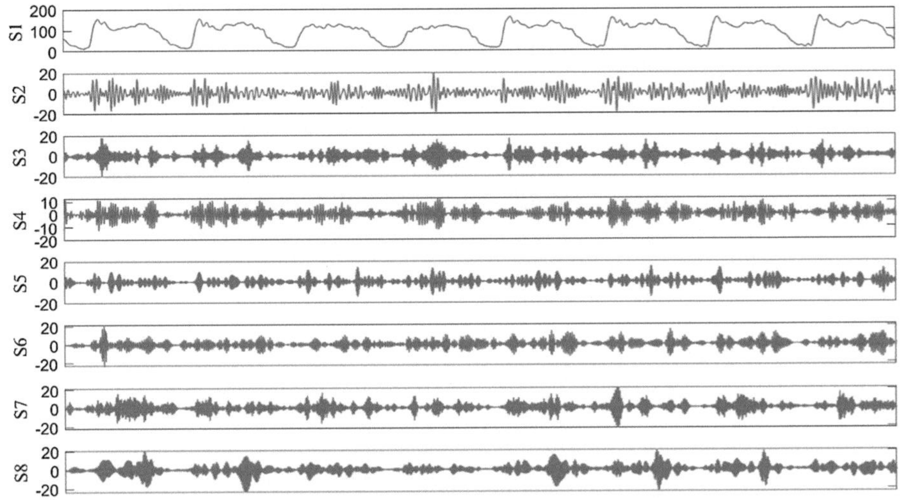

Fig. 7.21 The eight sub-series of Rd2 traffic flow series based on the WPD

output data structure of the 1-step prediction of the S1 data created by the three road segments.

(Step 4) Dividing the Training/Testing Set
Eight datasets have the same training/testing set, for each training/testing set, the first 6 days of data (1728 samples) are used as a training set to train the prediction model, and the last 2 days of data (567 samples) are used as a test set. A total of eight training/testing sets are built.

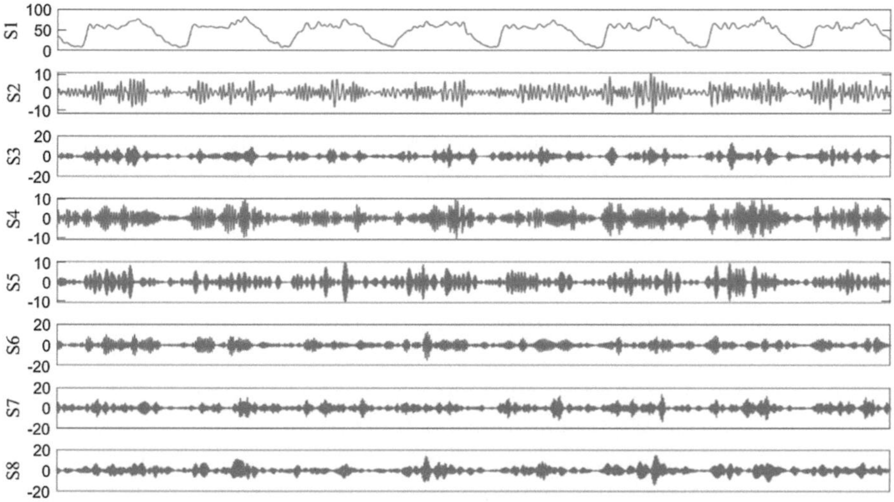

Fig. 7.22 The eight sub-series of Rd3 traffic flow series based on the WPD

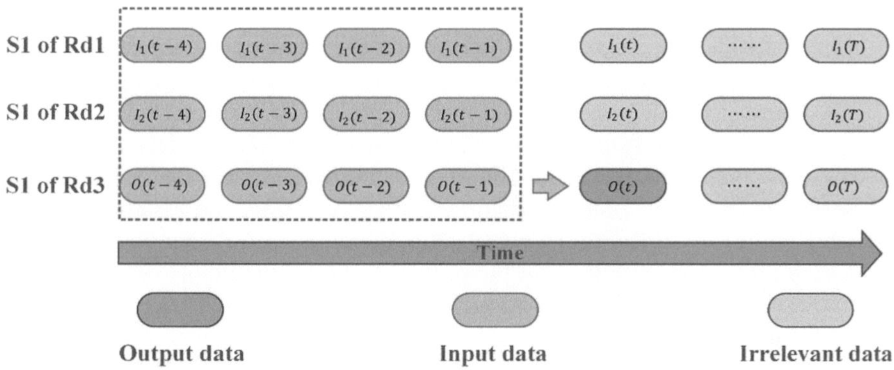

Fig. 7.23 The input/output data structure of 1-step prediction

(Step 5) Training Eight Prediction Models Based on the Elman Neural Network (or LSTM Network)

Eight prediction models based on Elman neural network (or LSTM network) are trained using eight training sets. Each prediction model will be used to predict the value of each sub-series. When training the prediction models based on Elman neural network, the maximum number of iterations is set to 3000 and the error prediction is 0.00001. When training the prediction models based on the LSTM network, the layer number of the LSTM is set to 300, the optimizer is "Adam," the "MaxEpochs" is set to 100, the "MiniBatchSize" is set to 70, and the "InitialLearnRate" is set to 0.005.

(Step 6) Testing the Prediction Performance
When inputting the input data of the test set into each prediction model, the predicted values of the eight sub-series of Rd3 can be obtained. The finally predicted traffic flow can be obtained by summing the eight predicted sub-series. The predicted traffic flow is compared with the actual traffic flow, and the five error evaluation index values of MAE, MAPE, RMSE, SDE, R and the training time of the eight prediction models are calculated to evaluate the prediction performance.

7.5.3 The Prediction Results

Figures 7.24, 7.25, 7.26, 7.27, and 7.28 show the prediction results from 1 to 5 steps of the vehicle volume obtained by using the WPD-Elman prediction model and the WPD-LSTM prediction model, respectively. Table 7.3 shows the predictive performance of the WPD-Elman prediction model and the WPD-LSTM prediction model based on different advanced prediction steps. Through analysis, the following conclusions can be obtained:

(a) As can be seen from Figures 7.24, 7.25, 7.26, 7.27, and 7.28, whether using Elman neural network as the predictor or LSTM network as the predictor, such WPD-prediction models can always predict traffic flow series well, and have good prediction effects on the jump points of traffic flow series. However, as the number of predicted steps increases, the prediction effect will decrease. Taking a sample point between 0 and 100 as an example, it is clear that the predicted series curves of the 1-step prediction, 2-step prediction, and 3-step prediction are significantly closer to the actual series curve than the 4-step prediction and 5-step prediction. In Fig. 7.27, the prediction of sample points between 0 and 100 is not very good in the first four steps.
(b) More detailed information about the predictive performance of the WPD-prediction model can be seen in Table 7.4. In terms of predictor selection, because the prediction performance of the WPD-Elman model is always better than the WPD-LSTM model, it is more sensible to use Elman neural network as predictor. Taking the 1-step prediction as an example, the MAE, MAPE, RMSE, SDE, R and the training time of the WPD-Elman prediction model are 0.76197, 2.0359%, 0.99159, 0.99153, 0.99902, and 32.508s, respectively. In comparison, the MAE, MAPE, RMSE, SDE, R and the training time of the WPD-LSTM prediction model are 0.90659, 3.0222%, 1.1191, 1.1187 0.99887, and 391.54s, respectively. In terms of the predicted steps, as the number of predicted steps increases, the predicted performance will also decrease. Take the WPD-Elman prediction model as an example, the MAE, MAPE, RMSE, SDE, R and the training time of the 1-step prediction model are 0.76197, 2.0359%, 0.99159, 0.99153, 0.99902, and 32.508s, respectively. The MAE, MAPE, RMSE, SDE,

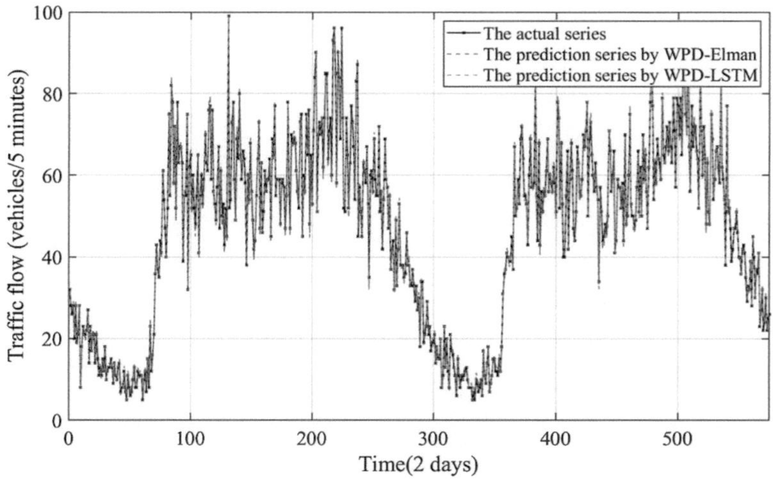

Fig. 7.24 The results of the 1-step prediction for the traffic flow series

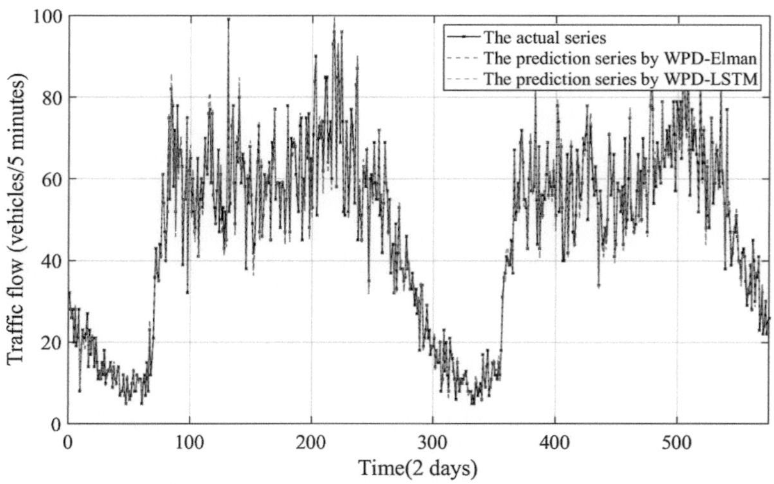

Fig. 7.25 The results of the 2-step prediction for the traffic flow series

R and the training time of the 2-step prediction model are 1.3068, 3.6089%, 1.6952, 1.6947 0.99715, and 32.421s, respectively. The MAE, MAPE, RMSE, SDE, R and the training time of the 5-step prediction model are 3.299, 8.8608%, 4.281, 4.2804, 0.98157, and 32.19s, respectively. As the number of predicted steps increases, the values of MAE, MAPE, RMSE, SDE, and R will increase, and only the training time will not be affected by the number of predicted steps.

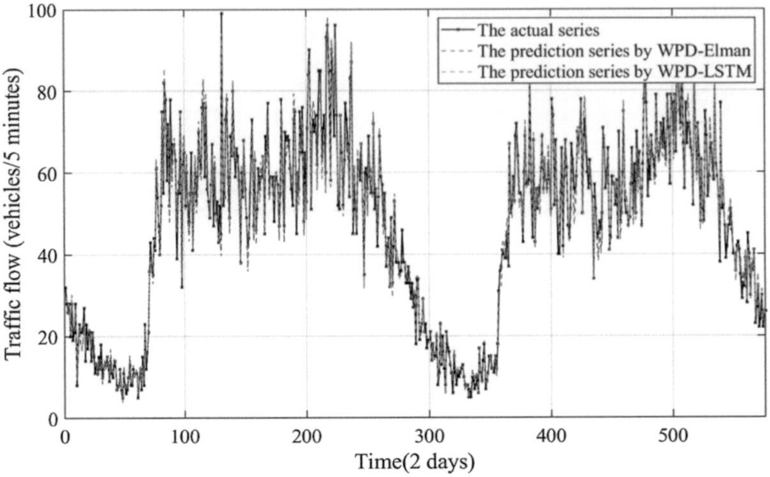

Fig. 7.26 The results of the 3-step prediction for the traffic flow series

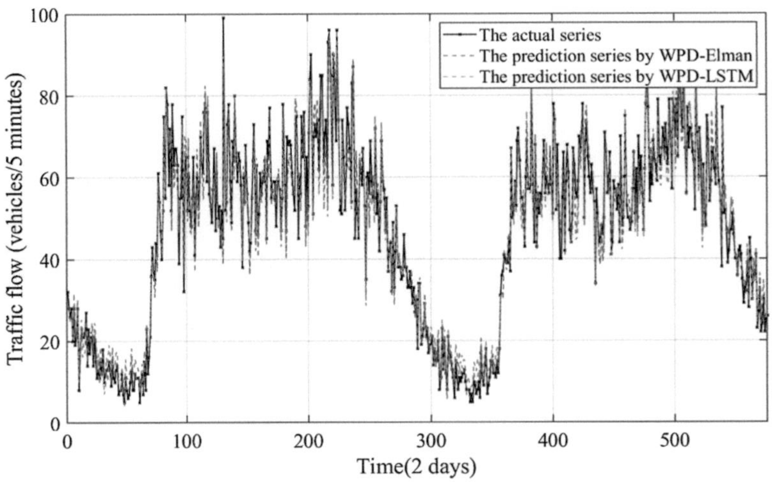

Fig. 7.27 The results of the 4-step prediction for the traffic flow series

7.6 Comparative Analysis of Forecasting Performance

In this chapter, four models are used to predict traffic flow: the Elman model, the LSTM model, the WPD-Elman model, and the WPD-LSTM model. In order to show the model prediction results more clearly, the prediction results of the first day of the test set (288 sample points) are plotted in Figs. 7.29, 7.30, 7.31, 7.32, and 7.33. The prediction performance indicators of the four models are shown in Table 7.5. From these figures and tables, the following conclusions can be drawn:

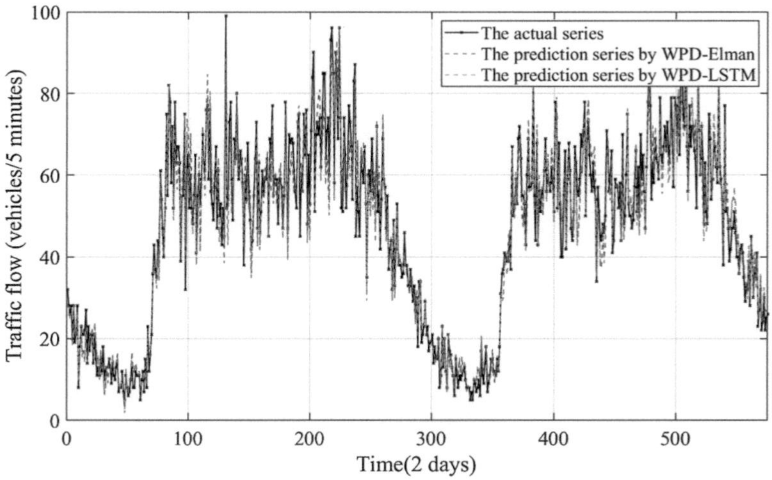

Fig. 7.28 The results of the 5-step prediction for the traffic flow series

Table 7.4 The performance estimating results of the WPD-prediction model

Model	Step	MAE	MAPE	RMSE	SDE	R	Training time(s)
WPD-Elman	1	0.76197	2.0359	0.99159	0.99153	0.99902	32.508
	2	1.3068	3.6089	1.6952	1.6947	0.99715	32.421
	3	2.1914	5.8263	2.8399	2.8393	0.99196	32.397
	4	2.7471	7.3865	3.511	3.5106	0.98765	31.448
	5	3.299	8.8608	4.281	4.2804	0.98157	32.19
WPD-LSTM	1	0.90659	3.0222	1.1191	1.1187	0.99887	391.54
	2	1.5677	4.0226	2.0545	2.0449	0.99584	381.33
	3	2.3387	5.971	3.0601	3.0427	0.99078	389.77
	4	3.478	11.711	4.2834	4.2804	0.98393	393.35
	5	3.6295	9.5912	4.7016	4.6602	0.97853	511.76

(a) The single prediction model of the Elman model and LSTM model cannot handle data jump points very well, and the traffic flow series has a strong jump ability. As shown in Fig. 7.29, the predicted series of the Elman model and the LSTM model are unable to predict the jump point in all sample points, so the prediction performance of the traffic flow is low. This may be because that the Elman model and the LSTM model filter out data jump points as data anomalies.

(b) Since the wavelet packet decomposition algorithm decomposes the traffic flow series into a series of sub-series according to different frequencies, the Elman model and the LSTM model of each sub-series consider the frequency of the entire series during training. Therefore, when a prediction model is established for each sub-series, the jump point is not filtered as an abnormal point. It can be seen from Figs. 7.29, 7.30, 7.31, 7.32, and 7.33 that the prediction model

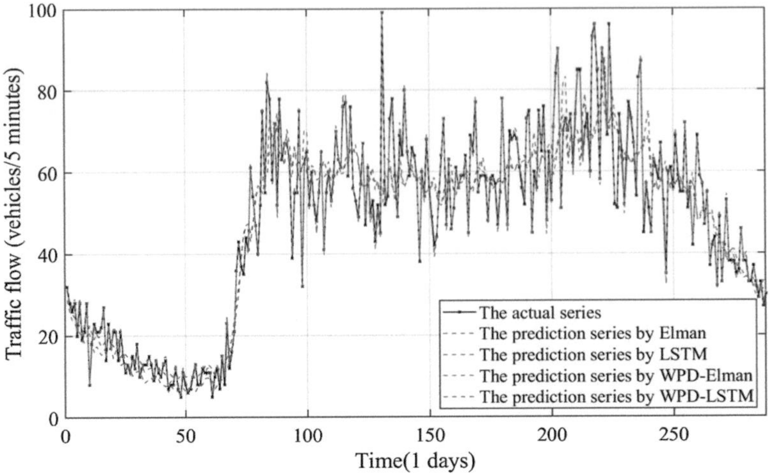

Fig. 7.29 The results of the 1-step prediction for the traffic flow series

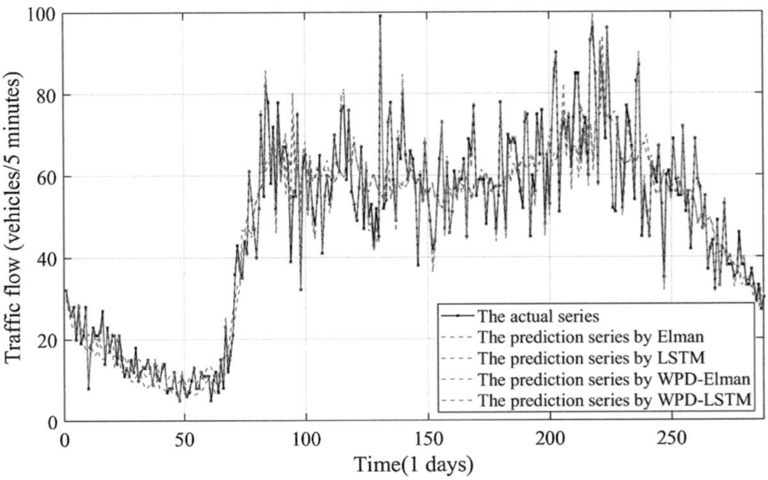

Fig. 7.30 The results of the 2-step prediction for the traffic flow series

combined with the wavelet packet decomposition algorithm has a good processing ability for the data jump point and a good prediction effect on the traffic flow series, and the prediction curve is close to the actual curve.

Comparing the specific prediction performance indicators, the prediction performance of the Elman model is similar to that of the LSTM model in a single model, and their prediction performance is not ideal. Moreover, the predicted performance indicators of the models are not affected by the number of predicted steps. Taking the Elman model as an example, the MAE, MAPE, RMSE, SDE, R and the training time of the 1-step prediction model are 7.0369, 18.043%, 9.3783, 0.90813, 5.593,

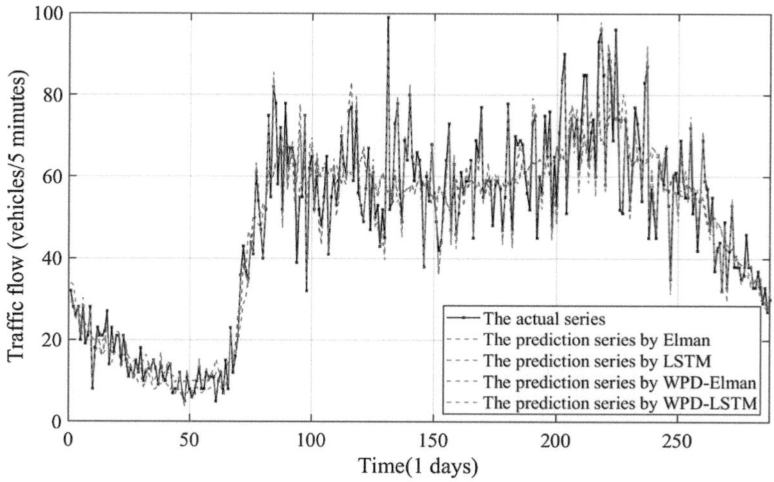

Fig. 7.31 The results of the 3-step prediction for the traffic flow series

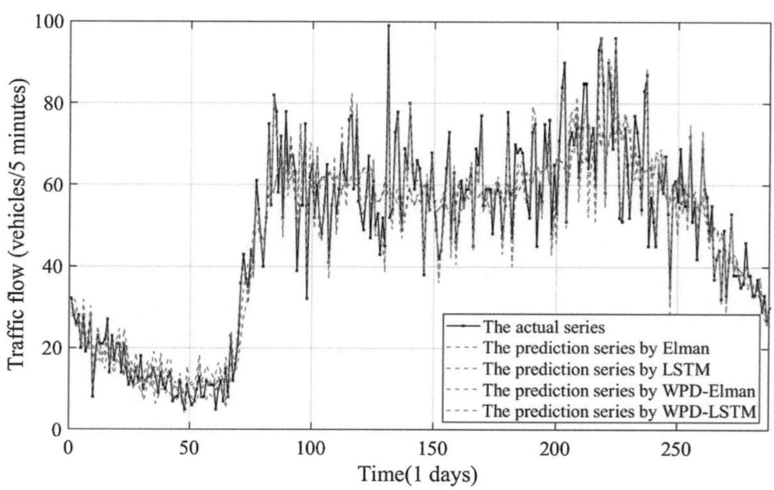

Fig. 7.32 The results of the 4-step prediction for the traffic flow series

and 9.3783s, respectively. The MAE, MAPE, RMSE, SDE, R and the training time of the 5-step prediction model are 7.2391, 19.211%, 9.5415, 0.90488, 4.478, and 9.5415s, respectively.

Comparing the specific prediction performance indicators, the prediction model combined with the wavelet packet decomposition algorithm has better prediction performance than the single model. Taking the 1-step prediction of the Elman model and WPD-Elman model as an example, the MAE, MAPE, RMSE, SDE, R and the training time of the Elman model are 7.0369, 18.043%, 9.3783, 0.90813 5.593, and 9.3783s, respectively. The MAE, MAPE, RMSE, SDE, R and the training time of

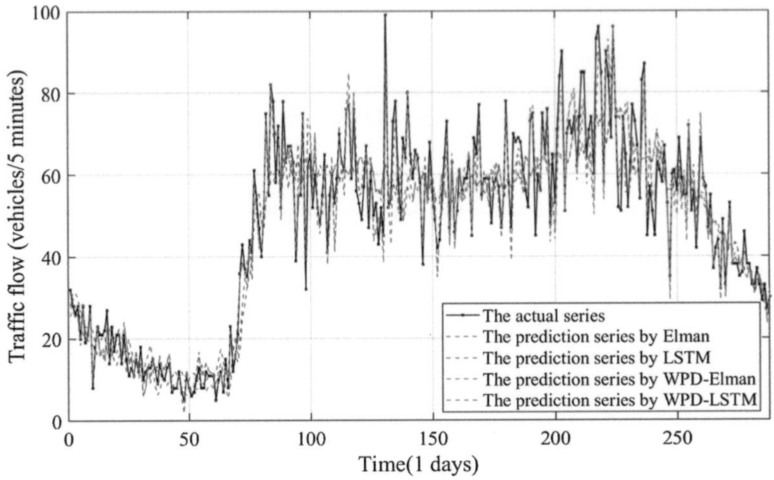

Fig. 7.33 The results of the 5-step prediction for the traffic flow series

the WPD-Elman model are 0.76197, 2.0359%, 0.99159, 0.99153, 0.99902, and 32.508s, respectively. It can be seen from these indicators that wavelet packet decomposition algorithm can greatly improve the prediction performance, but the training time of the model will increase.

Comparing the specific prediction performance indicators, although the prediction performance of the WPD-Elman model and WPD-LSTM are better, the prediction performance of the WPD-Elman model is superior. Taking the 1-step prediction as an example, the MAE, MAPE, RMSE, SDE, R and the training time of the WPD-Elman model are 0.76197, 2.0359%, 0.99159, 0.99153, 0.99902, and 32.508s, respectively. The MAE, MAPE, RMSE, SDE, R and the training time of the WPD-LSTM model are 0.90659, 3.0222%, 1.1191, 1.1187, 0.99887, and 391.54s, respectively.

Comparing the specific prediction performance indicators, the prediction performance of different single models, WPD-Elman model and WPD-LSTM model will decrease with the increase of predicted steps. The MAE, MAPE, RMSE, SDE, R and the training time of the 1-step prediction model are 0.76197, 2.0359%, 0.99159, 0.99153, 0.99902, and 32.508s, respectively. The MAE, MAPE, RMSE, SDE, R and the training time of the 2-step prediction model are 1.3068, 3.6089%, 1.6952, 1.6947 0.99715, and 32.421s, respectively. The MAE, MAPE, RMSE, SDE, R and the training time of the 5-step prediction model are 3.299, 8.8608%, 4.281, 4.2804, 0.98157, and 32.19s, respectively.

7.7 Conclusion

This chapter studies the traffic flow prediction models based on multi-dimensional data. During the running of the traffic vehicle, there is an uninterrupted flow of traffic from other road segments into the target road segment, and the traffic of the

Table 7.5 The performance estimating results of the involved prediction models

Model	Step	MAE	MAPE	RMSE	SDE	R	Training time(s)
Elman	1	7.0369	18.043	9.3783	0.90813	5.593	9.3783
	2	7.011	18.172	9.3797	0.90811	4.355	9.3797
	3	6.9137	18.045	9.263	0.91055	4.127	9.263
	4	7.141	18.582	9.4572	0.90655	4.478	9.4572
	5	7.2391	19.211	9.5415	0.90488	4.279	9.5415
LSTM	1	7.1817	17.219	9.5345	9.5283	0.9069	49.609
	2	7.1341	17.244	9.5016	9.4917	0.90786	46.895
	3	6.9526	16.714	9.3624	9.3589	0.90994	49.76
	4	7.1379	17.119	9.4919	9.4869	0.90727	50.486
	5	7.2984	17.779	9.6791	9.6707	0.90321	50.094
WPD-Elman	1	0.76197	2.0359	0.99159	0.99153	0.99902	32.508
	2	1.3068	3.6089	1.6952	1.6947	0.99715	32.421
	3	2.1914	5.8263	2.8399	2.8393	0.99196	32.397
	4	2.7471	7.3865	3.511	3.5106	0.98765	31.448
	5	3.299	8.8608	4.281	4.2804	0.98157	32.19
WPD-LSTM	1	0.90659	3.0222	1.1191	1.1187	0.99887	391.54
	2	1.5677	4.0226	2.0545	2.0449	0.99584	381.33
	3	2.3387	5.971	3.0601	3.0427	0.99078	389.77
	4	3.478	11.711	4.2834	4.2804	0.98393	393.35
	5	3.6295	9.5912	4.7016	4.6602	0.97853	511.76

adjacent road segments will affect the future traffic volume of the target road segment. Therefore, this chapter builds multi-dimensional datasets by using the traffic changes of adjacent road segments and the traffic changes of the target road segment, and establishes four traffic flow prediction models based on different model frameworks, namely the Elman model, LSTM model, WPD-Elman model, and WPD-LSTM model. Based on the experimental results, the following conclusions can be drawn:

(a) Among the four models, the traffic flow prediction performance of the WPD-Elman model using multi-dimensional data is the best, and the prediction performance of WPD-LSTM is second. The two models of the Elman model and the LSTM model have poor prediction performance.
(b) When building the traffic flow prediction model, the wavelet packet algorithm can deal with the data jump point very well, which can greatly improve the model prediction accuracy. The prediction performance of the Elman model and the LSTM model combined with the wavelet packet decomposition algorithm has been greatly improved.
(c) When the prediction performance is high, the number of prediction steps also has a certain influence on the prediction accuracy. As the number of prediction steps increases, the prediction accuracy will also decrease. This is the case with both the WPD-Elman model and the WPD-LSTM model. When the prediction performance is low, the prediction accuracy is not affected by the number of

prediction steps obviously. This is the case with both the Elman model and the LSTM model.

References

Ahn, J., Ko, E., & Kim, E. Y. (2016). Highway traffic flow prediction using support vector regression and Bayesian classifier. In *2016 International conference on big data and smart computing (BigComp)* (pp. 239–244). IEEE.

Chong S, Rui S, Jie L, Xiaoming Z, Jun T, Yunbo S et al (2016) Temperature drift modeling of MEMS gyroscope based on genetic-Elman neural network. Mechanical Systems and Signal Processing *72*:897–905

Duan Z, Yang Y, Zhang K, Ni Y, Bajgain S (2018) Improved deep hybrid networks for urban traffic flow prediction using trajectory data. IEEE Access 6:31820–31827

Goudarzi S, Kama M, Anisi M, Soleymani S, Doctor F (2018) Self-organizing traffic flow prediction with an optimized deep belief network for internet of vehicles. Sensors 18(10):3459

Graves A, Schmidhuber J (2005) Framewise phoneme classification with bidirectional LSTM and other neural network architectures. Neural Networks 18(5–6):602–610

Hochreiter S, Schmidhuber J (1997) Long short-term memory. Neural Computation 9(8):1735–1780

Kim Y-J, Hong J-S (2015) Urban traffic flow prediction system using a multifactor pattern recognition model. IEEE Transactions on Intelligent Transportation Systems 16(5):2744–2755

Koesdwiady A, Soua R, Karray F (2016) Improving traffic flow prediction with weather information in connected cars: A deep learning approach. IEEE Transactions on Vehicular Technology 65(12):9508–9517

Li PH, Li YG, Xiong QY, Chai Y, Zhang Y (2014) Application of a hybrid quantized Elman neural network in short-term load forecasting (Article). International Journal of Electrical Power & Energy Systems 55:749–759. https://doi.org/10.1016/j.ijepes.2013.10.020.

Li X, Zhang L, Wang Z, Dong P (2019a) Remaining useful life prediction for lithium-ion batteries based on a hybrid model combining the long short-term memory and Elman neural networks. Journal of Energy Storage 21:510–518

Li Z, Jiang S, Li L, Li Y (2019b) Building sparse models for traffic flow prediction: An empirical comparison between statistical heuristics and geometric heuristics for Bayesian network approaches. Transportmetrica B: Transport Dynamics 7(1):107–123

Liu H, Mi X, Li Y (2018) Smart deep learning based wind speed prediction model using wavelet packet decomposition, convolutional neural network and convolutional long short term memory network. Energy Conversion and Management 166:120–131. http://www.sciencedirect.com/science/article/pii/S019689041830356X

Luo X, Li D, Yang Y, Zhang S (2019) Spatiotemporal traffic flow prediction with KNN and LSTM. Journal of Advanced Transportation 2019

Qu L, Li W, Li W, Ma D, Wang Y (2019) Daily long-term traffic flow forecasting based on a deep neural network. Expert Systems with Applications 121:304–312

Sable AH, Talbar SN (2016) A Novel illumination invariant face recognition method based on PCA and WPD using YCbCr color space. Procedia Computer Science 92:181–187

Song X, Li W, Ma D, Wang D, Qu L, Wang Y (2018) A match-then-predict method for daily traffic flow forecasting based on group method of data handling. Computer-Aided Civil and Infrastructure Engineering 33(11):982–998

Tang K, Chen S, Khattak AJ (2018) A spatial–temporal multitask collaborative learning model for multistep traffic flow prediction. Transportation Research Record 2672(45):1–13

Zhang D, Kabuka MR (2018) Combining weather condition data to predict traffic flow: A GRU-based deep learning approach. IET Intelligent Transport Systems 12(7):578–585

Zhang Q, Gao J, Dong H, Mao Y (2018) WPD and DE/BBO-RBFNN for solution of rolling bearing fault diagnosis. Neurocomputing 312:27–33

Zheng Z, Yang Y, Liu J, Dai H-N, Zhang Y (2019) Deep and embedded learning approach for traffic flow prediction in urban informatics. IEEE Transactions on Intelligent Transportation Systems 20:1–13

Zhu Z, Peng B, Xiong C, Zhang L (2016) Short-term traffic flow prediction with linear conditional Gaussian Bayesian network. Journal of Advanced Transportation 50(6):1111–1123

Part IV
Smart Environment

In recent years, the attention of the public to environmental issues is increasing with the increasingly prominent environmental problems. Adopting reasonable and effective measures to deal with urban environmental problems has become a new challenge for urban managers. However, the formation of environmental problems is the result of a variety of complex factors. Traditional manual supervision is difficult to eliminate all kinds of problems that may threaten the urban environment from the source, nor can it achieve positive effects of urban environmental governance. Especially in the current era of rapid urban development, diversified governance bodies, and complex types of information, effective governance of urban environmental problems has encountered more difficulties.

Based on the above considerations, smart environment technology has been paid more and more attention to the construction process of smart cities. Based on the traditional urban governance model, smart environment integrates big data, cloud computing, and other new-generation information processing technologies into urban environmental governance, so as to improve the efficiency of the urban environmental monitoring system. Intelligent environment technologies can change the model of urban environmental governance and optimize the urban environment by using information platforms and intelligent systems.

This part introduces smart environmental technology from the perspective of big data prediction, including three aspects: urban air quality prediction, urban hydrological state prediction, and urban noise prediction. Chapter 8 introduces the application of the ELM in urban air prediction and discusses the influence of feature extraction on longitude prediction. Chapter 9 studies the application of the Bayesian model and the Elman model in hydrological status prediction, and then combines decomposition algorithm to optimize the single model. Chapter 10 takes noise prediction as the research object, explores the application of the RF algorithm, the BFGS algorithm, and the GRU model in the field of noise prediction, and introduces big data prediction combined with big data framework.

Chapter 8
Prediction Models of Urban Air Quality in Smart Environment

Abstract As an indispensable part of the environment, air quality has gradually become a research hotspot in recent years. The intelligent identification and prediction of atmospheric pollutants are to extract and identify the internal and external characteristics of atmospheric pollutant concentrations and integrate the influencing factors to realize the effective and stable prediction of air quality. To a certain extent, it can provide substantive reference for the control and early warning of regional atmospheric pollutant concentrations. In this chapter, single data-driven and multi-data-driven prediction models are constructed based on AQI historical data and six kinds of air pollutant concentrations data. The principal components are obtained by three feature extraction methods, the hybrid prediction model is constructed under the framework of feature extraction, and the proposed models are compared and analyzed. The results show that the prediction of pollutant concentrations after effectively extracting the main characteristics of air pollution is feasible. On this basis, this chapter also puts forward the big data calculation framework of two air pollution prediction models as a reference.

8.1 Introduction

As an important banner that leads to the healthy development of smart cities, environmental protection is also an important part of building a green city. Under the connotation of the construction of a smart city, the concept of *Smart Environmental Protection* came into being. *Smart Environmental Protection* has been extended and expanded on the basis of *Digital Environmental Protection*, which is regarded as *Environmental Protection 2.0* (Min and Sun 2011). The concept of *Smart Environmental Protection* is to use the new technologies such as the Internet of Things (IoT) and big data to monitor the regional environmental quality online in order to achieve comprehensive control of various pollution factors such as atmosphere, hydrology, and noise in the regional environment (Zhao et al. 2013; Zhang

H. Liu, *Smart Cities: Big Data Prediction Methods and Applications*,
https://doi.org/10.1007/978-981-15-2837-8_8

et al. 2015). The intelligent identification and prediction of atmospheric pollutants involved in this chapter are to extract and identify the internal and external characteristics of atmospheric pollutant concentrations, and to integrate the all-round influencing factors to achieve effective and stable prediction of air quality. And it can provide a substantial reference for the control and early warning of regional air pollutant concentrations to a certain extent.

The methods for forecasting the concentrations of atmospheric pollutants mainly include statistical methods (Li et al. 2018a), artificial neural networks and other intelligent methods (Yang and Wang 2017; Liu et al. 2019), physical process simulation methods (Jiachao et al. 2018; Zhu et al. 2017), etc., among which artificial neural networks are the most widely used. However, the traditional neural network prediction method has the defects of slow convergence speed and easy to fall into local minimization. As a generalized single-hidden layer feedforward neural network, the extreme learning machine (ELM) has fast learning speed and strong generalization ability, and has mature applications in the field of air pollutant concentrations prediction gradually. Zhang and Ding (2017) and Li et al. (2018b) constructed a single or integrated atmospheric pollutant concentration prediction model based on the extreme learning machine algorithm.

There are many kinds of atmospheric pollutants, including various gaseous pollutants and particulate pollutants, and there is an undescribed relationship between various pollutants. As a common measure of air quality, AQI is affected by the concentrations of various atmospheric pollutants. As a mature method for multi-factor analysis, the feature extraction method achieves the effect of reducing data complexity and processing difficulty under the premise of ensuring the accuracy of the data by reducing the dimension of multiple factors. Kalapanidas and Avouris (2003) used the improved genetic algorithm as a feature selection method to extract the main information components in the air pollution concentrations to predict the maximum daily concentrations of air pollutants.

Based on the physical relationship between the AQI index and the proposed concentrations of six atmospheric pollutants, this chapter uses AQI historical data and six atmospheric pollutant concentrations data to construct single data-driven and multiple data-driven prediction models, respectively. It uses three feature extraction methods to obtain the principal components, constructs the hybrid prediction models under the feature extraction framework, and compares and analyzes the proposed models. The schematic diagram in this chapter is shown in Fig. 8.1.

8.2 Original Air Pollutant Concentrations Series for Prediction

8.2.1 Original Sequence for Modeling

The time series in this chapter is derived from the original pollutant concentration time series of an atmospheric pollutant concentration monitoring site in Hefei, China, and the sampling interval of the original concentration data is 1 h. Figure 8.2

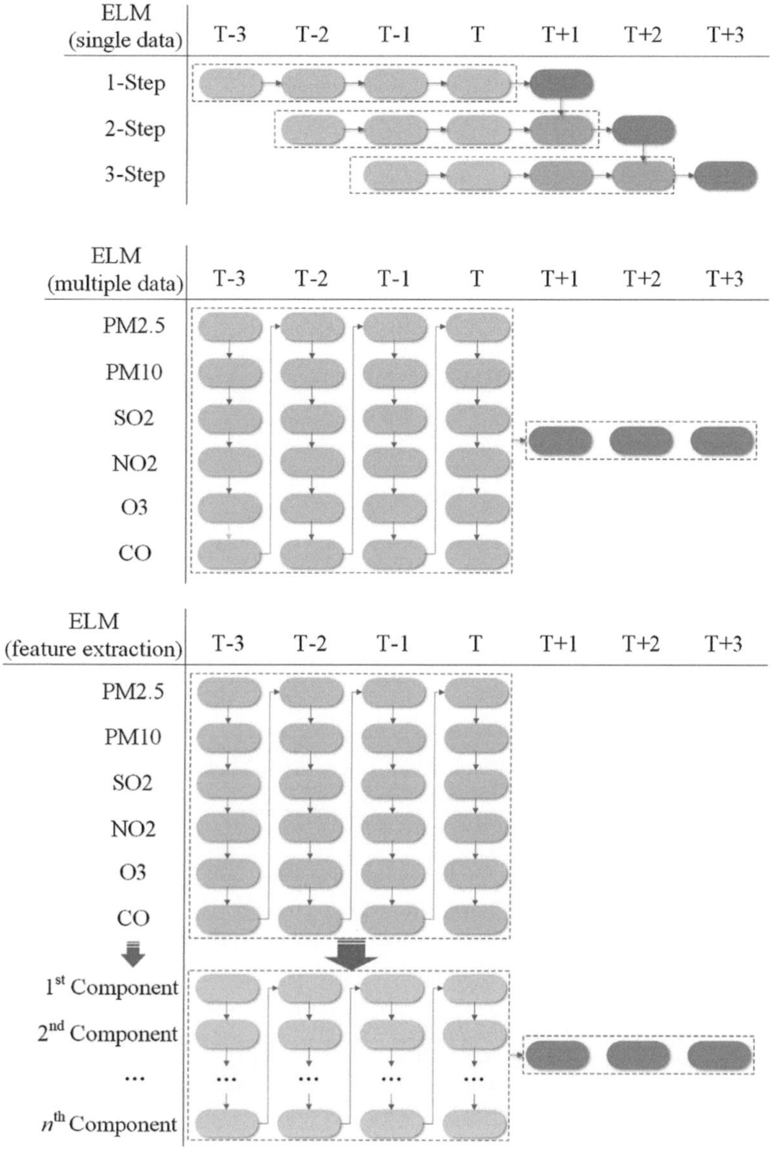

Fig. 8.1 Schematic diagram of the prediction process of the proposed prediction models

shows the original time series of AQI, denoted as $\{X_{\text{AQI}}\}$. Figure 8.3 shows the concentrations time series of $PM_{2.5}$, PM_{10}, SO_2, NO_2, O_3, and CO, denoted as $\{X_{\text{PM2.5}}\}$, $\{X_{\text{PM10}}\}$, $\{X_{\text{SO2}}\}$, $\{X_{\text{NO2}}\}$, $\{X_{\text{O3}}\}$, and $\{X_{\text{CO}}\}$, respectively. All of the air pollutant concentrations time series and AQI original series contain 800 sample points. From the waveform of the original sequences, it can be seen that the concentration series of

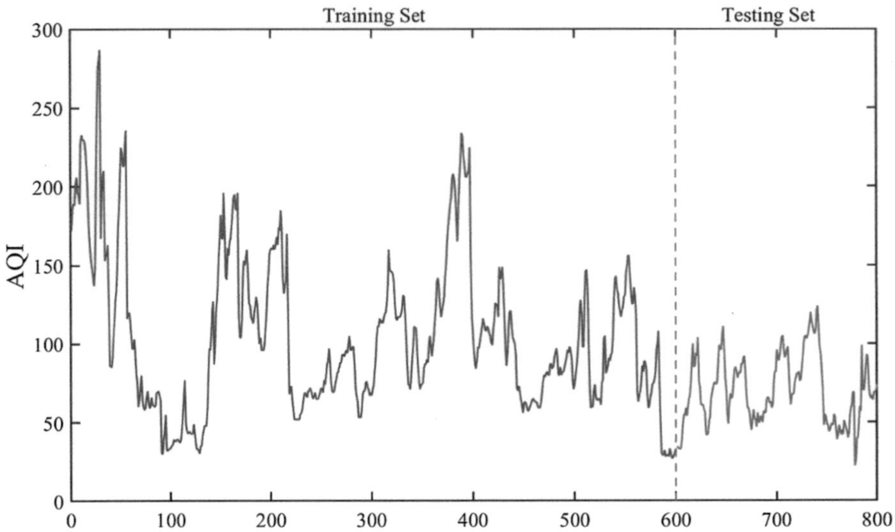

Fig. 8.2 Original AQI series

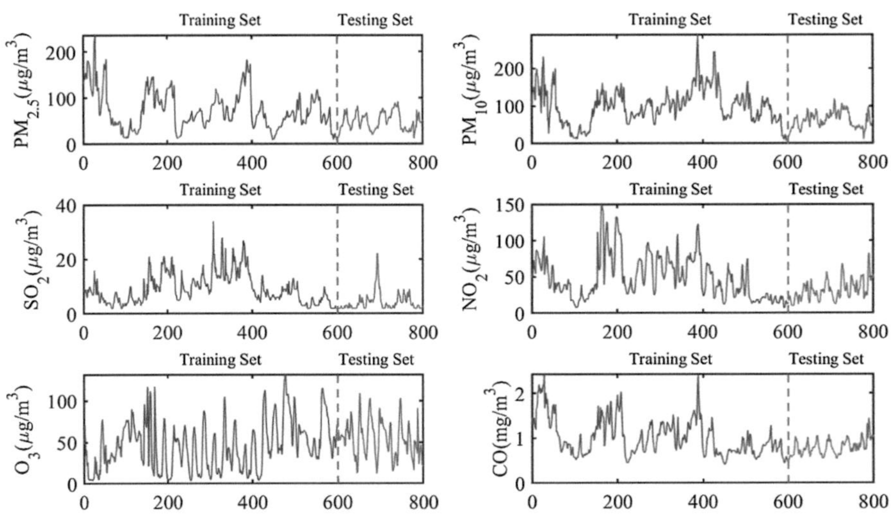

Fig. 8.3 Original air pollutant concentrations series

six kinds of air pollutant concentrations and the original AQI series all have strong randomness and non-stationarity.

In order to comprehensively analyze the characteristics of AQI series and the concentration time series of the six pollutants, and further calculate the common statistical characteristics of the seven groups of time series, such as extremum, mean value, standard deviation, skewness, and kurtosis, Table 8.1 is the statistical characteristics of the seven groups of time series adopted in this chapter, where, AQI

Table 8.1 Statistical characteristics of the original series

Original series	Minimum	Maximum	Mean value	Standard deviation	Skewness	Kurtosis
AQI	22	287	96.3750	46.8155	1.1033	4.0455
$PM_{2.5}$	9	236	67.7650	39.0126	1.0673	4.0910
PM_{10}	7	290	90.7638	45.5050	0.7481	3.7880
SO_2	2	34	7.8775	5.34703	1.1943	4.3100
NO_2	7	150	44.7000	27.5076	1.1288	4.0364
O_3	4	132	47.8288	27.6395	0.4206	2.6642
CO	0.4110	2.4400	0.9695	0.3791	1.2019	4.1944

series has no dimension, $PM_{2.5}$, PM_{10}, SO_2, NO_2, and O_3 have a dimension of μg/m^3, and CO has a dimension of mg/m^3. Where, skewness is the third-order ratio of the central moment and standard deviation of time series, which reflects the deviation of time series data distribution relative to symmetric distribution. When skewness value is negative, its distribution is skewed to the left, when skewness value is positive, its distribution is skewed to the right, and when skewness value is 0, its distribution is symmetrical. Kurtosis is the ratio of the fourth-order central moment of time series to the fourth power of standard deviation, which reflects the outliers of data. The kurtosis value adopted is the kurtosis value relative to the standard normal distribution. When the kurtosis value is positive, it indicates that the dispersion degree of the distribution is greater than the standard normal distribution.

The skewness values of the original AQI series and the concentration time series of the six air pollutants are all positive, indicating that their distributions are all skewed to the right relative to the symmetrical distribution, and their statistical distribution characteristics are closer to the symmetrical distribution. The kurtosis values of the seven groups of time series are all positive, indicating that the distribution of sample points relative to the mean is relatively scattered, and the probability of occurrence of extreme values in sample points is higher than that in normal distribution.

8.2.2 Separation of Sample

In order to construct the model and evaluate the models' performance, the original AQI series and six groups of air pollutant concentration time series for simulation need to be grouped. The seven groups of time series in this chapter all contain 800 sample points and are divided into training dataset and testing dataset. The 1st to 600th sample points are used as training samples to train the prediction model, and the 601st to 800th sample points are used as testing samples to test the obtained model after the model training is completed, so as to obtain the prediction output of the model, calculate the prediction error of the model, and evaluate the prediction performance of the model. The results of separation of sample are shown in Figs. 8.2 and 8.3.

8.3 Air Quality Prediction Model Driven by Single Data

8.3.1 Model Framework

Figure 8.4 shows the model framework for predicting the air quality coefficient AQI with a single data-driven ELM prediction model.

8.3.2 Theoretical Basis of ELM

Extreme learning machine (ELM) is a generalized single-hidden layer feedforward neural network (Huang et al. 2004, 2006), which has fast learning speed, good generalization ability, and a strong ability to search for global optimal solution (Dong et al. 2017). The extreme learning machine can randomly generate the connection weights of the input layer and the hidden layer and the bias of the hidden layer, and there is no need to adjust in the learning process (Huang and Chen 2007). Given N samples of $X_k = \{x_k, y_k\}$, $k = 1, 2, \ldots, N$, where x_k is the input data, y_k is the true value, activation function is $f(\cdot)$, and hidden layer nodes are n, the solution process of ELM is shown below (Dong et al. 2017).

$$o_j = \sum_{i=1}^{n} \lambda_i f\left(W_i \cdot X_k + b_i\right), \quad k = 1, 2, \ldots, N \tag{8.1}$$

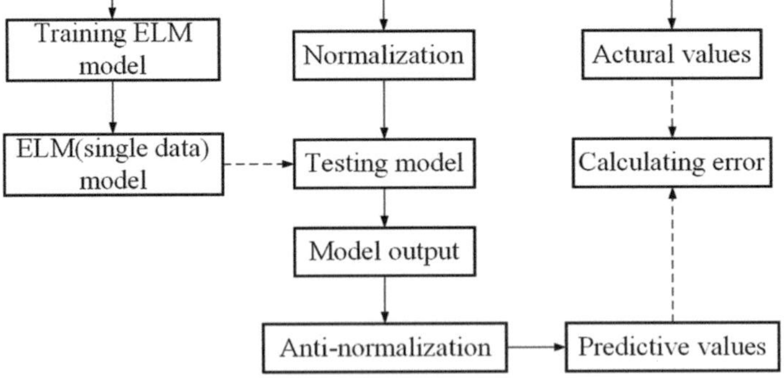

Fig. 8.4 Modeling flowchart of ELM prediction model driven by single data

where o_j is the output value of the ELM, $W_i = \{\omega_{i1}, \omega_{i2}, \ldots, \omega_{in}\}'$ is the connection weight between the input node and the ith hidden layer node, b_i is the bias of the ith node of the hidden layer, λ_i is the connection weight of the ith hidden layer node and the output node.

Therefore, only the number of hidden layer nodes and the activation function of the extreme learning machine need to be set to train it.

8.3.3 Steps of Modeling

a. Dividing the original AQI time series into a training set and a testing set, and normalizing the input data and the output data of the training set, respectively;
b. Training ELM after the initialization of parameters such as the number of hidden layer nodes of the model and the activation function, and obtaining the trained ELM model;
c. Performing the same normalization operation as the input data in step (a), and then input the ELM model obtained by training in step (b) for testing, and anti-normalize the model output to obtain the predicted value relative to the original data;
d. Comparing the predicted values of the model obtained in step (c) with the actual values, and calculating the four error evaluation index values of the MAE, MAPE, RMSE, and SDE.

8.3.4 Forecasting Results

The forecasting results of the ELM (single data) model are shown in Figs. 8.5, 8.6, and 8.7.

The evaluation indices of the ELM (single data) forecasting results are shown in Table 8.2.

The following conclusions can be drawn from Figs. 8.5, 8.6, 8.7 and Table 8.2:

a. The ELM (single data) model can achieve accurate predictions in AQI forecasting. It can well capture the variation trend of the original AQI series and the extreme change value in the peak and trough region. However, the prediction deviation is obvious at the 780–800 sample points.
b. With the increase of the advance prediction steps, the prediction accuracy of the ELM (single data) model decreases continuously, and the time delay phenomenon becomes more and more serious. This is because the ELM (single data) prediction model adopts a rolling prediction strategy, and the prediction errors of the previous step will be calculated in the prediction process of the later step, which will lead to the accumulation of prediction errors.

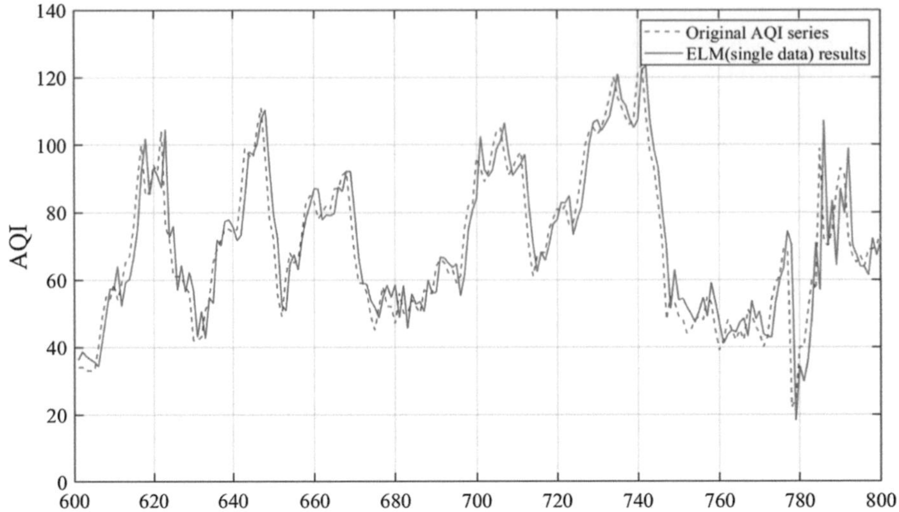

Fig. 8.5 The forecasting results of AQI time series by ELM (single data) in 1-step

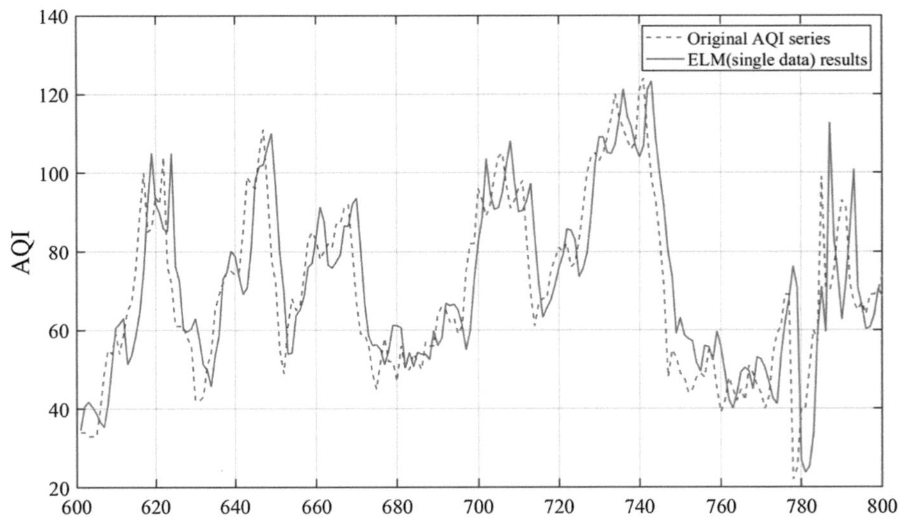

Fig. 8.6 The forecasting results of AQI time series by ELM (single data) in 2-step

8.4 Air Quality Mixture Prediction Model Driven by Multiple Data

8.4.1 Model Framework

Figure 8.8 shows the model framework for predicting the air quality coefficient AQI with the multiple data-driven ELM prediction model.

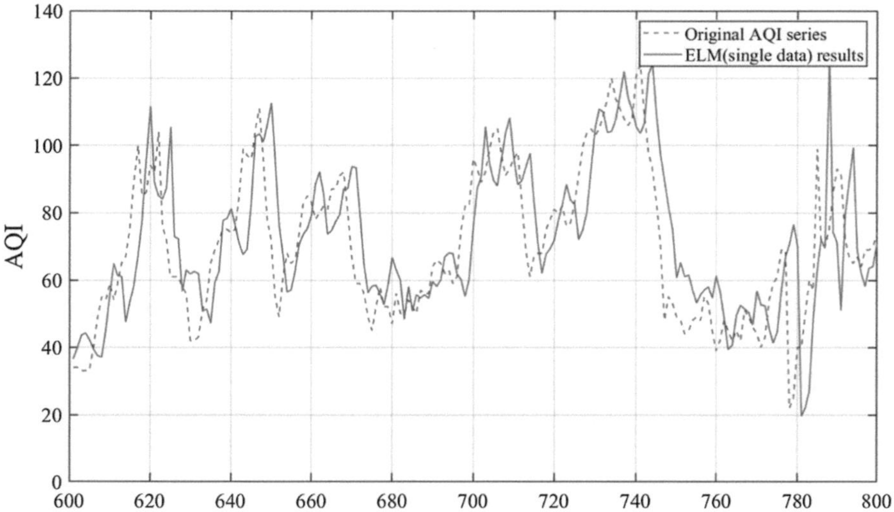

Fig. 8.7 The forecasting results of AQI time series by the ELM (single data) in 3-step

Table 8.2 The forecasting performance indices of the ELM (single data) model

Model	Step	MAE	MAPE (%)	RMSE	SDE
ELM (single data)	1	9.7579	6.4656	9.2485	9.2205
	2	15.3085	10.2725	13.6276	13.5702
	3	18.5984	12.4919	16.6371	16.5211

8.4.2 *Steps of Modeling*

a. The original concentration series of the six atmospheric pollutants and the original AQI time series are divided into the training set and the testing set in the same proportion, and the input data and the output data of the training set are, respectively, normalized;
b. Training the ELM after initialization of parameters such as the number of hidden layer nodes and activation function of the model, and the trained ELM model is obtained, where the input is six kinds of atmospheric pollutant concentration series data, and the output is AQI series data;
c. Performing the same normalization operation as the input data in step (a), and then input the ELM model obtained by training in step (b) for testing, and anti-normalize the model output to obtain the predicted value relative to the original data;
d. Comparing the predicted values of the model obtained in step (c) with the actual values, and calculate the four error evaluation index values of MAE, MAPE, RMSE, and SDE.

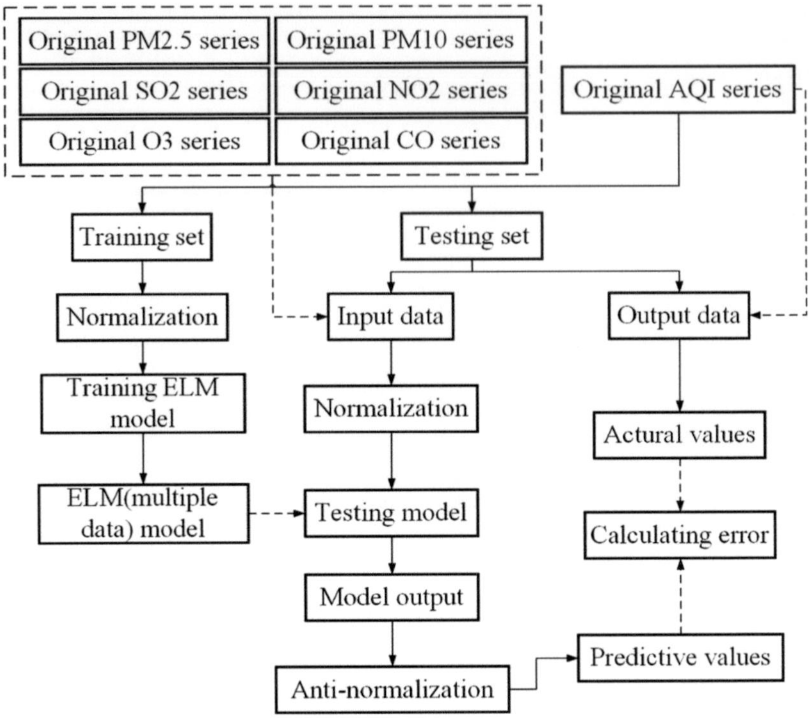

Fig. 8.8 Modeling flowchart of the ELM prediction model driven by multiple data

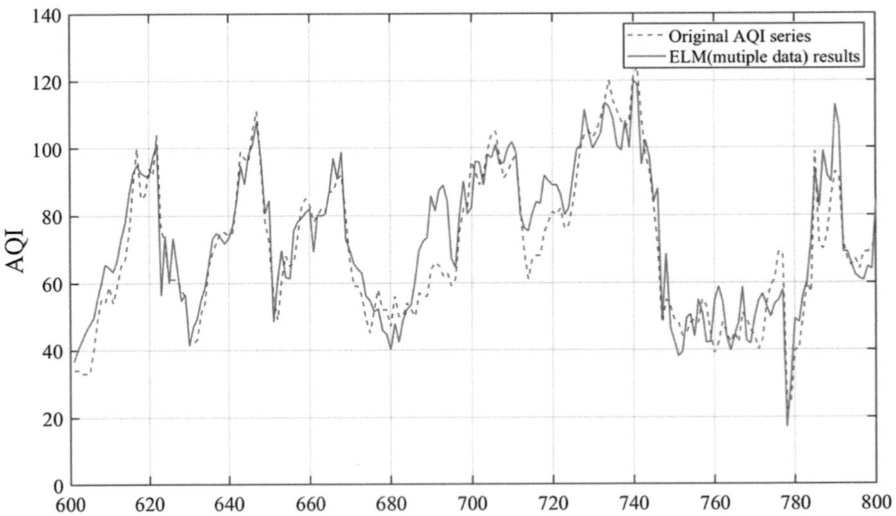

Fig. 8.9 The forecasting results of AQI series by the ELM (multiple data) in 1-step

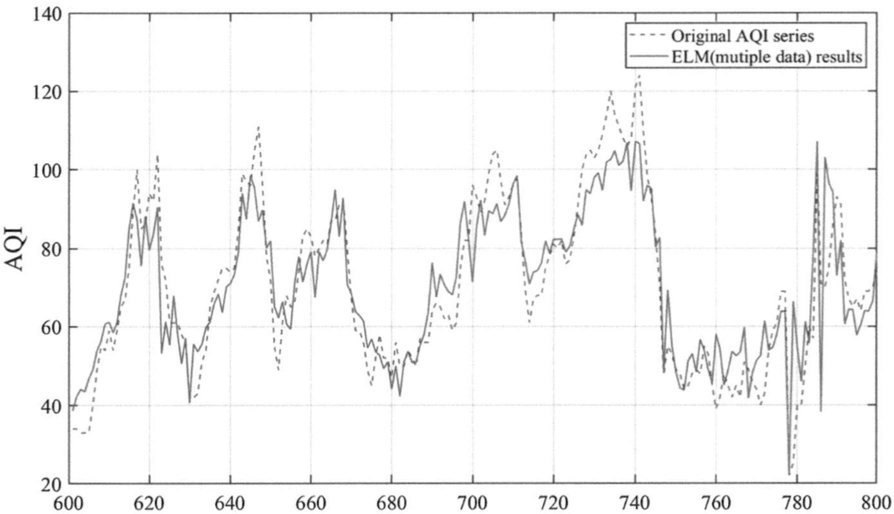

Fig. 8.10 The forecasting results of AQI series by the ELM (multiple data) in 2-step

8.4.3 Forecasting Results

The forecasting results of the ELM (multiple data) model are shown in Figs. 8.9, 8.10, and 8.11.

The evaluation indices of the ELM (multiple data) forecasting results are shown in Table 8.3.

The following conclusions can be drawn from Figs. 8.9, 8.10, 8.11 and Table 8.3:

a. The forecasting accuracy of ELM (multiple data) model for AQI series is obviously worse than ELM (multiple data) model, and its fitting ability for the variation trend of original series and the limit value of peak and trough section is also relatively weak, especially at the 780–800 sample points, the prediction deviation is particularly obvious.
b. With the increase of the advance prediction steps, the prediction accuracy of the ELM (multiple data) model decreases continuously, but the time delay phenomenon is not obvious. The reason of it might be the difference of the prediction strategy of ELM (multiple data) model, this model adopts multidimensional data fitting to driving forecasting model, thus there is no accumulation of errors, even occurs the case where the prediction accuracy of advanced high step is higher than that of advanced low step.

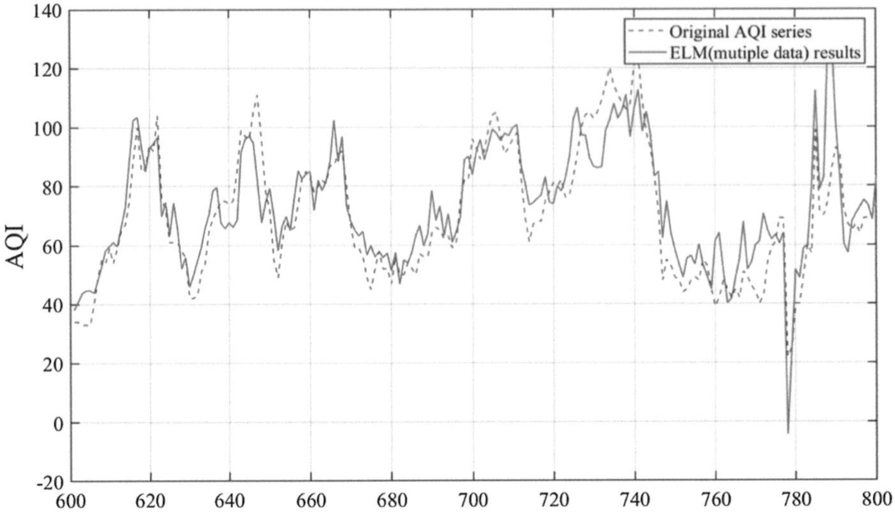

Fig. 8.11 The forecasting results of AQI series by the ELM (multiple data) in 3-step

Table 8.3 The forecasting performance indices of the ELM (multiple data) model

Model	Step	MAE	MAPE (%)	RMSE	SDE
ELM (multiple data)	1	11.2560	6.8925	8.7173	8.2431
	2	11.6617	7.3029	9.6265	9.6265
	3	12.9554	8.0589	10.7956	10.3399

8.5 Air Quality Prediction Under Feature Extraction Framework

8.5.1 Model Framework

Figure 8.12 shows the model framework for predicting the air quality coefficient AQI with a multiple data-driven ELM prediction model under the feature extraction framework.

8.5.2 Theoretical Basis of Feature Extraction Method

8.5.2.1 Principal Component Analysis

Principal Component Analysis (PCA) is also called principal component regression analysis. The main idea is to transform the multivariate data that may be correlated into the irrelevant data with few variables through orthogonal transformation, so as to decrease the dimension of the dataset and preserve the characteristic information as much as possible (Jolliffe and Cadima 2016).

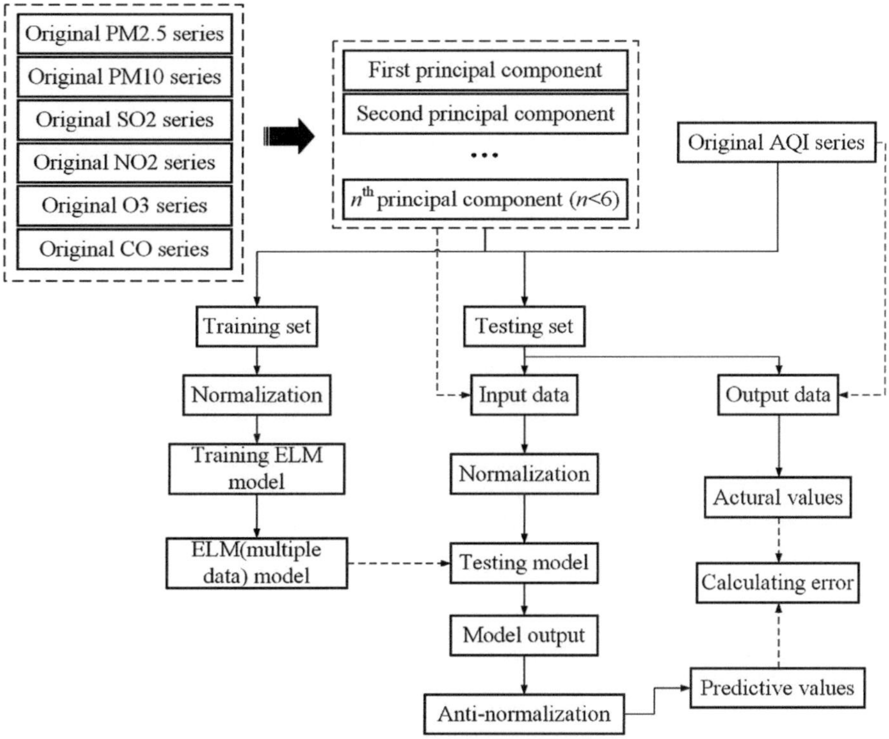

Fig. 8.12 Modeling flowchart of the ELM prediction model under feature extraction framework

Algorithm Principle

Suppose there is data $\mathbf{X}_{n \times d} = \{x_1, x_2, \cdots, x_n\}^T$ with n variables, and any variable x_i, $i = 1, 2, \ldots n$ represents a row vector with dimension $1 \times d$, and d represents the sample number of the variable, then the principal component analysis method function can be expressed as follows:

$$\begin{cases} x_1 = a_{11}S_1 + a_{12}S_2 + \cdots + a_{1m}S_m + \varepsilon_1 \\ x_2 = a_{21}S_1 + a_{22}S_2 + \cdots + a_{2m}S_m + \varepsilon_2 \\ \vdots \\ x_n = a_{n1}S_1 + a_{n2}S_2 + \cdots + a_{nm}S_m + \varepsilon_n \end{cases} \tag{8.2}$$

where $x_1, x_2, \ldots, x_n$ is n variable vectors with dimension $1 \times d$, $S_1, S_2, \ldots, S_m$ are the m principal components with dimensions $1 \times d$ extracted by PCA, which mean m eigenvectors of the sample covariance matrix, a_{ij} represents the load value of variable x_i in the principal component F_j. The larger the load value is, the higher the correlation between the variable and the principal component will be. $\varepsilon_1, \varepsilon_2, \ldots, \varepsilon_n$

is the n error vectors with dimension $1 \times d$, representing the error between each principal component and the actual variable vector.

The above formula can be expressed in the form of Eq. (8.3).

$$\begin{bmatrix} x_1 \\ x_2 \\ \vdots \\ x_n \end{bmatrix} = \begin{bmatrix} a_{11} & a_{12} & \cdots & a_{1m} \\ a_{21} & a_{22} & \cdots & a_{2m} \\ \vdots & \vdots & \ddots & \vdots \\ a_{n1} & a_{n2} & \cdots & a_{nm} \end{bmatrix} \begin{bmatrix} S_1 \\ S_2 \\ \vdots \\ S_m \end{bmatrix} + \begin{bmatrix} \varepsilon_1 \\ \varepsilon_2 \\ \vdots \\ \varepsilon_n \end{bmatrix} \tag{8.3}$$

That is,

$$\mathbf{X}_{n \times d} = \mathbf{A}_{n \times m} \mathbf{S}_{m \times d} + \boldsymbol{\varepsilon}_{n \times d} \tag{8.4}$$

where $\mathbf{X}_{n \times d}$ represents the original data matrix, $\mathbf{A}_{n \times m}$ represents the component coefficient matrix in principal components; $\mathbf{S}_{m \times d}$ represents the matrix composed of principal components; $\boldsymbol{\varepsilon}_{n \times d}$ is the error matrix.

The Identification Process

Data Standardization

The chapter introduces the six main influencing factors of AQI, namely $PM_{2.5}$, PM_{10}, SO_2, NO_2, O_3, and CO, and the original data has been introduced in Sect. 8.2. In order to analyze the data dispersion of the six characteristic variables, a box diagram as shown in Fig. 8.13 is drawn for analysis.

The value of CO is much smaller than that of other features, and the dispersion of PM_{10} features is the largest with great difference. The dispersion of $PM_{2.5}$, NO_2, and O_3 features are next. This is due to the differences between units of each feature data values. The feature variables with the same unit can be carried out by PCA in the original characteristics data, but the unit of various characteristics data is different, and there are large differences of various characteristics variance. Direct PCA will highlight the role of feature data with larger values and reduce the role of feature data with smaller values, so it is necessary to standardize the original data.

After standardization, the box diagram shown in Fig. 8.14 is analyzed again. It is clear that the dimensions of each characteristic data are eliminated, and the values are all within the range of [0, 7].

Principal Components

The main steps of extracting principal components can be divided into five steps.

a. Calculate the sample covariance matrix;
b. Calculate the eigenvalues and eigenvectors of the covariance matrix;
c. Descending order according to the size of the eigenvalues

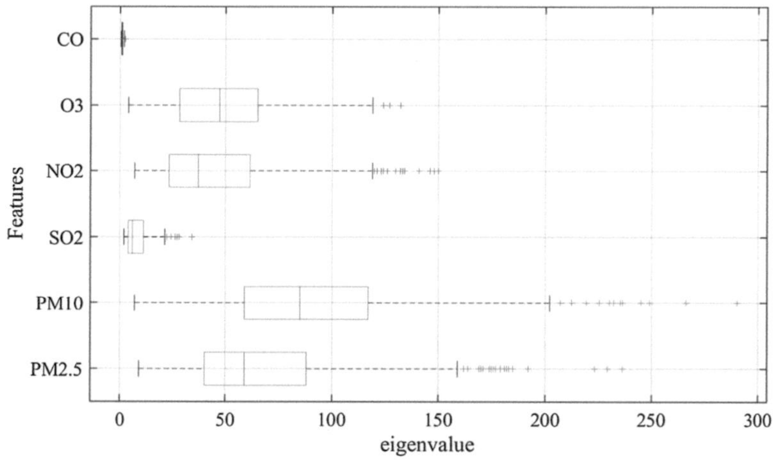

Fig. 8.13 Box diagram of original feature data

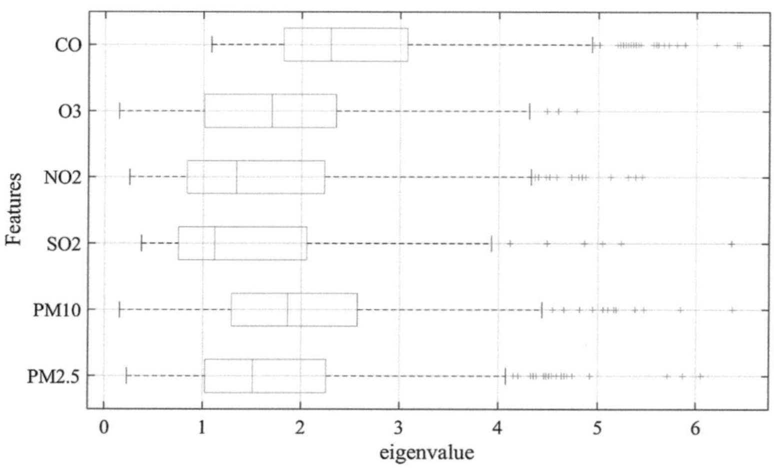

Fig. 8.14 Box diagram of standardized feature data

d. Select the number of principal components to be extracted, m;
e. Calculate the corresponding m principal components.

Information Contribution Rate and Cumulative Information Contribution Rate

In order to make the selected principal components contain more original sample information and eliminate unnecessary data noise, it needs to pay attention to the selection of the number of principal component m. It has been explained above that the larger the eigenvalue is, the more information the original sample data reflected by the eigenvector will have. Therefore, the information contribution rate and cumulative information contribution rate of each principal component can be calculated by the eigenvalue.

The information contribution rate of the *i*th principal component can be expressed as follows:

$$r_i = \frac{\lambda_i}{\sum_{k=1}^{n} \lambda_k} \times 100\% \tag{8.5}$$

where $\lambda_i \in \{\lambda_1, \lambda_2, \ldots, \lambda_n\}$ is the n eigenvalues.

The cumulative information contribution rate of the *i*th principal component can be expressed as follows:

$$R_i = \sum_{p=1}^{i} r_p = \sum_{p=1}^{i} \frac{\lambda_p}{\sum_{k=1}^{n} \lambda_k} \times 100\% \tag{8.6}$$

Identification Result

After standardization, principal component analysis is carried out for the six influencing factors of AQI. Table 8.4 shows the sample covariance characteristic values, information contribution rate, and cumulative information contribution rate of the six principal components extracted by PCA. It can be seen from Fig. 8.15 that the lower-ranking principal components hardly contain the information of the original sample data and belong to data noise, so it can be eliminated. In this chapter, 85% of the cumulative information contribution rate is selected as the threshold, and the principal components whose information contribution rate exceeds 85% are extracted, namely the first three principal components are taken as follow-up research objects. Figure 8.16 shows the data distribution of the three principal components extracted.

Table 8.4 Identification results of PCA

Principal component	Covariance eigenvalue of sample	Information contribution rate (%)	Cumulative information contribution rate (%)
1	3.8110	63.52	63.52
2	0.9327	15.55	79.07
3	0.6402	10.67	89.74
4	0.3187	5.31	95.05
5	0.2055	3.42	98.47
6	0.0919	1.53	100

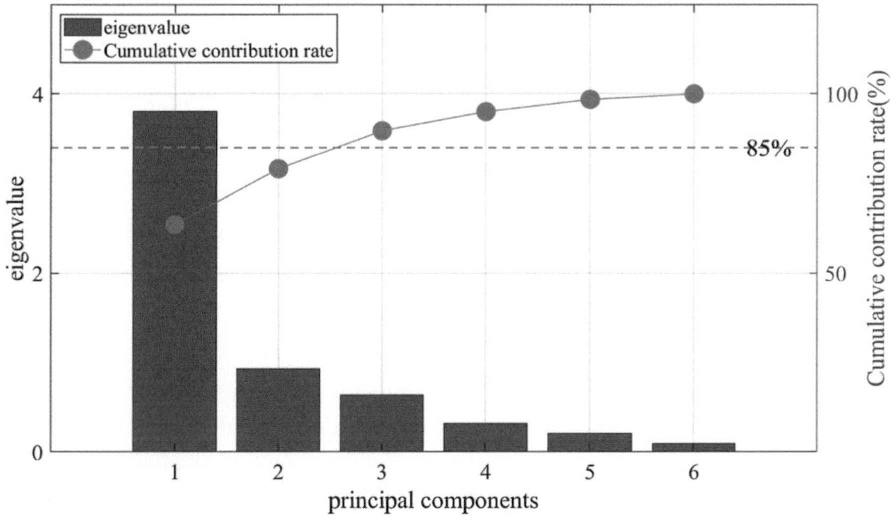

Fig. 8.15 Identification results of PCA

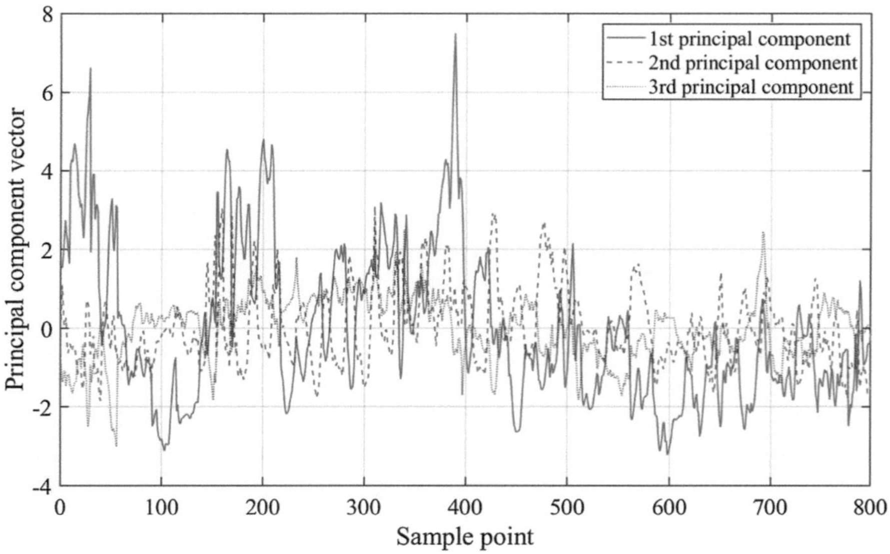

Fig. 8.16 Sample series of selected principal components

8.5.2.2 Kernel Principal Components Analysis

Kernel Principal Components Analysis (KPCA) is a nonlinear extension of the PCA. Different from the ability of the PCA to process linear data, the KPCA has better nonlinear data processing ability, can extract more significant features from nonlinear data, and is widely used in fault detection (Huang and Yan 2016; Liu et al. 2017; Navi et al. 2018), image recognition (Vinay et al. 2015; Li et al. 2008), signal denoising (Teixeira et al. 2008), and other fields. The main idea is to introduce a nonlinear mapping function, map the original sample data to a high-dimensional sample space, then use PCA to map the new sample data to another low-dimensional sample space, thereby realize the data dimensionality reduction (Jaffel et al. 2017). In this section, six kinds of AQI influencing factors are taken as the original sample set, and the KPCA method is used to achieve the dimensionality reduction.

Algorithm Principle

Suppose there is raw sample data $\mathbf{X}_{d \times n}$, d is the number of variable samples, and n is the number of variables, where $d > > n$. Introducing the nonlinear mapping function Φ, the original sample data $\mathbf{X}_{d \times n}$ of the $d \times n$ dimension is mapped to the $d \times d$ high-dimensional space, denoted as $\Phi(\mathbf{X}_{d \times n})\Phi(\mathbf{X}_{d \times n})^T$, and the PCA method is used in the high-dimensional space, which can be expressed as follows:

$$\Phi\left(\mathbf{X}_{d\times n}\right)\Phi\left(\mathbf{X}_{d\times n}\right)^T s_i = \lambda s_i \tag{8.7}$$

where $s_i \in \{s_1, s_2, \cdots, s_n\}$ represents n feature vectors of the high-dimensional spatial sample $\Phi(\mathbf{X}_{d \times n})\Phi(\mathbf{X}_{d \times n})^T$, and $\lambda_i \in \{\lambda_1, \lambda_2, \cdots, \lambda_n\}$ represents the corresponding n feature values.

Define the vector v_i to satisfy $s_i = \Phi(\mathbf{X}_{d \times n})v_i$ and bring it into the Eq. (8.7).

$$\Phi\left(\mathbf{X}_{d\times n}\right)\Phi\left(\mathbf{X}_{d\times n}\right)^T \Phi\left(\mathbf{X}_{d\times n}\right)v_i = \lambda\Phi\left(\mathbf{X}_{d\times n}\right)v_i \tag{8.8}$$

Multiply both sides by $\Phi(\mathbf{X}_{d \times n})^T$ to get the following formula:

$$\Phi\left(\mathbf{X}_{d\times n}\right)^T \Phi\left(\mathbf{X}_{d\times n}\right)\Phi\left(\mathbf{X}_{d\times n}\right)^T \Phi\left(\mathbf{X}_{d\times n}\right)v_i = \lambda\Phi\left(\mathbf{X}_{d\times n}\right)^T \Phi\left(\mathbf{X}_{d\times n}\right)v_i \tag{8.9}$$

Since the nonlinear mapping function Φ is a recessive function, define a high-dimensional kernel matrix $\mathbf{K}_{d \times d}$ as follows:

$$\mathbf{K}_{d\times d} = \Phi\left(\mathbf{X}_{d\times n}\right)^T \Phi\left(\mathbf{X}_{d\times n}\right) = \left(k_{ij}\right)_{d\times d} \tag{8.10}$$

where k_{ij} is represented as the element of the ith row and the jth column of the high-dimensional kernel matrix. The elements in the high-dimensional kernel matrix are calculated by kernel functions.

The Identification Process

Nonlinear Mapping Based on the Gaussian Kernel

The main steps of extracting principal components using kernel principal component analysis can be divided into seven steps:

a. Data normalization;
b. Compute the kernel matrix;
c. Kernel matrix for centralization;
d. Calculate the eigenvalues and eigenvectors;
e. Sort the eigenvalues in descending order;
f. Select the number of principal components to be extracted;
g. Calculate the corresponding *m* principal components.

Information Contribution Rate and Cumulative Information Contribution Rate

Using the KPCA method, the square root of the eigenvalues of the kernel matrix obtained in the foregoing can be used to indicate that the corresponding principal component reflects the amount of information.

The information contribution rate of the *i*th principal component can be expressed as follows:

$$r_i = \frac{\sqrt{\lambda_i}}{\sum_{k=1}^{n} \sqrt{\lambda_k}} \times 100\% \tag{8.11}$$

where λ_i represents the *i*th eigenvalue of the high-dimensional kernel matrix.

The cumulative information contribution rate of the *i*th principal component can be expressed as follows:

$$R_i = \sum_{p=1}^{i} r_p = \sum_{p=1}^{i} \frac{\sqrt{\lambda_p}}{\sum_{k=1}^{n} \sqrt{\lambda_k}} \times 100\% \tag{8.12}$$

Identification Result

The six influencing factors of AQI are reduced by using KPCA based on the Gaussian kernel function.

Table 8.5 and Fig. 8.17 show the kernel matrix eigenvalues, information contribution rates, and cumulative information contribution rates of the six principal components extracted by the Gaussian KPCA. The principal component of the cumulative information contribution rate of 85% is selected as the threshold value.

Therefore, the first five principal components are selected as the subsequent research objects.

8.5.2.3 Factor Analysis

Factor Analysis (FA) is one of several common methods for data dimensionality reduction, which is realized by using a few factors to describe the relationship among numerous factor indicators. The main idea of the dimension reduction is that every factor or indicator data of the original sample data is divided into two parts. One part can be represented by a linear combination by a handful of public factor and the other part can be denoted by the unique factors of corresponding factors or indicator data. Thus a small number of factors are substituted for factors or indicators to achieve the purpose of dimension reduction (Bandalos and Finney 2018).

Algorithm Principle

Factor Analysis Model

Suppose there is data $\mathbf{X}_{n \times d} = \{x_1, x_2, \cdots, x_n\}^T$ with n variables, and any variable x_i, $i = 1, 2, \ldots n$ represents a row vector with dimension $1 \times d$, and d represents the sample number of the variable, then the factor analysis model can be expressed as follows:

$$\begin{cases} x_1 = \overline{x}_1 + a_{11}S_1 + a_{12}S_2 + \cdots + a_{1m}S_m + \varepsilon_1 \\ x_2 = \overline{x}_2 + a_{21}S_1 + a_{22}S_2 + \cdots + a_{2m}S_m + \varepsilon_2 \\ \vdots \\ x_n = \overline{x}_n + a_{n1}S_1 + a_{n2}S_2 + \cdots + a_{nm}S_m + \varepsilon_n \end{cases} \tag{8.13}$$

where $x_1, x_2, \ldots, x_n$ are n variable vectors with dimension $1 \times d$, $\overline{x}_1, \overline{x}_2, \supset, \overline{x}_n$ are the mean of the variable vectors; $S_1, S_2, \ldots, S_m$ are m common factor vectors with dimension $1 \times d$, abbreviated as common factor; $\varepsilon_1, \varepsilon_2, \ldots, \varepsilon_n$ are n characteristic factor

Table 8.5 The identification results of the Gaussian KPCA

Principal component	Covariance eigenvalue of sample	Information contribution rate (%)	Cumulative information contribution rate (%)
1	8.8791	22.8984	22.8984
2	7.4057	19.0988	41.9972
3	6.6140	17.0571	59.0543
4	6.0301	15.5512	74.6055
5	4.9691	12.8149	87.4204
6	4.8779	12.5796	100

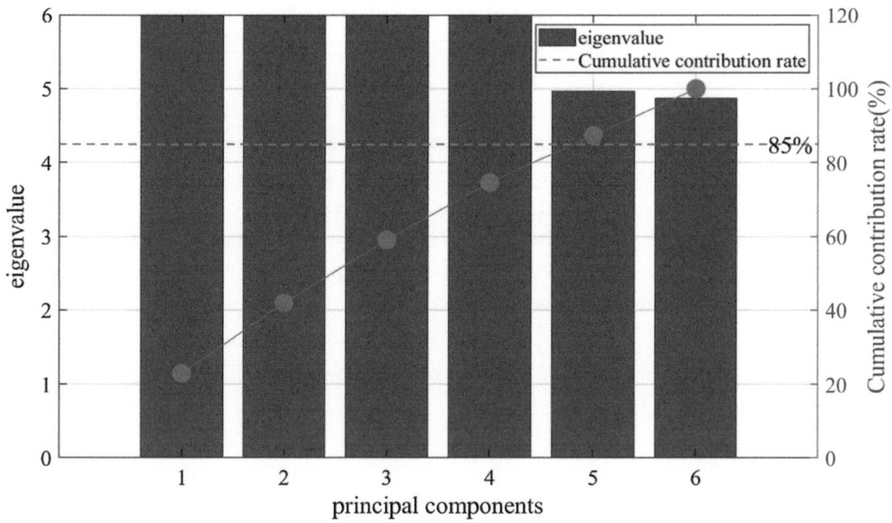

Fig. 8.17 The identification results of the Gaussian KPCA

vectors with dimension $1 \times d$, abbreviated as characteristic factor; a_{ij} is the factor loading value.

Contribution Degree

The factor loading value is the correlation coefficient between the variable and the common factor, indicating the information contribution degree of the common factor to the variable. The larger the factor loading value is, the higher the contribution to the corresponding variable will be. The contribution degree of the factor loading value to the overall sample data can be expressed as follows:

$$q_j = \sum_{i=1}^{n} a_{ij}^2 \tag{8.14}$$

where q_j represents the contribution degree of the jth factor to the overall sample data.

Heywood Case

Variances of variables can be expressed by factor loading values.

$$\mathrm{Var}\left(x_i\right) = \sigma^2 + \xi_i^2 = a_{i1}^2 + a_{i2}^2 + \cdots + a_{im}^2 + \xi_i^2 \tag{8.15}$$

where $\mathrm{Var}(x_i)$ is the variance of the variable x_i; σ^2 is the generic variance, expressing the contribution of all common factors to the ith variable; ξ_i^2 is special variance.

If the variance of all variables is equal to 1 after standardization of the original sample data, which means $\mathrm{Var}(x_i) = 1, \quad \forall\, i \in [1, n]$, the sum of the generic variance and the special variance will be equal to 1, meaning that the values of the generic variance and the special variance are between 0 and 1. But in the actual calculation, the generic variance will be greater than or equal to 1. If the generic variance is equal to 1, it will be called the Heywood case; if the generic variance is greater than 1, it will be called ultra-Heywood case (Savalei and Kolenikov 2008). Heywood and ultra-Heywood case mean that there are abnormal phenomena in which the special variance is zero or even negative. And if this happens, the parameters should be adjusted and the data should be reprocessed.

The Identification Process

Data Standardization

Before the principal component extraction of six influencing factors of AQI by the FA method, the data should be dedimensionalized and centralized to eliminate the difference of unit dimension between different influencing factors of sample data and to make the mean value of each influencing factor data zero.

The dedimensionalization formula is as follows:

$$x = x / \mathrm{Std}(x) \tag{8.16}$$

where x represents the data of AQI influencing factors, and $\mathrm{Std}(x)$ denotes the standard deviation of x.

The centralization formula is as follows:

$$x = x / \overline{x} \tag{8.17}$$

where $\overline{x}$ is the mean value of AQI influencing factor data x.

The Applicability of FA

After standardizing and centralizing the characteristic data, it is necessary to test whether the factor index meets the requirements of the FA. The main test methods include the KMO test and the Bartlett sphericity test (Cecchetto and Pellanda 2014). This chapter uses the KMO test for applicability analysis. KMO test statistics are commonly utilized to compare variable correlation coefficient and partial correlation coefficient. The value range is 0–1. The KMO statistics can be calculated as follows:

$$\mathrm{KMO} = \frac{\sum\sum r_{ij}^2}{\sum\sum r_{ij}^2 + \sum\sum p_{ij}^2} \tag{8.18}$$

where r_{ij}^2 represents the correlation coefficient between the ith variable and the jth variable, and p_{ij}^2 represents the partial correlation coefficient between the ith variable and the jth variable.

When the value of the KMO is larger than or equal to 0.5, the FA is considered suitable for the sample data. The KMO of the data used in this chapter is 0.5900, so the FA method can be used for the sample data.

Information Contribution Rate and Cumulative Information Contribution Rate

q_j can reflect the contribution degree of the jth common factor to the whole sample data when using the FA method.

The information contribution rate of the jth common factor is defined as follows:

$$f_j = \frac{q_j}{\sum_{k=1}^{m} q_k} \times 100\% \tag{8.19}$$

The cumulative information contribution rate of the jth common factor is defined as follows:

$$F_j = \sum_{p=1}^{j} f_p = \sum_{p=1}^{j} \frac{q_p}{\sum_{k=1}^{m} q_k} \times 100\% \tag{8.20}$$

Identification Result

Through calculation, when the number of common factors exceeds 3, the subsequent factor load value is zero, so the factor number is set to 3 for the FA method.

Table 8.6 shows the contribution degree, information contribution rate, and cumulative information contribution rate of the three factors. Figure 8.18 shows that the low-ranking factor has a low contribution rate to the sample data and has almost no impact on the sample data. It may even be data noise, so it can be eliminated.

In this chapter, 85% of the cumulative information contribution rate is selected as the threshold, and the principal components whose contribution rate is more than 85% are extracted, namely the three factors whose contribution rate ranks high, as the follow-up research objects. Figure 8.19 shows the data distribution of these three factors.

Table 8.6 Identification results of the FA

Common factor	Contribution degree	Information contribution rate (%)	Cumulative information contribution rate (%)
1	2.0204	42.3954	42.3954
2	1.5462	32.4455	74.8409
3	1.1990	25.1591	100

8.5.3 *Steps of Modeling*

a. The original concentration series of the six atmospheric pollutants concentrations and the original AQI time series are divided into the training set and the testing set in the same proportion, and the input data and the output data of the training set are, respectively, normalized;
b. Using the PCA, KPCA, and FA methods to extract the principal components of the training set and the testing set, respectively, as new model inputs;
c. Training the ELM after initialization of parameters such as the number of hidden layer nodes of the model and the activation function, and obtaining the trained ELM model, where the input is the principal components data obtained in the step (b), and the output is the AQI series data;
d. The input data of the testing set is subjected to the same normalization operation as in the step (a), and then the ELM model obtained by the training in the step (c) is input for testing, and the model output is anti-normalized to obtain the predicted value relative to the original data;

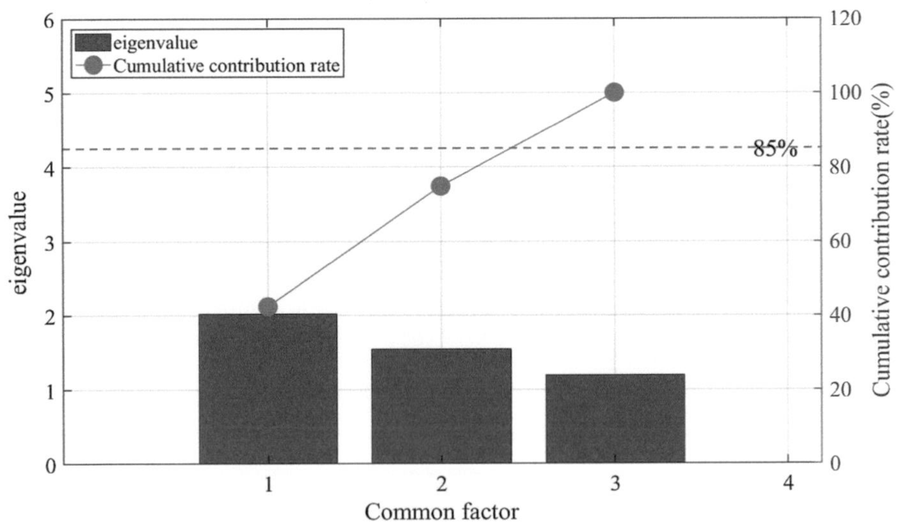

Fig. 8.18 Identification results of the FA

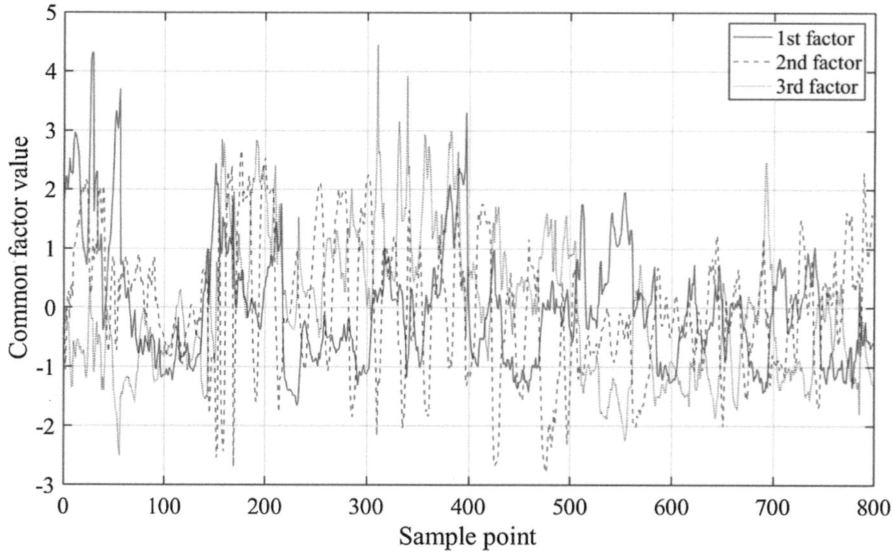

Fig. 8.19 Sample series of selected common factors

e. Comparing the predicted values of the model obtained in step (d) with the actual values, and calculating the four error evaluation index values of MAE, MAPE, RMSE, and SDE.

8.5.4 Forecasting Results

The forecasting results of the ELM model under the feature extraction framework are shown in Figs. 8.20, 8.21, and 8.22.

The evaluation indices of the ELM model under feature extraction framework results are shown in Table 8.7.

The following conclusions can be drawn from Figs. 8.20, 8.21, 8.22 and Table 8.7:

a. In this chapter, feature extraction is adopted as the pre-processing method of the prediction process, which does not improve the prediction accuracy of the model, on the contrary, it further reduces the accuracy of the ELM prediction model. The prediction model under PCA and FA feature extraction framework still has some curve fitting ability, while the KPCA-ELM model basically loses its prediction ability and its prediction accuracy is very poor. The possible reason for this phenomenon is that the physical interpretation of AQI is obtained by using the complex calculation of the six atmospheric pollutant concentrations data in this chapter. Feature extraction to achieve data dimensionality may destroy this

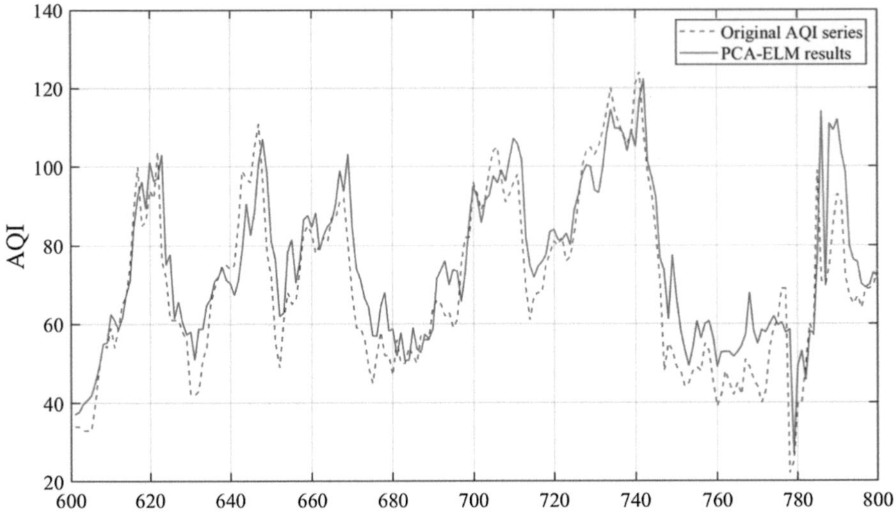

Fig. 8.20 The forecasting results of AQI time series by the PCA-ELM in 1-step

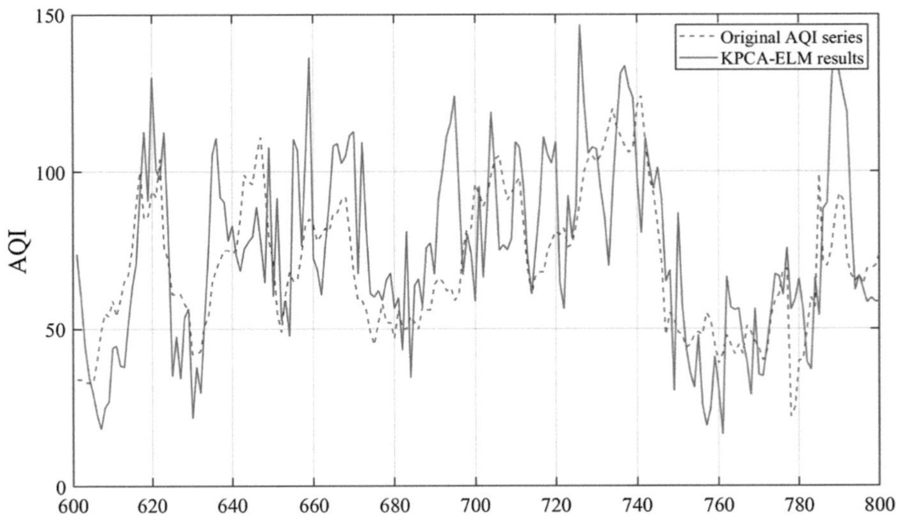

Fig. 8.21 The forecasting results of AQI time series by the KPCA-ELM in 1-step

physical relationship. As the contribution threshold of principal component, the 85% may be lower, losing some of the main information in the original series.

b. The prediction model under the feature extraction framework also does not have the phenomenon of time delay, which is consistent with the ELM (multiple data) prediction model. It uses the historical data of other multiple data to fit with the future data of AQI so as to realize advance prediction without the gradual accumulation of prediction errors.

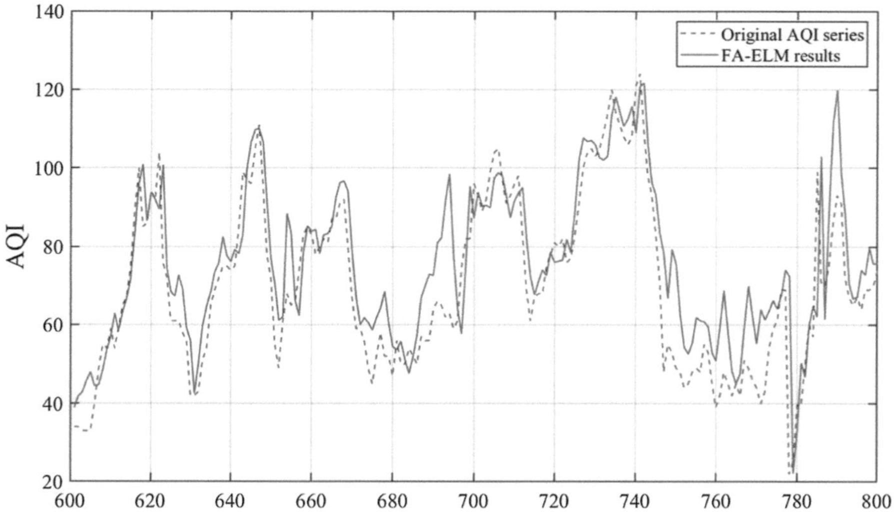

Fig. 8.22 The forecasting results of AQI time series by the FA-ELM in 1-step

Table 8.7 The forecasting performance indices under feature extraction framework

Model	Step	MAE	MAPE (%)	RMSE	SDE
PCA-ELM	1	13.6767	8.2044	11.0127	9.4615
	2	14.5743	8.6259	11.6723	11.0014
	3	15.5078	9.1660	12.2135	10.7769
KPCA-ELM	1	28.3962	18.2526	22.8717	22.3643
	2	30.3466	18.8829	24.3160	23.7980
	3	31.6014	19.0784	24.9286	24.5529
FA-ELM	1	14.9122	8.6400	11.5609	10.0151
	2	15.1325	8.7326	11.1800	9.1924
	3	16.4206	9.3897	12.6611	10.6174

8.6 Big Data Prediction Architecture of Urban Air Quality

8.6.1 The Idea of Urban Air Quality Prediction Based on Hadoop

With the advent of the era of big data, massive air quality monitoring data are no longer suitable for processing on the traditional single machine. In the face of massive air quality monitoring and related impact factor data prediction and processing problems, a single predictor will lead to long training time and a lack of memory and other problems. Therefore, on the basis of study of the chapter, it has become a new trend to study how to realize the parallelization of the ELM algorithm in the environment of big data platforms. In the paper, an urban air quality prediction

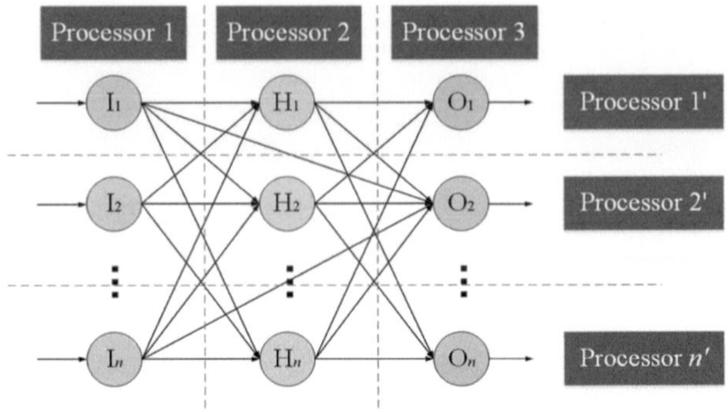

Fig. 8.23 Structure-based parallelized ELM

algorithm based on MR-ELM is proposed by combining the MapReduce distributed computing framework of Hadoop platform with the ELM algorithm.

8.6.2 Parallelization Framework of the ELM

The idea of parallel computing of the ELM algorithm can be divided into two kinds: the one is structure-based parallelization as shown in Fig. 8.23, and the other is data-based parallelization as shown in Fig. 8.24. The former divides the nodes of the ELM into multiple processors, and its division mode is divided into horizontal division and vertical division. The latter divides the sample data into different processors, each processor will have a complete ELM model. Structure-based parallelization is to divide the amount of computation, but the traffic is increased, so it is not suitable for cluster computers. Based on the parallelization of data, the amount of computation is divided, and the traffic is less, so it is more suitable for cluster computers.

8.6.3 The Parallelized ELM Under the MapReduce Framework

The ELM parallelization under the MapReduce framework combines the parallelization idea of the ELM with the MapReduce parallel computing framework, makes full use of the advantages of MapReduce, and greatly reduces the difficulty of ELM parallelization. The parallelization of MapReduce is divided into two stages: the Map phase and the Reduce phase, and the datasets processed must have the following characteristics: (1) the dataset can be divided into multiple small datasets; (2) each small dataset can be processed independently.

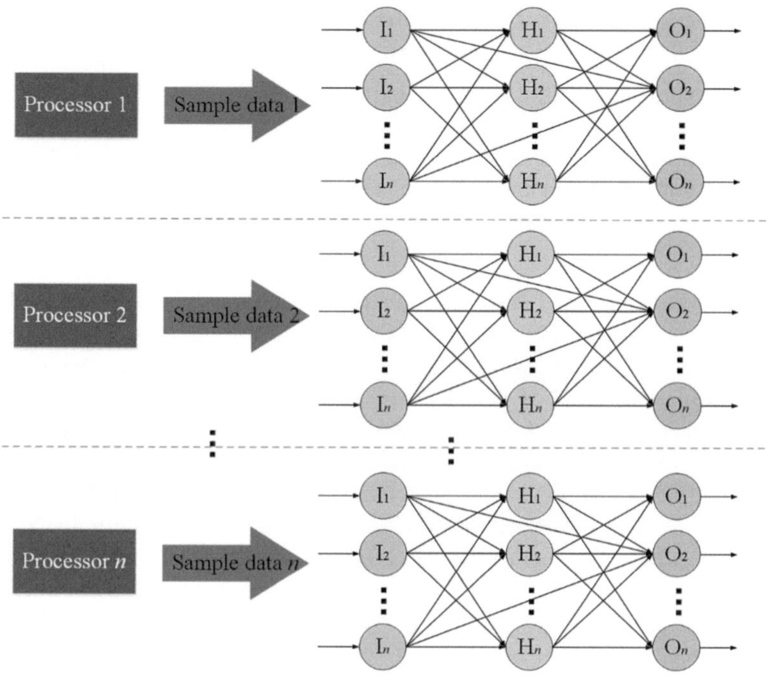

Fig. 8.24 The data-based parallelized ELM

The parallel design idea of the ELM based on the MapReduce is as follows:

a. Sample dataset partition

Dataset partition is to divide the whole sample dataset into multiple datasets and store them in different data nodes, which is carried out automatically by the Hadoop platform. The number of datasets is determined by the size of HDFS blocks in the Hadoop platform and the size and number of sample data files. Each data block will correspond to a map task.

b. Map function and reduce function

After the preparation of the above data, the sample data is read through the map function, the training process of the ELM is realized in the map function, and the batch mode is used as the training mode of each ELM model. Taking the number of rows and columns of the two core matrices of the ELM as the output (that is, the input of the reduce function), the output value of each map function is accumulated in the reduce function, and the cumulative value is used as the output core matrix of the ELM in the map function. Judge whether or not to meet the end conditions of the training model.

8.7 Comparative Analysis of Forecasting Performance

The prediction results of all the models in the chapter are shown in Figs. 8.25, 8.26, and 8.27.

The comprehensive forecasting performance indices of all the models in this chapter are shown in Table 8.8.

The following conclusions can be drawn from Figs. 8.25, 8.26, 8.27 and Table 8.8:

a. For the prediction of the AQI series, the ELM (single data) model has the highest prediction accuracy, followed by the ELM (multiple data) model, and the prediction accuracy of the PCA-ELM, FA-ELM, and KPCA-ELM models decreases. For example, the MAE, MAPE, RMSE, and SDE predicted in 1-step by the ELM (single data) model are 9.7579, 6.4656%, 9.2485, and 9.2205, respectively; the MAE, MAPE, RMSE, and SDE of the ELM (multiple data) model are 11.2560, 6.8925%, 8.7173, and 8.2431, respectively; the MAE, MAPE, RMSE, and SDE of the PCA-ELM model are 13.6767, 8.2044%, 11.0127, and 9.4615, respectively; the MAE, MAPE, RMSE, and SDE of the FA-ELM model are 14.9122, 8.6400%, 11.5609, and 10.0151, respectively; the MAE, MAPE, RMSE, and SDE of the FA-ELM model are 28.3962, 18.2526%, 22.8717, and 22.3643, respectively.
b. From the prediction results figures, the ELM (single data) model using AQI's own raw data for advance prediction is undoubtedly the best, and the ELM (multiple data) model has relatively good fitting and predictive ability. The predictive model under the KPCA framework completely lost its predictive power and the

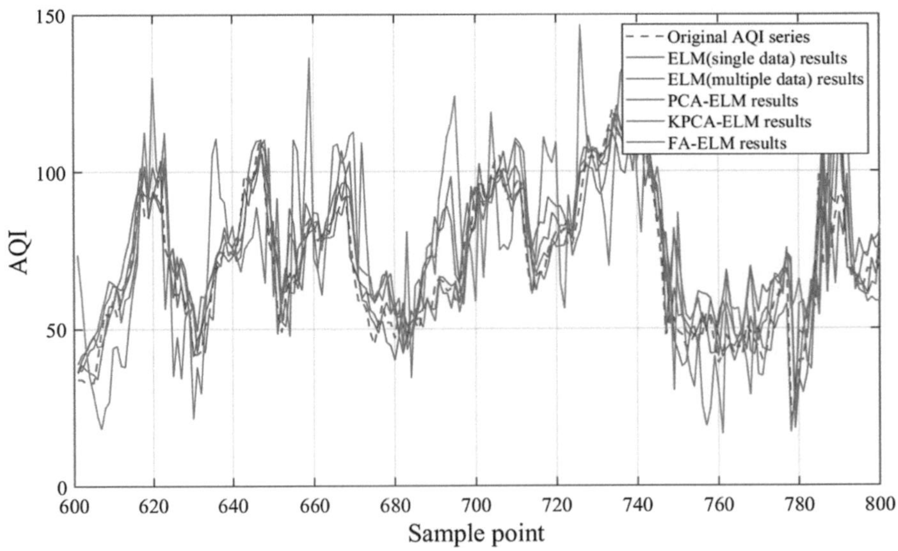

Fig. 8.25 The forecasting results of AQI time series by proposed models in 1-step

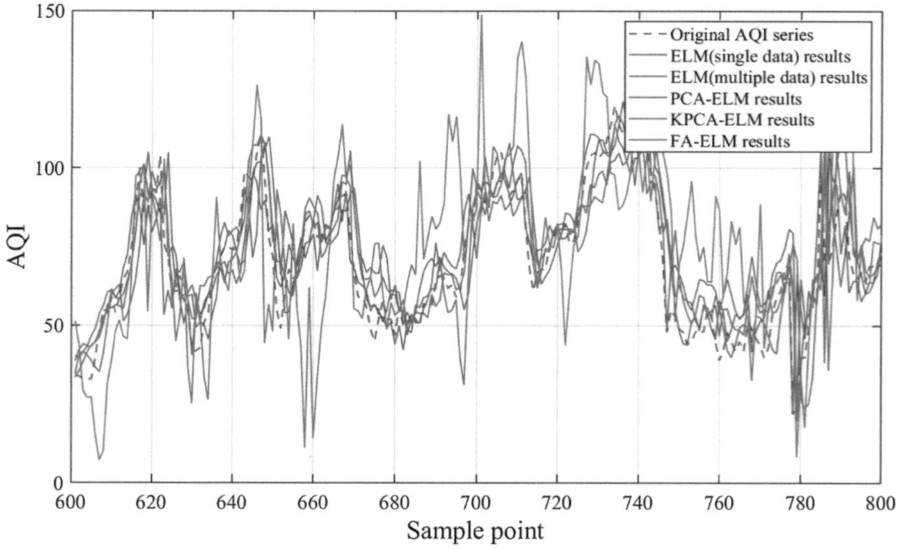

Fig. 8.26 The forecasting results of AQI time series by proposed models in 2-step

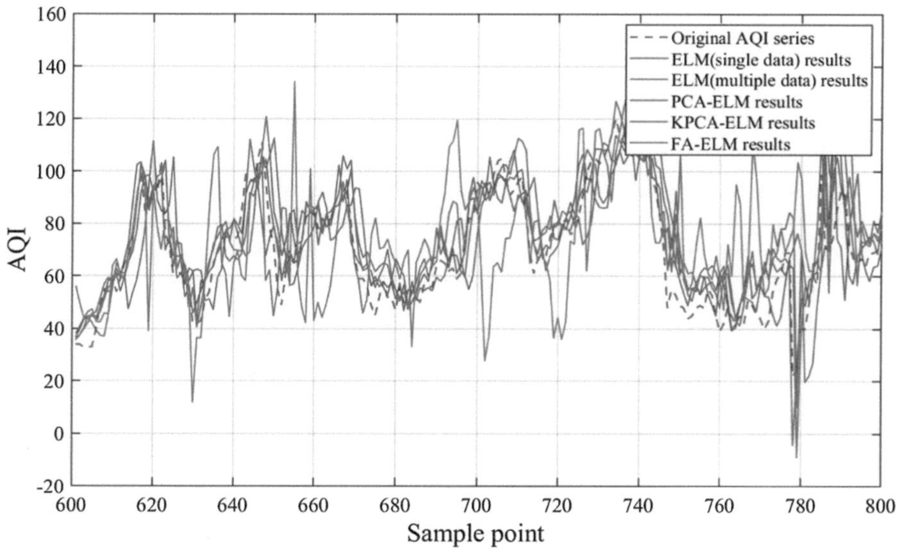

Fig. 8.27 The forecasting results of AQI time series by proposed models in 3-step

prediction effect is extremely poor. The prediction effect of the hybrid models under the PCA and FA frameworks is between them.

c. All models show significant prediction biases at the 780–800 sample points, which may be related to the sharp trend effect in the original AQI series.
d. Combined with the timeliness of model prediction, the ELM (single data) model is still the optimal prediction model, and the ELM (multiple data) model is the

Table 8.8 The comprehensive forecasting performance indices of proposed models

Model	Step	MAE	MAPE (%)	RMSE	SDE	Time (s)
ELM (single data)	1	9.7579	6.4656	9.2485	9.2205	0.1240
	2	15.3085	10.2725	13.6276	13.5702	
	3	18.5984	12.4919	16.6371	16.5211	
ELM (multiple data)	1	11.2560	6.8925	8.7173	8.2431	0.1840
	2	11.6617	7.3029	9.6265	9.6265	
	3	12.9554	8.0589	10.7956	10.3399	
PCA-ELM	1	13.6767	8.2044	11.0127	9.4615	0.3617
	2	14.5743	8.6259	11.6723	11.0014	
	3	15.5078	9.1660	12.2135	10.7769	
KPCA-ELM	1	28.3962	18.2526	22.8717	22.3643	0.5810
	2	30.3466	18.8829	24.3160	23.7980	
	3	31.6014	19.0784	24.9286	24.5529	
FA-ELM	1	14.9122	8.6400	11.5609	10.0151	0.6168
	2	15.1325	8.7326	11.1800	9.1924	
	3	16.4206	9.3897	12.6611	10.6174	

second. The hybrid prediction models under the three feature extraction frameworks are due to the existence of the feature extraction process, resulting in a decrease in timeliness.

e. The adoption of the three feature extraction methods leads to the possibility that the accuracy of the hybrid model does not rise and even fall. The main reason is that the number of original input data type is too small. Compared with the tens of dimensional data reduction, the six factors in this chapter are not counted. Moreover, feature extraction may destroy the main information in the original data, and fail to extract the main components. In addition, in the case where the main information is extracted without results, the feature extraction process also consumes the time of the hybrid model forecasting, making the overall timeliness of the model worse.
f. In this chapter, before using the FA method, the test value obtained by the KMO test is 0.5900. However, the KMO value is usually set at 0.7 or even 0.85 or more (Ebrahimy and Osareh 2014; Nair et al. 2016). Therefore, the factor analysis method is not absolutely applicable as a feature extraction method in this chapter.

8.8 Conclusion

The chapter considers the relationship between the concentrations of six common atmospheric pollutants and the air quality coefficient AQI. The AQI historical time data and the six atmospheric pollutant concentrations are used as input to predict the future time value of AQI. The main components of atmospheric pollutants are analyzed by three feature extraction methods, and multiple data-driven prediction models are constructed. The following conclusions are obtained:

a. The single data-driven ELM model is the optimal prediction model with the highest prediction accuracy and the best prediction timeliness.
b. The prediction models driven by the six kinds of atmospheric pollution concentrations data also have practical application value to some extent, and the comprehensive prediction performance is not much different from the ELM (single data) model, which proves the correctness of the method using the relevant data as the input of the prediction model.
c. The comprehensive prediction performance of the hybrid prediction models under the framework of the three feature extraction methods in the chapter is generally poor, especially the KPCA-ELM model, which completely loses the predictive ability. This issue needs to be improved and optimized in the follow-up work.

References

Bandalos DL, Finney SJ (2018) Factor analysis: Exploratory and confirmatory. Routledge, London

Cecchetto FH, Pellanda LC (2014) Construction and validation of a questionnaire on the knowledge of healthy habits and risk factors for cardiovascular disease in schoolchildren. Jornal de Pediatria 90(4):415–419

Dong H, Li M, Zhang S, Han L, Li J, Su X (2017) Short-term power load forecasting based on kernel principal component analysis and extreme learning machine. Journal of Electronic Measurement and Instrument 32:188–193

Ebrahimy S, Osareh F (2014) Design, validation, and reliability determination a citing conformity instrument at three levels: Normative, informational, and identification. Scientometrics 99(2):581–597

Huang G-B, Chen L (2007) Convex incremental extreme learning machine. Neurocomputing 70(16-18):3056–3062

Huang J, Yan X (2016) Related and independent variable fault detection based on KPCA and SVDD. Journal of Process Control 39:88–99

Huang G-B, Zhu Q-Y, Siew C-K (2004) Extreme learning machine: A new learning scheme of feedforward neural networks. Neural Networks 2:985–990

Huang G-B, Zhu Q-Y, Siew C-K (2006) Extreme learning machine: Theory and applications. Neurocomputing 70(1-3):489–501

Jaffel I, Taouali O, Harkat MF, Messaoud H (2017) Kernel principal component analysis with reduced complexity for nonlinear dynamic process monitoring. The International Journal of Advanced Manufacturing Technology 88(9-12):3265–3279

Jiachao S, Kun L, Jianren F, Junxi Z, Qing W, Xiang G et al (2018) Rapid response model of regional air pollutant concentration based on CMAQ and feed forward neural network. Acta Scientiae Circumstantiae 38(11):4480–4489

Jolliffe IT, Cadima J (2016) Principal component analysis: A review and recent developments. Philosophical Transactions of the Royal Society A: Mathematical, Physical and Engineering Sciences 374(2065):20150202

Kalapanidas E, Avouris N (2003) Feature selection for air quality forecasting: A genetic algorithm approach. AI Communications 16(4):235–251

Li J, Li X, Tao D (2008) KPCA for semantic object extraction in images. Pattern Recognition 41(10):3244–3250

Li, J., Shao, X., & Zhao, H. (2018a). An online method based on random forest for air pollutant concentration forecasting. in *2018 37th Chinese Control Conference (CCC)* (pp. 9641–9648). IEEE.

Li Y, Jiang P, She Q, Lin G (2018b) Research on air pollutant concentration prediction method based on self-adaptive neuro-fuzzy weighted extreme learning machine. Environmental Pollution 241:1115–1127

Liu Z, Guo W, Hu J, Ma W (2017) A hybrid intelligent multi-fault detection method for rotating machinery based on RSGWPT, KPCA and Twin SVM. *ISA transactions* 66:249–261

Liu H, Wu H, Lv X, Ren Z, Liu M, Li Y et al (2019) An intelligent hybrid model for air pollutant concentrations forecasting: Case of Beijing in China. Sustainable Cities and Society 47:101471

Min XU, Sun HL (2011) From "digital environmental protection" to "smart environmental protection". Administration & Technique of Environmental Monitoring 23(4):5–7

Nair SC, Satish KP, Sreedharan J, Ibrahim H (2016) Assessing health literacy in the eastern and middle-eastern cultures. BMC Public Health 16(1):831

Navi M, Meskin N, Davoodi M (2018) Sensor fault detection and isolation of an industrial gas turbine using partial adaptive KPCA. Journal of Process Control 64:37–48

Savalei V, Kolenikov S (2008) Constrained versus unconstrained estimation in structural equation modeling. Psychological Methods 13(2):150

Teixeira AR, Tomé AM, Stadlthanner K, Lang EW (2008) KPCA denoising and the pre-image problem revisited. Digital Signal Processing 18(4):568–580

Vinay A, Shekhar VS, Murthy KB, Natarajan S (2015) Face recognition using Gabor wavelet features with PCA and KPCA-a comparative study. Procedia Computer Science 57:650–659

Yang Z, Wang J (2017) A new air quality monitoring and early warning system: Air quality assessment and air pollutant concentration prediction. Environmental Research 158:105–117

Zhang J, Ding W (2017) Prediction of air pollutants concentration based on an extreme learning machine: The case of Hong Kong. International Journal of Environmental Research and Public Health 14(2):114

Zhang Y, Li N, Zhang Y (2015) The research of China's urban smart environmental protection management mode. Springer, Berlin

Zhao YJ, Du B, Liu BK (2013) Smart environmental protection: The new pathway for the application of the internet of things in environmental management. Applied Mechanics & Materials 411-414:2245–2250

Zhu J, Wei-Xin DU, Jun-Ying MA, Dong N, Lin WY, Deng WY (2017) Spatial changes of air pollutants distributions before and after the "coal to gas" project in Urumqi based on numerical simulation with MM5 and CALPUFF models. Arid Land Geography 40(1):165–171

Chapter 9
Prediction Models of Urban Hydrological Status in Smart Environment

Abstract Urban river water level early warning is not only an important means to ensure the normal operation of the urban system but also an integral part of the intelligent embodiment of the intelligent city. In addition to paying attention to its fluctuation state, the river water level is also very important for the accurate prediction of water level height in the future. For this reason, this chapter first constructs the prediction model of water level fluctuation state based on Naive Bayesian classifier, and on this basis, establishes the deterministic prediction model of water level height, and integrates the decomposition algorithm into the hybrid model. Have finally achieved good prediction results.

9.1 Introduction

Since the beginning of this century, the phenomena of a large amount of stagnant water in low-lying roads, down-through overpasses and tunnels caused by heavy rainfall have occurred from time to time, which have brought great inconvenience to people's travel, and even caused great losses of people's lives and property. The fluctuation of the water level of the main rivers in the city also closely affects the normal operation of the city and the normal living conditions of the residents. Smart cities integrate a variety of new technologies such as big data, Internet of Things, artificial intelligence, etc., to achieve comprehensive sensing capabilities in multiple fields, and to establish a spatial monitoring network for hybrid monitoring of water, land, and air. Many countries and cities have realized the intelligent monitoring of urban water level and laid the road for the city to develop intelligently (Krajewski et al. 2017; Kirby 2015; Werner et al. 2010).

As a common form of time series, the water level height series is still the most commonly used method system for statistical methods, machine learning methods and deep neural network methods in the field of time series prediction. In addition to the simple one-dimensional data-driven model prediction, the multi-dimensional

H. Liu, *Smart Cities: Big Data Prediction Methods and Applications*,
https://doi.org/10.1007/978-981-15-2837-8_9

data-driven hybrid forecasting model is constructed by using various related influencing factors such as meteorology, which is the mainstream methods of water level intelligent prediction. Zhen et al. used the four-year water level monitoring data of the Yangtze River basin to construct the ARIMA model, which basically realized the short-term prediction function of water level height. Based on this research, the influence of the prediction period on the forecasting performance of the model is analyzed (Zhen et al. 2017). Palash et al. cleverly combined with natural factors such as river flow persistence and basin precipitation to construct a data-driven prediction model. The experimental results achieved a precise prediction of the downstream water level of the river for many days in the future under the premise of limited data acquisition, which had strong practical value (Palash et al. 2018). Yang et al. also considered the problem of missing values of water level monitoring data due to sensor failures. Combining water level data with atmospheric data, using Random Forest as a predictor for water level prediction and played a good predictive effect. The experiment also proved that the feature extraction method could improve the performance of the multi-dimensional feature-driven forecasting model (Jun-He et al. 2017). In addition, some other machine learning methods such as the NARX neural networks (Wunsch et al. 2018) and the extreme learning machines (Yadav et al. 2017) have been applied in the field of water level prediction.

As a common pre-processing method, the decomposition algorithm has also applied maturely in the processing of water level data. Seo et al. used the Wavelet Decomposition algorithm as the pre-processing of ANN and ANFIS, which proved that the hybrid prediction model of fusion decomposition algorithm was better than single predictor (Seo et al. 2015). Abrokwah et al. used the discrete wavelet transform (DWT) to decompose multi-dimensional input data into multiple sub-sequences, and constructed an ANN model for each sub-sequence, which had a great prediction effect (Abrokwah and O'Reilly 2017). In addition to the decomposition algorithms mentioned above, the SD (Shuai et al. 2011), the EMD (Kisi et al. 2014) and other decomposition methods are also commonly used in the construction of water level hybrid forecasting models.

The chapter uses the water level monitoring data to construct a single Elman forecasting model and a hybrid forecasting model based on the decomposition framework. Based on this, the experiment and analysis of the influence of different decomposition parameters on the prediction performance of the hybrid model are carried out. In addition, the chapter also uses the Bayesian classification predictor to predict the trend of water level fluctuations and achieves good results.

9.2 Original Hydrological State Data for Prediction

9.2.1 Original Sequence for Modeling

The hydrological state data used in this chapter is mainly the water level data of river waters, which is derived from the average water level of several water level monitoring stations in river basins in two certain cities of China. The collection time

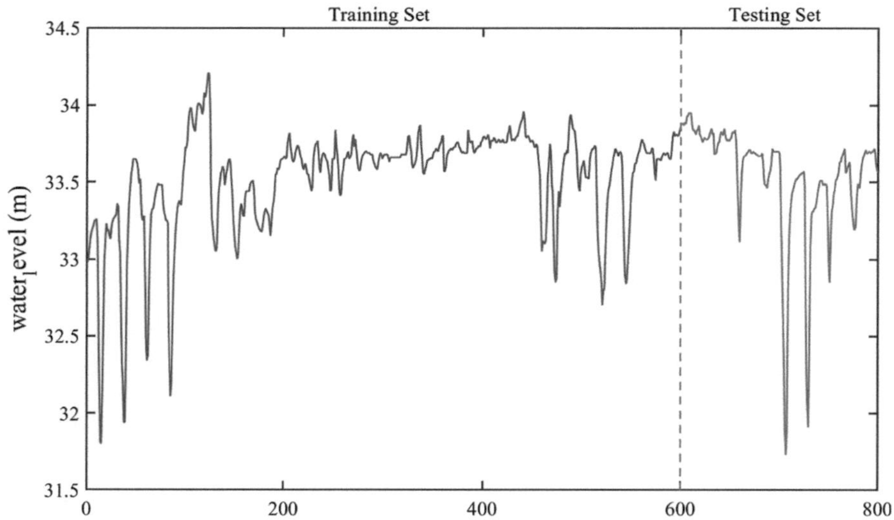

Fig. 9.1 Original water level height series $\{X_1\}$

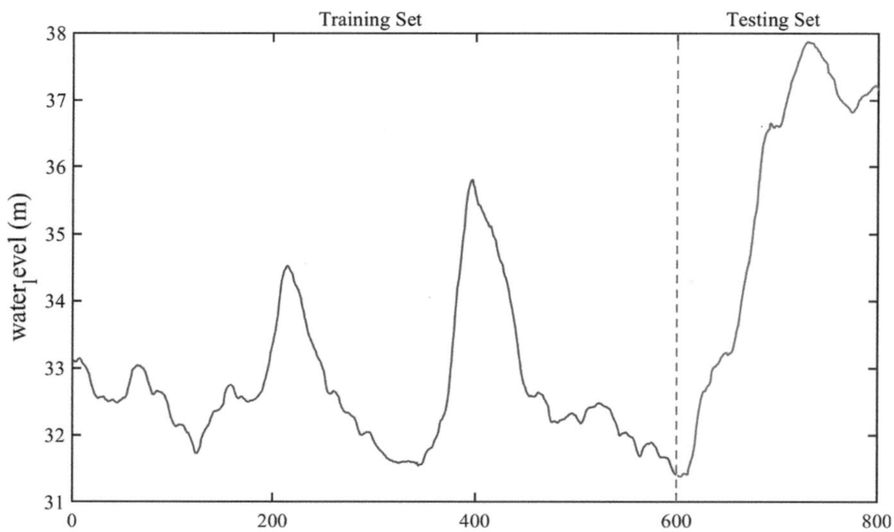

Fig. 9.2 Original water level height series $\{X_2\}$

interval of the water level data are both 1 h. Figures 9.1 and 9.2 show the two groups of original water level height series of 800 sample points, denoted by $\{X_1\}$ and $\{X_2\}$, respectively, and based on this, the water level fluctuation state diagrams of two groups of raw series are shown in Figs. 9.3 and 9.4. As can be seen from Figs. 9.1 and 9.2, the water level height of the water level series $\{X_1\}$ varies between 31.5 m and 34.5 m, and the water level height of the water level series $\{X_2\}$ varies between

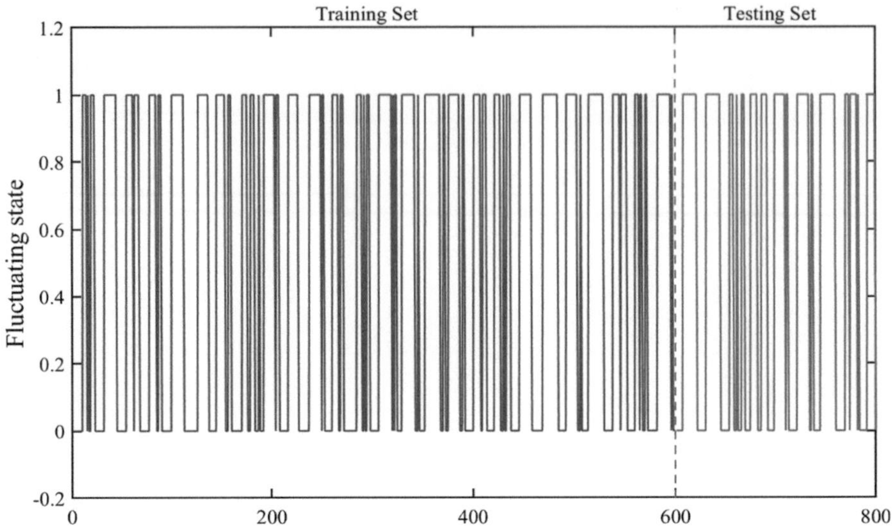

Fig. 9.3 Fluctuation state of water level series $\{X_1\}$

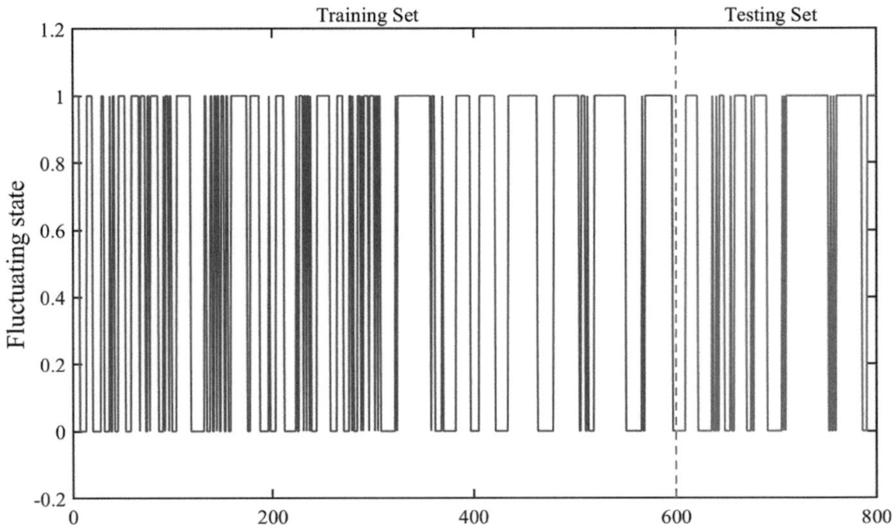

Fig. 9.4 Fluctuation state of water level series $\{X_2\}$

31.2 m and 37.9 m. Both sequences show obvious randomness and non-stationarity.

In order to comprehensively analyze the characteristics of original water level time series, and further calculate the common statistical characteristics of the two groups of the water level data, such as extremum, mean value, standard deviation, skewness and kurtosis, Table 9.1 is the statistical characteristics of the two groups of

Table 9.1 Statistical characteristics of the original water level series

Original series	Minimum	Maximum	Mean value	Standard deviation	Skewness	Kurtosis
$\{X_1\}$	31.7300	34.2100	33.5294	0.3521	−2.2006	9.4028
$\{X_2\}$	31.3800	37.8700	33.3897	1.8267	1.2095	3.1586

time series adopted in this chapter. The meaning of these statistical features has been described in detail in the previous chapters, and will not be repeated in the chapter.

The skewness value of the original water level time series $\{X_1\}$ is negative, indicating that its distributions are skewed to the left relative to the symmetrical distribution, and their statistical distribution characteristics are closer to the symmetrical distribution. The skewness value of the original water level time series $\{X_2\}$ is positive, and the meaning of the representative is just the opposite. The kurtosis values of the two groups of time series are all positive, indicating that the distribution of sample points relative to the mean is relatively scattered, and the probability of occurrence of extreme values in sample points is higher than that in normal distribution. In addition, the standard deviations of the two water level time series are small, indicating that the amplitude changes of the water level are relatively small.

9.2.2 *Separation of Sample*

In order to construct the model and evaluate the models' performance, the original water level time series for simulation need to be grouped. The two groups of water level series in this chapter contain 800 sample points, and divide both the two groups of series data into training dataset and testing dataset. For the two groups of original series, the 1st to 600th sample points are used as training samples to train the prediction model, and the 601st to 800th sample points are used as testing samples to test the obtained model after the model training is completed, so as to obtain the prediction output of the model, calculate the prediction error of the model, and evaluate the prediction performance of the model. The results of separation of sample are shown in Figs. 9.1 and 9.2.

9.3 Bayesian Classifier Prediction of Water Level Fluctuation

9.3.1 *Model Framework*

In this section, the Bayesian prediction model is used to predict the fluctuation trend of the selected water level, and calibrate the predicted water level fluctuation state as rising and falling, which are represented by the numbers 1 and 0, respectively. This is actually a supervised classification problem in machine learning. It uses a manually calibrated sample set to predict future data labels, that is, to classify the target problem status. Based on this consideration, the Bayesian theory is used, which is the most widely studied, simple and efficient Naive Bayesian classifier for experimental simulation. Figure 9.5 shows the model framework for predicting the fluctuation trend of the original water level with the Naive Bayesian classifier model.

9.3.2 *Theoretical Basis of the Bayesian Classifier*

Suppose the training sample set is (*A*, *B*), where *A* belongs to the attribute of the sample to be classified. Generally, *A* has multiple dimensions, that is, $A = \{a_1, a_2, \ldots, a_n\}$, *B* belongs to the manually calibrated sample label, and the dimension of *B* is the total number of labels, that is, $B = \{b_1, b_2, \ldots, b_m\}$. The basic idea of classification is to calculate the probability that a sample point belongs to each type of label, and then select the label category with the highest probability of attribution as the classification result.

Assuming that the classification result to be determined is *b*, the above process can be expressed as follows:

$$b = \arg\max\{P(b_1|A), P(b_2|A), \ldots, P(b_m|A)\} \tag{9.1}$$

The formula for calculating $P(b_m|A)$ in Eq. (9.1) is shown as follows:

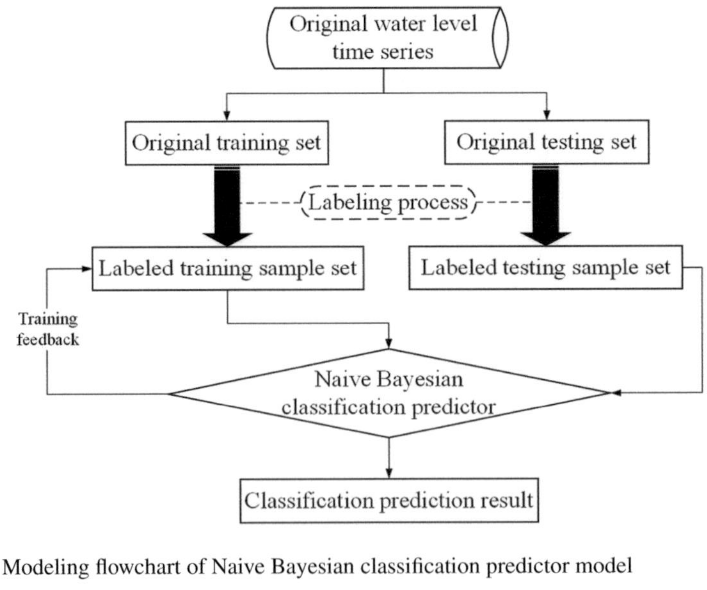

Fig. 9.5 Modeling flowchart of Naive Bayesian classification predictor model

$$P\left(b_m|A\right)=\frac{P\left(A|b_m\right)P\left(b_m\right)}{P\left(A\right)} \tag{9.2}$$

where $P(b_m)$ belongs to the prior probability and can be calculated according to the training dataset. Since $P(A)$ is constant for all classes, Eq. (9.2) can be transformed into the following form:

$$P\left(b_m|A\right)=P\left(b_m\right)P\left(a_1,a_2,\ldots,a_n|b_m\right) \tag{9.3}$$

where if each dimension feature in A is not independent of each other, assuming that the ith dimension a_i has p_j values, and each value corresponds to m label types, then the number of parameters in Eq. (9.3) is $m\prod_{j=1}^{n}p_j$. If the attribute independence assumption is carried out, then $P(a_1, a_2, \cdots, a_n| b_m)$ can be converted to $\prod_{i=1}^{n}P\left(a_n|b_m\right)$ and the number of parameters will be reduced to $\prod_{j=1}^{n}mp_j$.

9.3.3 *Steps of Modeling*

9.3.3.1 Dataset Preparation

The experimental dataset is constructed according to the method in Sect. 9.2.2. The input feature dimension of the sample set is the historical 10 time points of water level data, and the output label is the state of the water level rising or falling at the next time point, respectively. It is represented by the numbers 1 and 0. The training sample set and the testing sample set are then divided.

9.3.3.2 Classifier Training

At this time, the input dimension is 10, and the label type is 2, and the frequency of occurrence of each category in the training sample set and the conditional probability estimation of the state classification feature pair category are calculated according to Eqs. (9.1)–(9.3). It is assumed that the probability of the water level rising and falling state at the next moment is equal, then the problem of maximizing the solution of $P(b_m| A)$ becomes the problem of maximizing the solution of $P(a_1, a_2, \ldots, a_n| b_m)$. Then it is simplified according to the independence hypothesis. The solution problem of the final training process is the maximum solution of $\prod_{i=1}^{n}P\left(a_n|b_m\right)$. The probability problem can be solved with the evaluation of the training sample set data.

9.3.3.3 Self-Predicting Water Level Fluctuation Trend

When classifying the testing sample set, $P(a_1, a_2, \ldots, a_n| b_m)$ is calculated for each sample point in turn, and if and only if $P(a_1, a_2, \ldots, a_n| b_m)$ satisfies the following formula, it will consider that the sample point belongs to class m:

$$P\left(a_1, a_2, \ldots, a_n | b_m\right) > P\left(a_1, a_2, \ldots, a_n | b_q\right), 1 \le q \le m, q \ne m \tag{9.4}$$

9.3.4 *Forecasting Results*

The forecasting results of the Naive Bayesian classification prediction model are shown in Figs. 9.6 and 9.7. The accuracy of the Bayesian prediction for the fluctuation trend of water level series $\{X_1\}$ is 73.63%, and the accuracy of water level series $\{X_2\}$ is 77.61%.

As can be seen from Figs. 9.6 and 9.7, the Naive Bayesian classification predictor can capture the trend of water level fluctuations to a certain extent, and the prediction accuracy is also high. However, there are still two problems. First, it is impossible to accurately predict the sample points where the water level continuously changes sharply. This is the main factor for the prediction error. Second, there is a certain delay in the prediction process, which will affect the forecasting effectiveness of the method in practical applications.

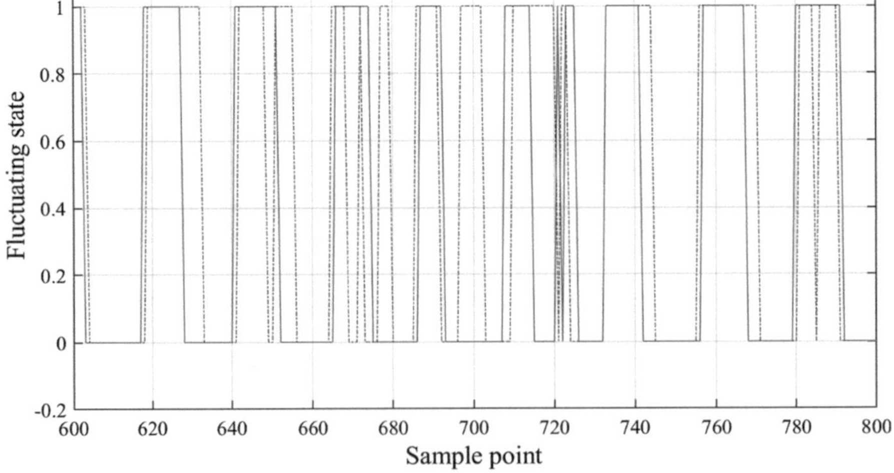

Fig. 9.6 Fluctuation trend prediction result of original water level series $\{X_1\}$

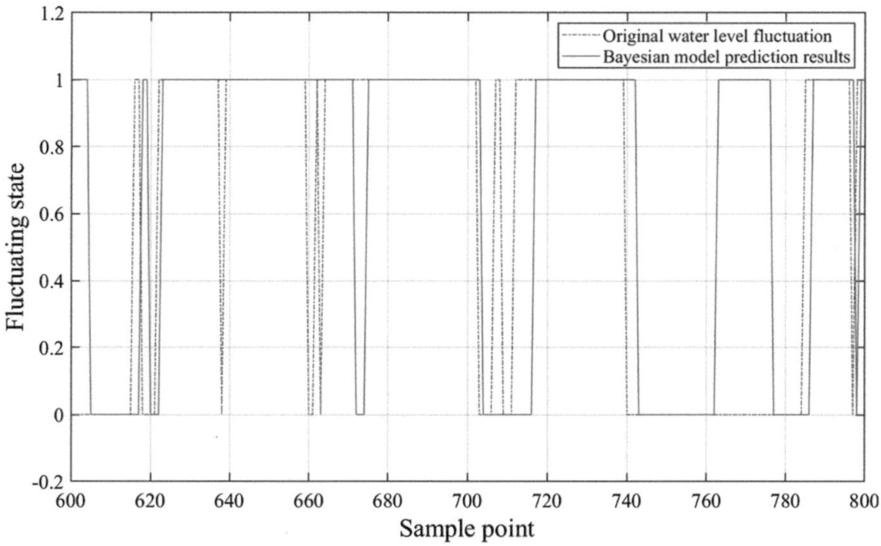

Fig. 9.7 Fluctuation trend prediction result of original water level series $\{X_2\}$

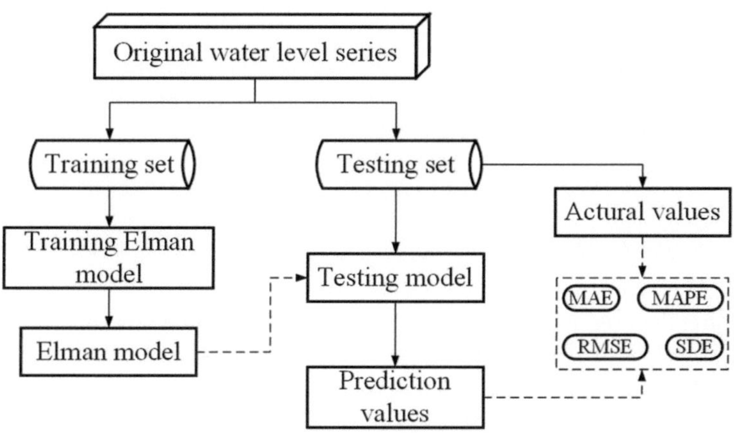

Fig. 9.8 Modeling flowchart of the Elman water level prediction model

9.4 The Elman Prediction of Urban Water Level

9.4.1 Model Framework

Figure 9.8 shows the model framework for predicting the original water level series by the single Elman prediction model.

9.4.2 The Theoretical Basis of the Elman

Compared with the BP neural network, the Elman neural network adds a receiving layer, also called a state layer, forming a network structure base on the framework including an input layer, a hidden layer, a receiving layer and an output layer. The output of the hidden layer is not only transmitted to the output layer, but also transmitted to the receiving layer for memory and storage. It acts as a delay operator with a 1-step delay effect and an input of the hidden layer after a certain delay through internal feedback. This structure increases the sensitivity of the Elman neural network to the historical state, so that the network has better processing ability for dynamic information and achieves the effect of dynamic modeling (Köker 2005).

The calculation method of the Elman neural network can be expressed as follows:

$$h(t) = f\left(\omega_1\left(x(t)\right) + \omega_2\left(h(t-1)\right) + a\right) \tag{9.5}$$

$$y(t) = g\left(\omega_3\left(h(t)\right) + b\right) \tag{9.6}$$

where $x(t)$ is the historical input data, $h(t)$ is the output data of hidden layer, and $y(t)$ is the output data; ω_1, ω_2, and ω_3 are the weight vectors of hidden layer connected to input layer, the receiving layer connected to hidden layer, and the output layer connected to hidden layer, respectively; a and b are the threshold vectors of the hidden layer and the output layer, respectively; $f(\cdot)$ and $g(\cdot)$ represent the activation functions of the hidden layer and the output layer, respectively.

9.4.3 Steps of Modeling

a. Dividing the original water level time series into a training dataset and a testing dataset, and normalize the input data and the output data of the training dataset, respectively;
b. Training the Elman neural network after the parameters initialized such as the maximum number of iterations of the model and the network learning rate, and obtain the trained Elman model;
c. The input data of the testing dataset are subjected to the same normalization operation as in step (a), then are input to the Elman model obtained by the training in step (b) for testing, and the model output result is inversely normalized to obtain predicted value relative to the original data;
d. Comparing the predicted values of the model obtained in step (c) with the actual values, and calculate the four error evaluation index values of the mean absolute error (MAE), the mean absolute percent error (MAPE), the root mean square error (RMSE) and the standard deviation error (SDE).

9.4.4 Forecasting Results

The forecasting results of the Elman prediction model are shown in Figs. 9.9 and 9.10.

The evaluation indices of the Elman forecasting results are shown in Table 9.2.

The following conclusions can be drawn from Figs. 9.9 and 9.10 and Table 9.2:

a. For the water level time series $\{X_1\}$, the single Elman neural network model has high prediction accuracy and great prediction effect, and can basically fit the change trend of the water level series, only appearing obvious prediction bias at some peak points with sharply changes, such as at the 705th and 718th sample points. This prediction bias becomes more severe with the increase of the number of advance forecasting steps.
b. For the water level time series $\{X_2\}$, the overall prediction effect of the Elman model is general. After the 684th sample point, a serious prediction trend deviation occurs, the amplitude of the change of the original sequence is not fitted, and the deviation also increases with the increase of the advanced prediction steps.
c. As the number of advanced prediction steps increases, the prediction accuracy and fitting ability of the single Elman model are declining, and the delay phenomenon begins to appear, which also becomes more serious as the number of prediction steps increases. The reason for this problem is mainly because the prediction process of the Elman model in this chapter adopts the rolling recursive prediction strategy. The prediction error of the previous step will accumulate in the prediction result of the latter step. Therefore, the larger the prediction step is, the larger the prediction error will be.

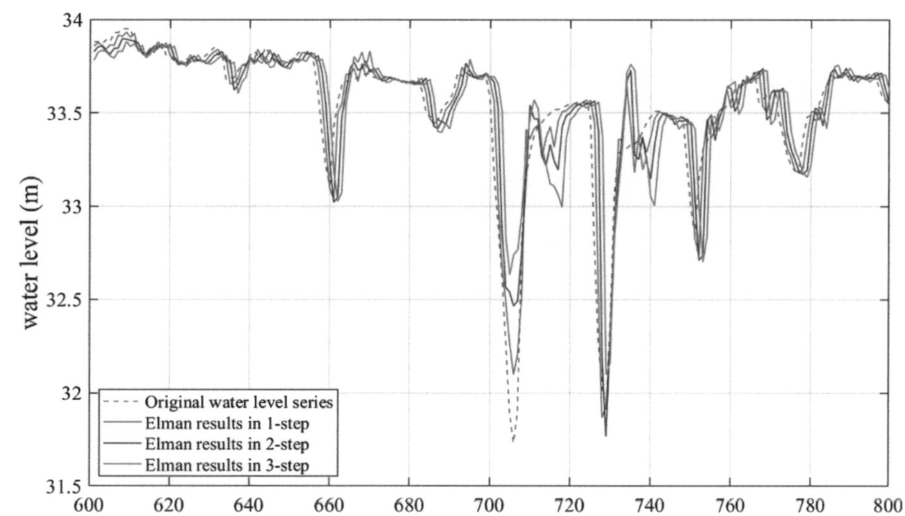

Fig. 9.9 The forecasting results of water level time series $\{X_1\}$ by the Elman

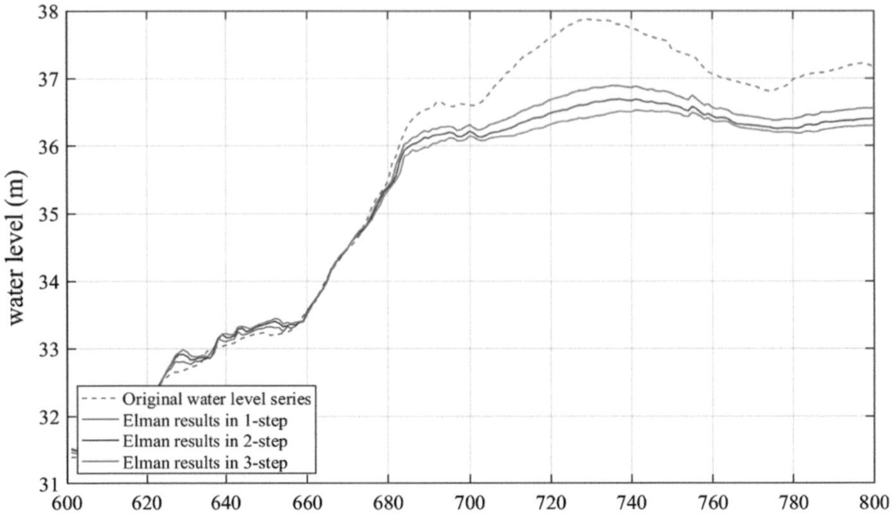

Fig. 9.10 The forecasting results of water level time series $\{X_2\}$ by the Elman

9.5 Urban River Water Level Decomposition Hybrid Prediction Model

9.5.1 Model Framework

The section uses the decomposition algorithm as a pre-processing method for water level prediction. The decomposition algorithms used include the MODWT, EMD and SSA. The framework of the constructed model is shown in Fig. 9.11.

9.5.2 The Theoretical Basis

9.5.2.1 Maximal Overlap Discrete Wavelet Transform

The Wavelet Decomposition (WD) is developed on the basis of the short-time Fourier Transform. Through the definition of the scale factor and the translation factor, a change window that can be translated in the time axis is obtained to realize the observation of the time–frequency characteristics of the time series. The change window varies with the frequency of the proposed frequency, and the sampling step size used in the time domain for different frequency components is variable. The Wavelet Decomposition is usually divided into the continuous wavelet transform (CWT) and discrete wavelet transform (DWT). Since the time series in the actual engineering application is usually a discrete point series, the DWT is more effective.

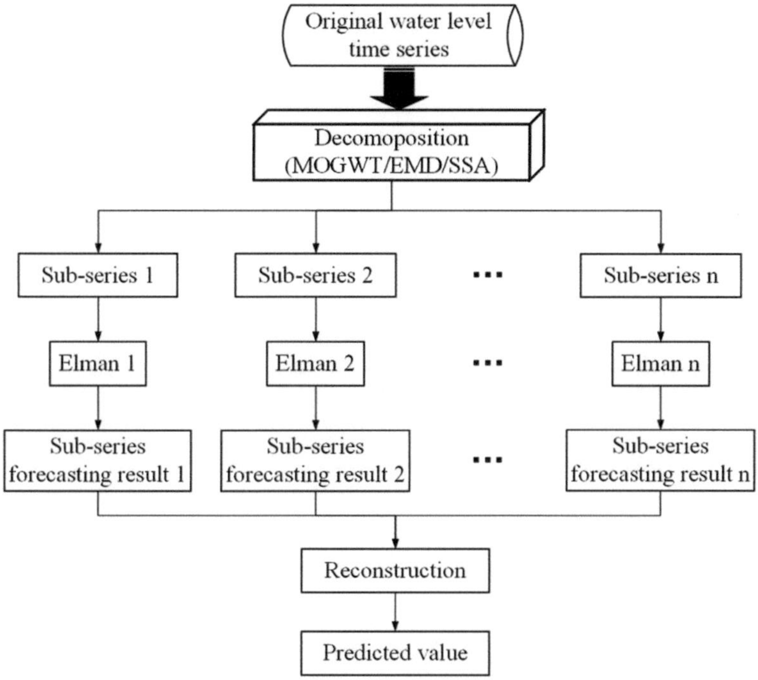

Fig. 9.11 Modeling flowchart of the Elman water level prediction model under decomposition framework

Table 9.2 The forecasting performance indices of the Elman model

Data series	Step	MAE (m)	MAPE (%)	RMSE (m)	SDE (m)
$\{X_1\}$	1	0.1823	0.0606	0.1093	0.1091
	2	0.3196	0.1064	0.1893	0.1888
	3	0.4633	0.1545	0.2662	0.2651
$\{X_2\}$	1	1.0834	0.3939	0.5122	0.3627
	2	1.3787	0.4983	0.6331	0.4507
	3	1.5862	0.5714	0.7238	0.5169

Unlike orthogonal transform of DWT, the Maximal Overlap Discrete Wavelet Transform (MODWT) is a non-orthogonal wavelet transform that overcomes some of the drawbacks of the DWT. First, the length of processed data is required to be divisible by the exponent of 2. Second, it is sensitive at both ends of the time series values. Since the extraction mechanism of the orthogonal wavelet transform is sensitive to cyclic motion, the wavelet coefficients and scale coefficients obtained by the decomposition are not shiftable. Third, carry out the orthogonal wavelet transform each time, the lengths of the obtained wavelet and scale coefficients are 1/2 of the previous scale, which makes the coefficients obtained at high resolution unsuitable for further statistical analysis (Li et al. 2014). In this section, to ensure the consistency of the result comparison, the related parameter of the MODWT in this model is: the number of decomposition layers is 3, and the mother wavelet is "db7." The original water level series $\{X_1\}$ is taken as an example, and the sub-component results obtained by the MODWT are shown in Fig. 9.12.

It can be concluded from Fig. 9.12 that the component S1 is the main component of the original water level series. The amplitude is basically consistent with the original series. It contains most of the energy of the original series, indicating the trend of the original series, and is the low-frequency component of the original series. The amplitudes of S3, S2, and S4 are sequentially decreased, and the frequencies are sequentially increased to be high-frequency components of the original series. According to the principle of the MODWT, every time the wavelet decomposition is performed on the time series, only the low-frequency components are decomposed, and the high-frequency components are not decomposed. Therefore, a three-layer MODWT is performed on the original water level data to obtain a low-frequency component and three high-frequency components.

9.5.2.2 Empirical Mode Decomposition

The Empirical Mode Decomposition (EMD) is a pre-processing method that uses the Hilbert-Huang Transform (HHT) to adaptively decompose nonlinear and non-stationary time series data (Huang et al. 1998). The basic principle is to decompose the original data into a sequence of the Intrinsic Mode Function (IMF) (Bi et al. 2018). In the EMD, each IMF must meet the following two prerequisites (Zhu et al. 2018):

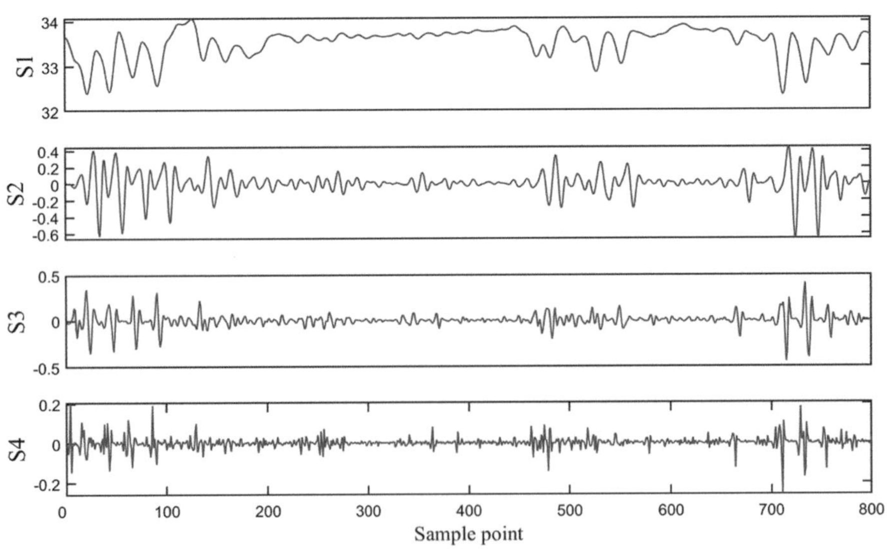

Fig. 9.12 MODWT results of the original water level series $\{X_1\}$

a. In the entire time series, the number of extremum and zero-crossing point must be equal, or different up to 1.
b. The average value of the envelope curve determined by the local maximum and minimum must be zero at all points.

Compared to the traditional Fourier decomposition and the WD, the EMD has many advantages. First, it is better interpreted and easier to implement. Secondly, the EMD can adaptively select the volatility features from the original time series, which is robust to nonlinear and non-stationary time series decomposition. Third, the EMD can decompose the original time series into several independent IMFs and a residue adaptively. The linearity and non-linearity of all IMF components and their residual components only depend on the characteristics of the time series under study. Finally, in the WD, the mother wavelet basis must be predetermined, and the EMD does not need to determine the basic function before decomposition. Figure 9.13 shows the modal signals corresponding to each Intrinsic Mode Function obtained by the EMD using the original water level series $\{X_1\}$ as example.

It can be seen from Fig. 9.13 that the modal signals IMF1 to IMF7 and residue are, respectively, corresponding from the low-frequency band to the high-frequency band. Among them, Residue has the largest amplitude, and concentrates most of the energy in the original water level time series, reflecting the overall trend of the water level series. The amplitude of IMF2 is second, the amplitudes of IM1, IMF3, and IMF4 are roughly equal, and the amplitude of IMF5 to IMF7 is the smallest, which is the high-frequency component in the original water level series. Obviously, compared with the MODWT, the EMD decomposition sub-sequence trend information is more detailed, and the difference between the low-frequency component and high-frequency component is more significant.

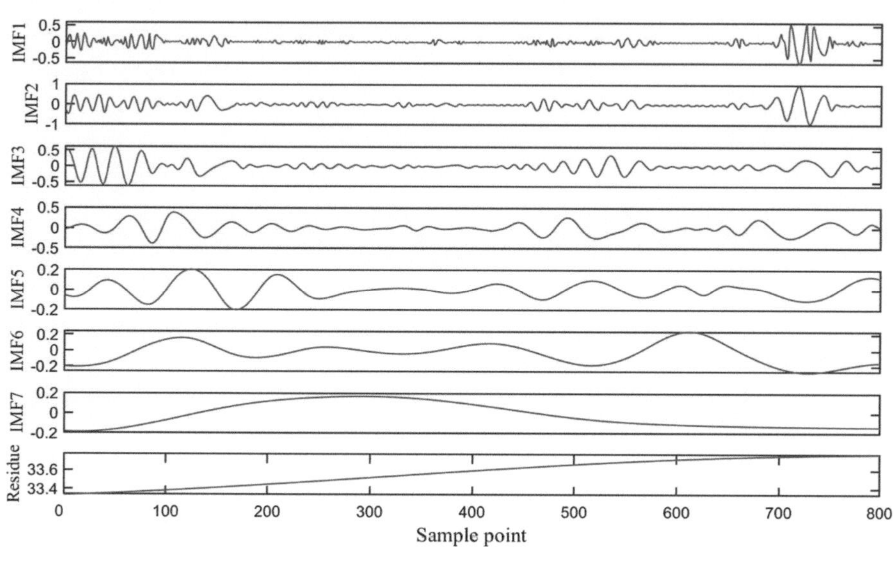

Fig. 9.13 The EMD results of the original water level series $\{X_1\}$

9.5.2.3 Singular Spectrum Analysis

The Singular Spectrum Analysis (SSA) analyzes the structure of time series and divides it into the trend component and residual component (Shen et al. 2018). In the SSA process, the selection of the window length L is very important. It is necessary to continuously try to select the appropriate window length. To ensure the consistency of the result comparison, the related parameter of singular spectrum in this model is: the window length $L = 120$. Figure 9.14 shows the singular spectrum analysis result of the original water level height time series $\{X_1\}$. The upper picture shows the original time series and the reconstructed sequence. It can be seen that the reconstructed time series occupies most of the energy of the original time series and represents the trend of the water level data. The residual component is a high-frequency signal, and the amplitude fluctuates in the range [−0.2, 0.2]. The singular spectrum analysis can decompose the original signal into a trend component and a residual component, thereby reducing the difficulty of subsequent prediction.

9.5.3 *Steps of Modeling*

a. Dividing the original water level time series into a training dataset and a testing dataset. Decompose the training dataset and the testing dataset into a plurality of sub-sequences by the decomposition algorithm, and normalize all of the sub-sequence, respectively;

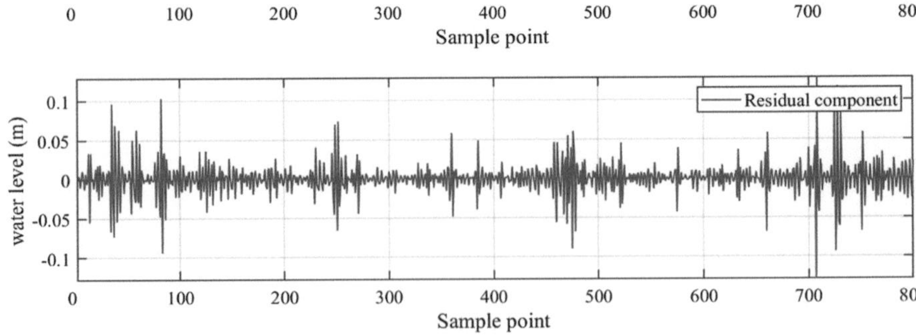

Fig. 9.14 The SSA results of the original water level series $\{X_1\}$

b. Training the Elman neural network after the parameters initialized such as the maximum number of iterations of the model and the network learning rate with the input normalized sub-sequences, and obtain the trained Elman model;
c. The input data of the testing dataset are subjected to the same normalization operation as in step (a), then are input to the Elman model obtained by the training in step (b) for testing, and the model output result is inversely normalized to obtain predicted value of every sub-sequence, then fuse the sub-value to the final predicted value relative to the original data;
d. Comparing the final predicted values of the model obtained in step (c) with the actual values, and calculate the four error evaluation index values of the mean absolute error (MAE), the mean absolute percent error (MAPE), the root mean square error (RMSE) and the standard deviation error (SDE).

9.5.4 Forecasting Results

Figures 9.15, 9.16, and 9.17 show the prediction results of the original water level series $\{X_1\}$ for the three hybrid prediction models of the MODWT-Elman, EMD-Elman and SSA-Elman based on the decomposition framework.

Table 9.3 lists the evaluation indices of the three hybrid models for the prediction of the two groups of water level series.

The following conclusions can be drawn from Figs. 9.15, 9.16, and 9.17 and Table 9.3:

a. When the Elman neural network is used as the predictor of hybrid models, the hybrid model under the MODWT decomposition framework has the best prediction performance, the SSA-Elman hybrid model has the second performance, and the EMD-Elman hybrid model has the worst prediction performance. For example, when the water level time series $\{X_1\}$ is predicted in 1-step, the MAE, MAPE, RMSE, and SDE of the MODWT-Elman model are 0.0757 m, 0.0253%, 0.0381 m, and 0.0380 m, respectively; the MAE, MAPE, RMSE, and SDE of the SSA-Elman model are 0.1440 m, 0.0480%, 0.0710 m, and 0.0710 m, respectively; the MAE, MAPE, RMSE, and SDE of the EMD-Elman model are 0.1822 m, 0.0607%, 0.1008 m, and 0.0996 m, respectively.
b. With the increase of the number of advanced prediction steps, the prediction accuracy of the three hybrid prediction models is declining. This is because the hybrid model advance prediction process in this chapter adopts the rolling prediction strategy, and the error caused by the advanced low-step prediction will accumulate to the advanced high-step prediction. For example, when the water level time series $\{X_2\}$ is predicted by the SSA-Elman hybrid model, the MAE, MAPE, RMSE, and SDE in 1-step forecasting are 0.5614 m, 0.2047%, 0.2664 m, and 0.1915 m, respectively; the MAE, MAPE, RMSE, and SDE in 2-step forecasting are 0.7252 m, 0.2636%, 0.3470 m, and 0.2540 m, respectively; the MAE,

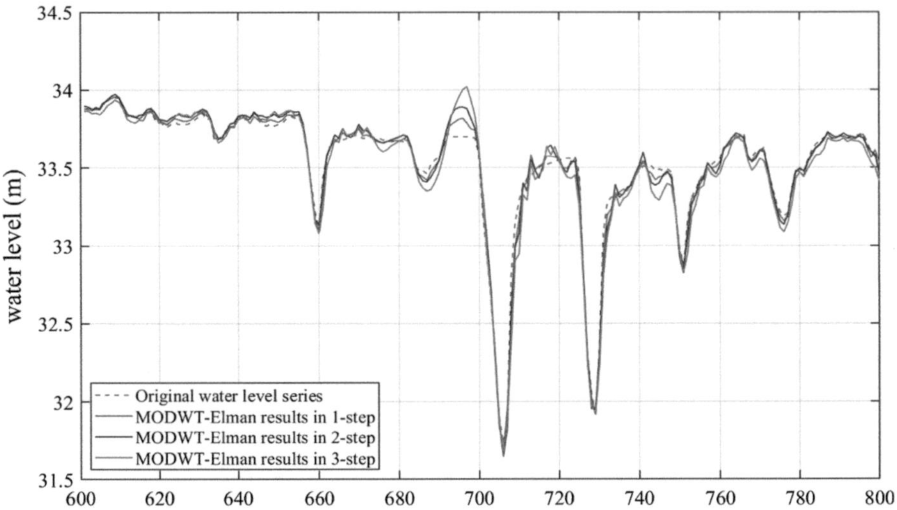

Fig. 9.15 The forecasting results of water level series $\{X_1\}$ by the MODWT-Elman

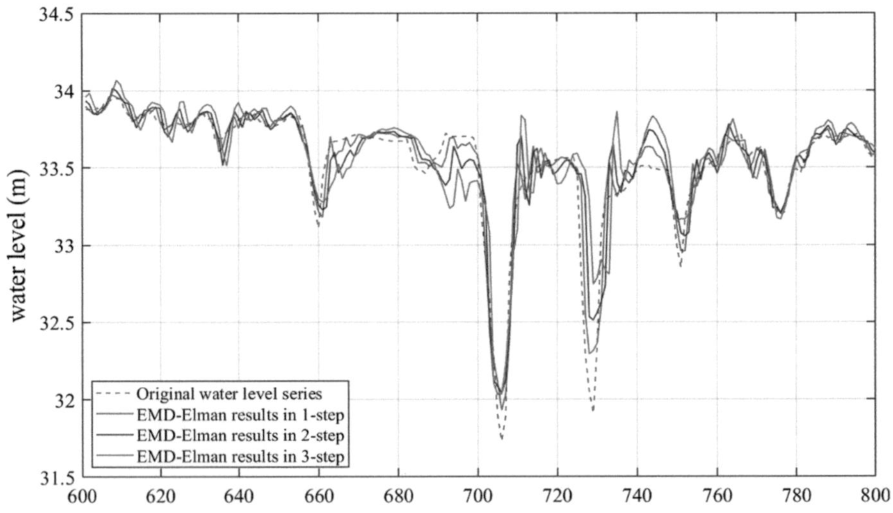

Fig. 9.16 The forecasting results of water level series $\{X_1\}$ by the EMD-Elman

MAPE, RMSE, and SDE in 3-step forecasting are 0.9685 m, 0.3508%, 0.4539 m, and 0.3253 m, respectively.

c. The MODWT-Elman hybrid prediction model has the best fitting effect on the original water level time series, and still has good fitting ability in the peaks and troughs region, while the deviations of the EMD-Elman and SSA-Elman in the peak and trough sections have been large, and overall fluctuations are relatively serious, and the fitting effect is very poor, such as the 685th to 700th sample

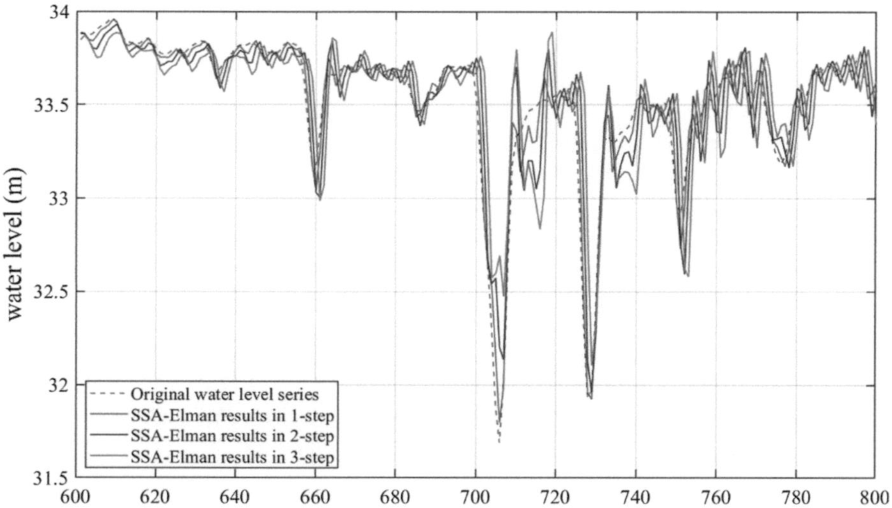

Fig. 9.17 The forecasting results of water level series $\{X_1\}$ by the SSA-Elman

Table 9.3 The forecasting performance indices of three hybrid models under decomposition framework

Data series	Model	Step	MAE (m)	MAPE (%)	RMSE (m)	SDE (m)
$\{X_1\}$	MODWT-Elman	1	0.0757	0.0253	0.0381	0.0380
		2	0.1284	0.0429	0.0636	0.0635
		3	0.1922	0.0642	0.0931	0.0887
	EMD-Elman	1	0.1822	0.0607	0.1008	0.0996
		2	0.3058	0.1019	0.1626	0.1616
		3	0.4206	0.1406	0.2195	0.2164
	SSA-Elman	1	0.1440	0.0480	0.0710	0.0710
		2	0.3384	0.1129	0.1684	0.1684
		3	0.5009	0.1670	0.2551	0.2550
$\{X_2\}$	MODWT-Elman	1	0.3991	0.1461	0.1996	0.1990
		2	0.3883	0.1414	0.1866	0.1864
		3	0.4460	0.1609	0.2084	0.1993
	EMD-Elman	1	1.5373	0.5531	0.7739	0.5664
		2	2.1878	0.7770	1.1012	0.8059
		3	2.5654	0.9025	1.3009	0.9594
	SSA-Elman	1	0.5614	0.2047	0.2664	0.1915
		2	0.7252	0.2636	0.3470	0.2540
		3	0.9685	0.3508	0.4539	0.3253

points in Fig. 9.16. This phenomenon is mainly because the non-stationary performance of the original water level series is not particularly strong, and excessive decomposition may destroy the trend information of the original series. In addition, the performance of the MODWT algorithm is limited by the number of decomposition layers (Zhai 2011) and the mother wavelet. The performance of the SSA algorithm is limited by the window length (Fenghua et al. 2014). These parameters will affect the prediction performance of the hybrid prediction model under decomposition framework. For this reason, the chapter will compare and analyze the performance of the decomposition algorithm based on the analysis of these three parameters.

9.5.5 *Influence and Analysis of Decomposition Parameters on Forecasting Performance of Hybrid Models*

9.5.5.1 Effect Analysis of the Decomposition Layers on the Performance of the MODWT

In order to study the effect of the decomposition layer on the performance of the MODWT algorithm, this section compares and analyzes the four forecasting performance indices of the MODWT-Elman model with the decomposition layer number 2–6, respectively. The results are shown in Fig. 9.18.

It can be seen from the analysis of Fig. 9.18 that when the mother wavelet of the MODWT is consistent, as the number of decomposition layers increases, the prediction accuracy of the hybrid prediction model first rises and then falls, and when the number of decomposition layer equals to 5, the model has the highest prediction accuracy and the best prediction performance. When the number of decomposition layer is 2, the model prediction performance is the worst. The main reason is when the number of decomposition layers is low, the internal feature information in the original water level time series is not completely excavated, and some features in the high-frequency components are not learned by the predictive model. However, when the number of decomposition layers increases, although the high-frequency components are gradually decomposed and the internal feature information is continuously stripped out, excessive decomposition will destroy the integrity and time-varying characteristics of the original series, it will not improve the forecasting performance of the hybrid model.

9.5.5.2 Effect Analysis of the Mother Wavelet on the Performance of the MODWT

In the chapter, the MODWT algorithm uses the "Daubechies" wavelet family. Therefore, in addition to the decomposition layer number, the mother wavelet (i.e. wavelet base) has a certain influence on the performance of the decomposition algo-

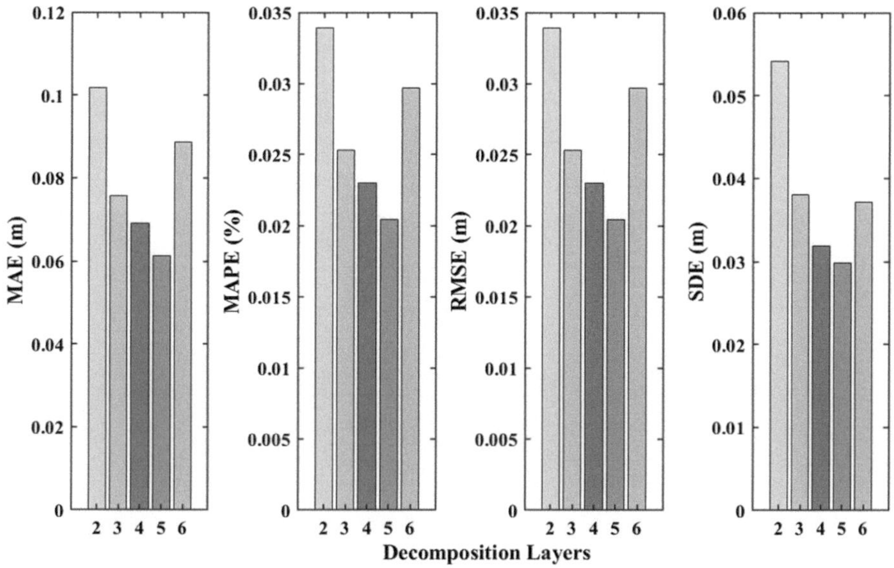

Fig. 9.18 Forecasting performance indices of different decomposition layer of the MODWT

rithm. In the section, the forecasting performance indices of the MODWT-Elman model with "Daubechies" wavelet base from "db2" to "db10" and five different types of mother wavelet are analyzed, where, in order to ensure the consistency of the comparison process, the five wavelet bases select "haar," "db5," "coif5," "sym5," and "meyr," respectively. The results are shown in Figs. 9.19 and 9.20, respectively.

By analyzing Figs. 9.19 and 9.20, the following conclusions can be drawn:

a. When the number of decomposition layers of the MODWT remains the same, as the mother wavelet of the "Daubechies" wavelet family transforms from "db2" to "db10," the prediction accuracy of the hybrid prediction model first rises and then falls. When the wavelet base is "db5," the model has the highest prediction accuracy and the best prediction performance. When the wavelet base is "db10," the model predicts the worst performance. The main reason is when the vanishing moment of the wavelet base is too large, the support length of the wavelet base becomes longer, which will generate more high-amplitude wavelet coefficients, and the boundary problem may occur, which is also disadvantageous to the model prediction performance. However, when the vanishing moment of the wavelet function is small, there are fewer wavelet coefficients equal to zero, and more series noise is not eliminated, which leads to the poor forecasting performance of hybrid model. Therefore, it is necessary to compromise the wavelet base vanishing moment. The best performance of the "db5" wavelet base in the study is verified by the argument.
b. For the water level data used in the chapter, the "Daubechies" wavelet family is undoubtedly the most suitable, and the performance of the corresponding decomposition algorithm is far superior to the other four wavelet bases. Of course, the

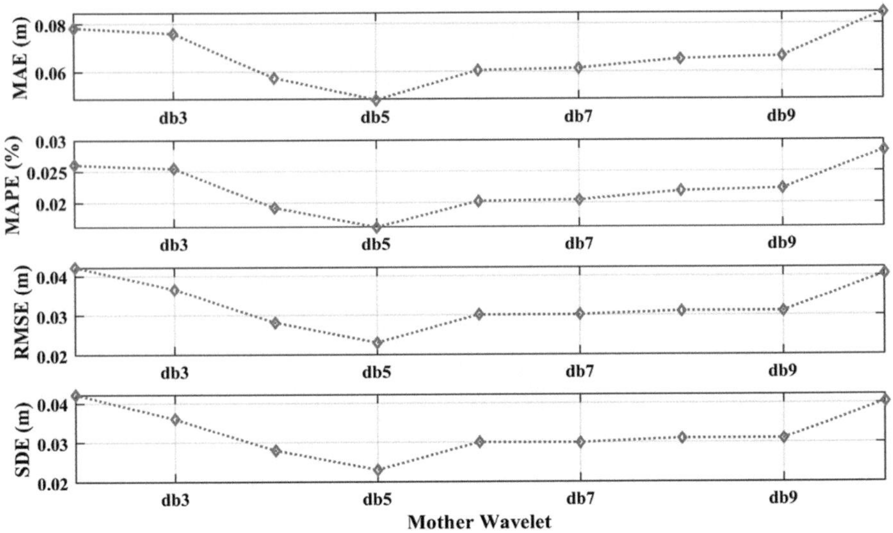

Fig. 9.19 Forecasting performance indices of different mother wavelet of the MODWT

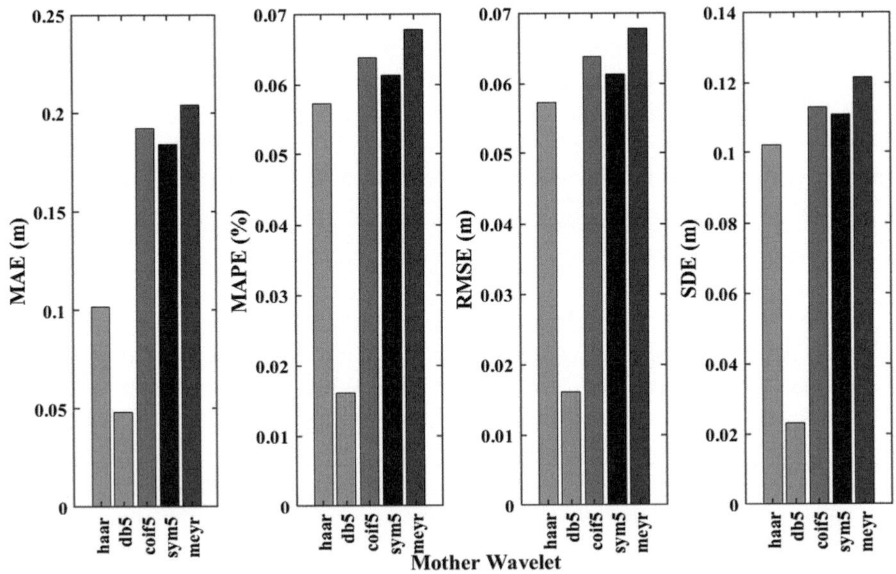

Fig. 9.20 Forecasting performance indices of different types of mother wavelet of the MODWT

decomposition algorithms corresponding to several other wavelet families are also limited by the influence of various factors such as the type of samples and vanishing moments, which also provides ideas for the development of related research work.

9.5.5.3 Effect Analysis of the Window Length on the Performance of the SSA

The length of the sample data in the chapter is 800. In general, the window length should not exceed 1/3 of the data length (Fenghua et al. 2014). That is, the length of the window in the SSA algorithm in the chapter is less than 267. In order to study the effect of window length on the performance of the SSA algorithm, this section compares and analyzes the four forecasting performance indices of the SSA-Elman models with window lengths of 300, 200, 150, 120, 100, 75, 60, and 24, respectively. The result is shown in Fig. 9.21.

It can be seen from the analysis of Fig. 9.21 that as the length of the SSA window decreases, the prediction accuracy of the hybrid prediction model generally rises and then decreases, and when the window length equals to 100, the hybrid model enjoys the highest accuracy and the best predictive performance. When the window length is 300 and 24, the model prediction performance is very poor. This phenomenon indicates that a reasonable window length is beneficial to convert the original data into a multi-dimensional series and form an appropriate trajectory matrix.

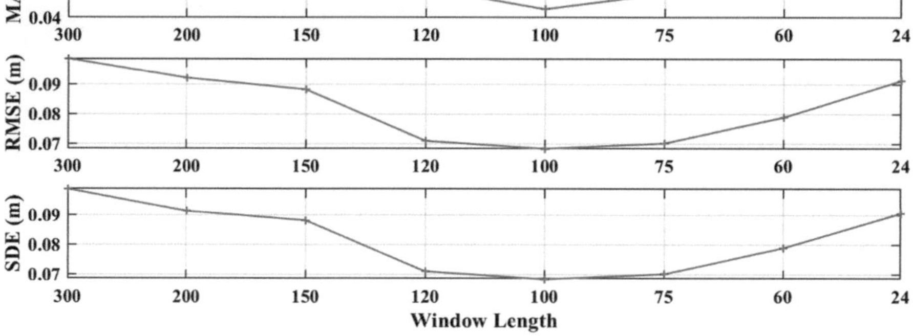

Fig. 9.21 Forecasting performance indices of different window length of the SSA

9.6 Comparative Analysis of Forecasting Performance

According to the research and analysis in the previous sections, the prediction performance will be optimal when the number of decomposition layers of the MODWT equals to 5 and the wavelet base takes the "db5" of the "Daubechies" wavelet family. The optimal window length of the SSA is 100. The section will comprehensively compare the predictions of the proposed models with the corresponding optimal models taking the original water level series $\{X_1\}$ as an example. Figures 9.22, 9.23, and 9.24 show the prediction results in 1-step to the 3-steps of the water level series $\{X_1\}$ for all models in the chapter. Table 9.4 shows the corresponding model comprehensive forecasting performance indices.

The following conclusions can be drawn from Figs. 9.22, 9.23, and 9.24 and Table 9.4:

a. The decomposition algorithm can significantly improve the prediction performance of the model and greatly improve the prediction accuracy of the model. Taking the forecasting results in 1-step of the original water level series $\{X_1\}$ as example, the MAE of the Elman model is 0.1823 m, the MAE of the MODWT-Elman model is 0.0757 m, the MAE of the EMD-Elman model is 0.1822 m, the MAE of the SSA-Elman model is 0.1440 m. It can be found that the EMD algorithm has no significant effect on the improvement of model prediction performance, which can only be seen in the prediction in 2-step and 3-step ahead.
b. The prediction accuracy of the hybrid model has greatly improved compared with the previous models after selecting the optimal decomposition layer and the optimal mother wavelet by the MODWT algorithm. The prediction accuracy of

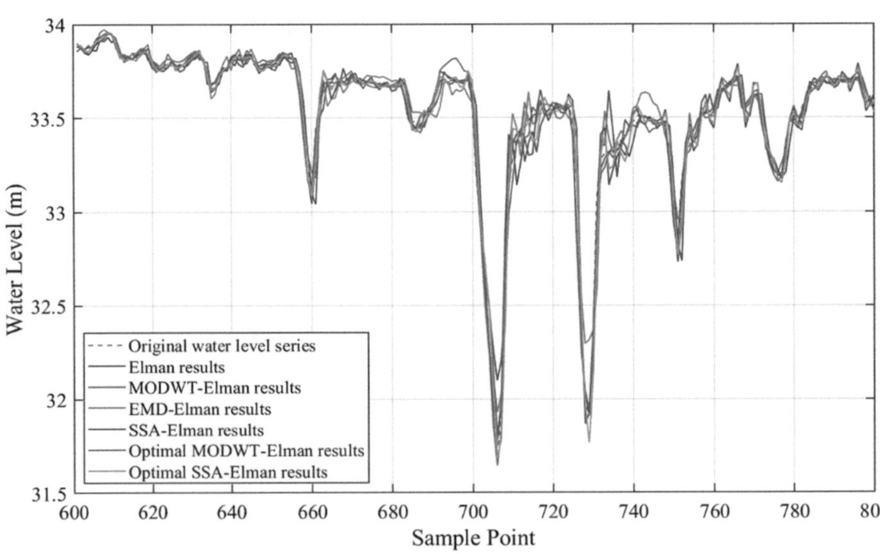

Fig. 9.22 Forecasting results of water level series $\{X_1\}$ by optimal models in 1-step

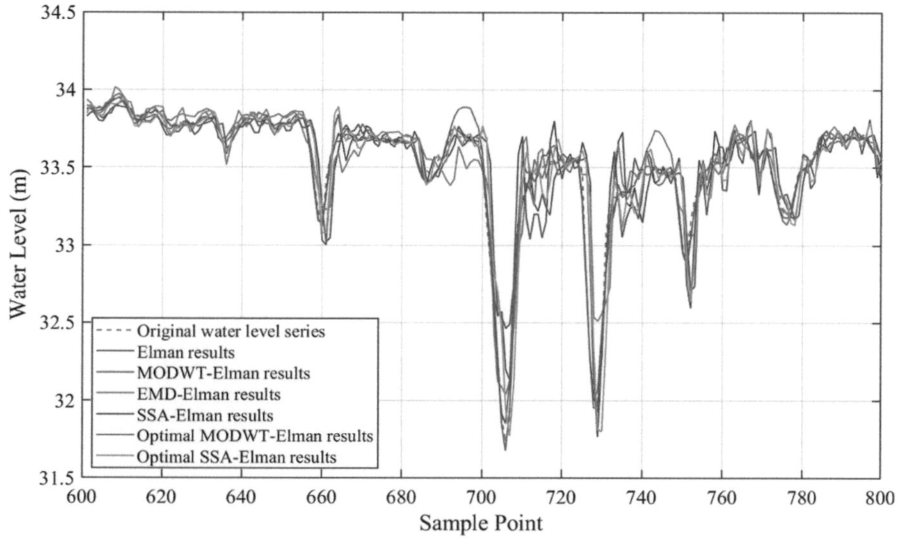

Fig. 9.23 Forecasting results of water level series $\{X_1\}$ by optimal models in 2-step

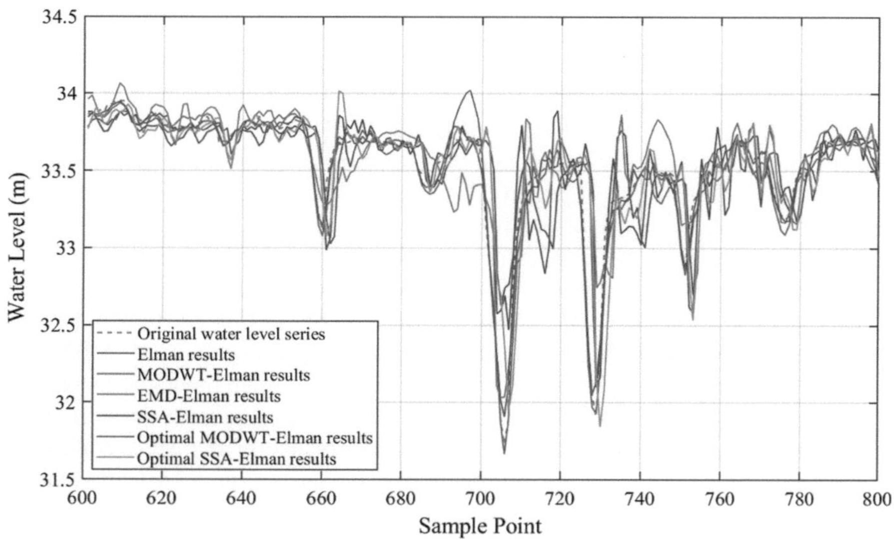

Fig. 9.24 Forecasting results of water level series $\{X_1\}$ by optimal models in 3-step

the hybrid model has also improved to some extent after selecting the optimal window length by the SSA algorithm. For example, when the water level time series $\{X_1\}$ is forecasting by hybrid model in 1-step, the MAE, MAPE, RMSE, and SDE of the MODWT-Elman model are 0.0757 m, 0.0253%, 0.0381 m, and 0.0380 m, respectively; the MAE, MAPE, RMSE, and SDE of the optimal

Table 9.4 The forecasting performance indices of optimal hybrid models

Data series	Model	Step	MAE (m)	MAPE (%)	RMSE (m)	SDE (m)	Time (s)
$\{X_1\}$	Elman	1	0.1823	0.0606	0.1093	0.1091	1.2049
		2	0.3196	0.1064	0.1893	0.1888	
		3	0.4633	0.1545	0.2662	0.2651	
	MODWT-Elman	1	0.0757	0.0253	0.0381	0.0380	2.6538
		2	0.1284	0.0429	0.0636	0.0635	
		3	0.1922	0.0642	0.0931	0.0887	
	EMD-Elman	1	0.1822	0.0607	0.1008	0.0996	3.4853
		2	0.3058	0.1019	0.1626	0.1616	
		3	0.4206	0.1406	0.2195	0.2164	
	SSA-Elman	1	0.1440	0.0480	0.0710	0.0710	1.2478
		2	0.3384	0.1129	0.1684	0.1684	
		3	0.5009	0.1670	0.2551	0.2550	
	Optimal MODWT-Elman	1	0.0462	0.0154	0.0224	0.0224	3.2625
		2	0.0910	0.0303	0.0455	0.0455	
		3	0.1210	0.0404	0.0562	0.0546	
	Optimal SSA-Elman	1	0.1233	0.0410	0.0664	0.0664	1.2268
		2	0.3163	0.1052	0.1641	0.1641	
		3	0.5096	0.1696	0.2616	0.2616	

MODWT-Elman model are 0.0462 m, 0.0154%, 0.0224 m, and 0.0224 m, respectively. The MAE, MAPE, RMSE, and SDE of the SSA-Elman model are 0.1440 m, 0.0480%, 0.0710 m, and 0.0710 m, respectively; the MAE, MAPE, RMSE, and SDE of the optimal SSA-Elman model are 0.1233 m, 0.0410%, 0.0664 m, and 0.0664 m, respectively.

c. With the increase of the advance prediction steps, the prediction accuracy of the six prediction models continues to decline. The reason has been analyzed in the previous part of the chapter, which is caused by the rolling prediction strategy adopted in the chapter. In the 1-step advance prediction, the MODWT-Elman model has the best prediction performance, the SSA-Elman model ranks second, and the EMD-Elman model is worst. In the case of 2-step advance prediction and 3-step advance prediction, the prediction performance of the MODWT-Elman model is still the best, but the prediction performance of the SSA-Elman model is worse than that of the EMD-Elman model.
d. Selecting the optimal parameters of the decomposition algorithm can further improve the prediction effect of hybrid model, which is also the main purpose of parameter optimization in time series prediction. In this chapter, parameter optimization belongs to manual parameter tuning. In the follow-up work, an integrated optimization algorithm can be used to search the optimal parameters of the hybrid model adaptively.
e. From the Figs. 9.22, 9.23, and 9.24 as you can see, with the increase of advanced prediction steps, the Optimal MODWT-Elman model and Optimal SSA-Elman

model always has a good fitting capability and high prediction precision in most of the peaks and troughs of the tendency of the segment, which still can accurately capture the sequence and state, in addition to the serious deviation in the 706th and 728th sample points, this may be due to abnormal data of testing sample set. In the subsequent research work, the two models can be further optimized, such as integrated optimization algorithm or pre-processing and post-processing algorithms such as anomaly detection and error modeling, so as to improve the prediction accuracy of the hybrid model and enhance the practicability of the prediction model.

f. The SSA-Elman model in 1-step to predict in advance still have good curve fitting ability, but the prediction deviation is larger in 2-steps and 3-step, the EMD-Elman model loses the ability to predict effectively in both advanced steps, and have taken place in certain time delay phenomenon, the practical performance is poor.
g. Combined with the computational time required for prediction, the EMD-Elman model has the worst prediction accuracy and the longest computation time, which is the worst model. The MODWT-Elman model has the highest prediction accuracy, while the optimal SSA-Elman model has the fastest prediction speed, and the two models have better prediction indices. Therefore, the two models have stronger practical value, and the follow-up work can be expanded based on these two models.

9.7 Conclusion

The chapter is based on the deterministic prediction of urban water level height time series. The water level state trend and the water level height change sequence are predicted based on various models, and combined with the prediction results for comparative analysis, including the following conclusions:

a. The Bayesian classification predictor can predict the trend of water level state to a certain extent, but it cannot adapt to the situation of sharp changes in water level, and needs further improvement;
b. The decomposition algorithm can improve the prediction performance of the model and the feature capture capability to the original series.
c. Under the MODWT framework, when the number of decomposition layers is 5 and the wavelet function selects "db5," the prediction performance of the hybrid model is optimal. Under the SSA decomposition framework, when the window length is 100, the predictive performance of the model is optimal.
d. When using the Elman neural network model as the predictor, comprehensively considering the prediction performance of the hybrid prediction model in 1-step and multi-step ahead, the MODWT-Elman model and the optimal SSA-Elman model have the highest prediction accuracy and the fastest prediction speed, and these two hybrid models are more practical.

References

Abrokwah, K., & O'Reilly, A. M. (2017). *Comparison of hybrid spectral-decomposition artificial neural network models for understanding climatic forcing of groundwater levels*. AGU Fall Meeting.

Bi S, Bi S, Chen X, Ji H, Lu Y (2018) A climate prediction method based on EMD and ensemble prediction technique. Asia-Pacific Journal of Atmospheric Sciences 54:1–12

Fenghua W, Jihong X, Zhifang H, Xu G (2014) Stock price prediction based on SSA and SVM. Procedia Computer Science 31:625–631

Huang NE, Shen Z, Long SR, Wu MC, Shih HH, Zheng Q et al (1998) The empirical mode decomposition and the Hilbert spectrum for nonlinear and non-stationary time series analysis. Proceedings of the Royal Society of London Series A: Mathematical, Physical and Engineering Sciences 454(1971):903–995

Jun-He Y, Ching-Hsue C, Chia-Pan C (2017) A time-series water level forecasting model based on imputation and variable selection method. Computational Intelligence & Neuroscience 2017(3):1–11

Kirby, D. (2015). Flood integrated decision support system for Melbourne (FIDSS). in *Proceedings of the 2015 Floodplain Management Association National Conference, Brisbane, Australia* (pp. 19–22).

Kisi O, Latifoğlu L, Latifoğlu F (2014) Investigation of empirical mode decomposition in forecasting of hydrological time series. Water Resources Management 28(12):4045–4057

Köker R (2005) Reliability-based approach to the inverse kinematics solution of robots using Elman's networks. Engineering Applications of Artificial Intelligence 18(6):685–693

Krajewski WF, Ceynar D, Demir I, Goska R, Kruger A, Langel C et al (2017) Real-time flood forecasting and information system for the state of Iowa. Bulletin of the American Meteorological Society 98(3):539–554

Li Z, Wang Y, Fan Q (2014) MODWT-ARMA model for time series prediction. Applied Mathematical Modelling 38(5-6):1859–1865

Palash W, Jiang Y, Akanda AS, Small DL, Nozari A, Islam S (2018) A streamflow and water level forecasting model for the Ganges, Brahmaputra and Meghna Rivers with requisite simplicity. Journal of Hydrometeorology 19(1):201–225

Seo Y, Kim S, Kisi O, Singh VP (2015) Daily water level forecasting using wavelet decomposition and artificial intelligence techniques. Journal of Hydrology 520(520):224–243

Shen Y, Guo J, Liu X, Kong Q, Guo L, Li W (2018) Long-term prediction of polar motion using a combined SSA and ARMA model. Journal of Geodesy 92(3):333–343

Shuai, W., Ling, T., & Yu, L. (2011). SD-LSSVR-based decomposition-and-ensemble methodology with application to hydropower consumption forecasting. in *Fourth International Joint Conference on Computational Sciences & Optimization.*

Werner M, Cranston M, Harrison T, Whitfield D, Schellekens J (2010) Recent developments in operational flood forecasting in England, Wales and Scotland. Meteorological Applications 16(1):13–22

Wunsch A, Liesch T, Broda S (2018) Forecasting groundwater levels using nonlinear autoregressive networks with exogenous input (NARX). Journal of Hydrology 567:734–758

Yadav B, Ch S, Mathur S, Adamowski J (2017) Assessing the suitability of extreme learning machines (ELM) for groundwater level prediction. Journal of Water & Land Development 32(1):103–112

Zhai B (2011) Financial high frequency time sequence MODWT fluctuation analysis. Computer Knowledge & Technology 7(10):2454–2455

Zhen, Y., Lei, G., Jiang, Z., & Liu, F. (2017). ARIMA modelling and forecasting of water level in the middle reach of the Yangtze River. in *International Conference on Transportation Information & Safety.*

Zhu J, Panpan S, Yuan G, Pengfei Z (2018) Clock differences prediction algorithm based on EMD-SVM. Chinese Journal of Electronics 27(1):128–132

Chapter 10
Prediction Model of Urban Environmental Noise in Smart Environment

Abstract With the continuous progress of modern urbanization, urban noise pollution is also increasing. Noise pollution has been disturbing the normal life of people, and serious noise pollution may even affect people's health. For this reason, it is necessary to make an effective noise prediction. Noise prediction uses historical data from noise monitoring points to predict future noise values, helping to provide effective noise regulation. In this chapter, the RF, BFGS, and GRU models are used to conduct feasibility studies for noise prediction. These three models are used to predict the public noise, traffic noise, and neighborhood noise. Through comprehensive comparative analysis of the experimental results, it can be concluded that the noise prediction performance of the BFGS model is the best in this experiment. Neighborhood noise is the most predictable among the three types of noise data.

10.1 Introduction

10.1.1 Hazard of Noise

In recent years, environmental noise has become a serious public hazard (Javaherian et al. 2018). At present, the main noise pollution is traffic noise, industrial noise, construction noise, and social noise (Purwaningsih et al. 2018). Noise pollution seriously affects people's normal life and causes potential harm to human health. Noise from various sources appears around people all the time (Khan et al. 2018). These noises will not only affect people's physical and mental health, but also interfere with normal living habits, making the incidence of neurological diseases high.

If people are exposed to excessive noise for a long time, they will not only suffer from hearing loss (Dewey et al. 2018) but also be easily affected by emotions. The data suggest that in a noisy environment, people's brain waves become disturbed by noise, showing an irregular state. These changes will show up in residential

H. Liu, *Smart Cities: Big Data Prediction Methods and Applications*,
https://doi.org/10.1007/978-981-15-2837-8_10

emotional control ability, irritability, insomnia, depression (Singh et al. 2018). Symptoms such as dizziness and difficulty concentrating can seriously interfere with residential normal work and quality of life.

Noise has a huge effect on the adult hearing system, which is already relatively mature. In daily work, the office buildings around the main road are often the areas with relatively serious noise pollution. Noise from urban traffic can easily penetrate the building's exterior walls and glass into the interior, even at great distances (Auger et al. 2018). These noises will make it difficult for people to concentrate on their work and reduce their work efficiency. For a long time in the past, it would inevitably lead to the body sub-health due to a long-term lack of rest.

The harm of noise to human body is systemic, which can not only cause the change of hearing system, but also have influence on non-hearing system (Kumar et al. 2018). The early stage of these effects is mainly physiological changes, long-term exposure to relatively strong noise, can cause pathological changes. In addition, noise in the workplace can interfere with language communication, affect work efficiency, and even cause accidents.

10.1.2 The Significance of Noise Prediction for Smart City

Currently, with the urbanization process, cities are expanding in size and quantity, and the air, water, and energy in cities all affect the overall development of cities. Ecological environment pollution is accompanied by economic development (Sengers et al. 2018). In the process of urbanization, various kinds of ecological environment pollution occur in many countries, including water resource shortage and serious pollution, air quality problems, industrial solid waste and household waste, urban noise pollution, traffic congestion, and urban layout disorder (Anthopoulos et al. 2019).

Urban noise pollution is divided into traffic noise pollution, industrial noise pollution, domestic noise pollution, and noise pollution in other noise areas. Noise pollution has a great impact on the environment and human body. According to the inspection of the environmental protection department, the proportion of cities with good acoustic environment quality is low.

In general, monitoring the noise work in cities usually has a certain complexity, to realize effective detection of noise in the city, it is necessary to use advanced detection equipment and detection technologies in monitoring city noise. However, urban noise monitoring technology started relatively late, at the same time, the detection system was not complete. The main reason is there is no standard of noise pollution inspection to adapt to relatively advanced information technology. In addition, noise monitoring equipment is so old that cannot achieve good detection result.

In this context, urban noise prediction plays a crucial role. Noise prediction uses the historical noise time series of noise monitoring stations to predict future noise value. By establishing the corresponding time-series model, the mapping relationship between the historical data and the noise points at the future moment is obtained. Noise monitoring through noise prediction is an effective method to manage noise pollution in smart city.

10.1.3 Overall Framework of Model

As a kind of time series, the common prediction models of noise data include statistical model, machine learning method, neural network, and deep network. The three models selected in this chapter are the random forest (RF) algorithm model (Liaw and Wiener 2002), BFGS (BFGS is made up of the initials *C. G. Broyden*, *R. Fletcher*, *D. Goldfarb,* and *D. F. Shanno*) algorithm model (Yuan 1991), and Gated Recurrent Units (GRU) deep neural network (Krizhevsky et al. 2012).

The RF is a flexible and widely machine learning algorithm, which can get good results in most cases even without hyperparametric tuning (Zhu et al. 2018). It is also one of the most commonly used algorithms because it is simple and can be used for both classification and regression tasks (Grange et al. 2018). The BFGS algorithm is an iterative optimization algorithm applied in the training iteration process of the neural network, so it can be regarded as a kind of neural network model (Boggs and Byrd 2019). As a quasi-Newton method, the BFGS is often used to solve nonlinear optimization problems (Chang et al. 2019). The GRU is based on multilayer neural network. It is a differentiated network structure mainly designed for image processing. The GRU contains two unique cell layers, namely convolution layer and pooling layer. In the convolution layer, each neuron is locally connected with the anterior layer and data features are extracted. The pooling layer is the feature dimension reduction without damaging the internal features of data. The GRU is composed of input layer, one or more groups of convolution-pooling layers, full connection layer, and output layer (Rastegari et al. 2016). The GRU mines the correlation between local information in data space through the form of local connection between adjacent nodes (Acharya et al. 2018).

In the chapter, three models as based-predictors are used to conduct prediction experiments on different types of noise datasets. The components of the framework are described below and Fig. 10.1 shows the general framework of the chapter.

a. Dataset: The noise dataset in this chapter contains three kinds of noise, which are namely public noise, neighborhood noise, and traffic noise. Each dataset is divided into training set and test set according to the method described in Sect. 10.2.2.
b. Based-predictor: In the chapter, the RF model, BFGS model, and GRU model are selected as based-predictors. Based-predictors are the core components of the model framework, which is utilized for training and testing.
c. Results and analysis: In this part, the prediction results and corresponding evaluation indicators of each predictor will be obtained. Then, the experimental forecasting results are compared and analyzed to explore the difference of prediction

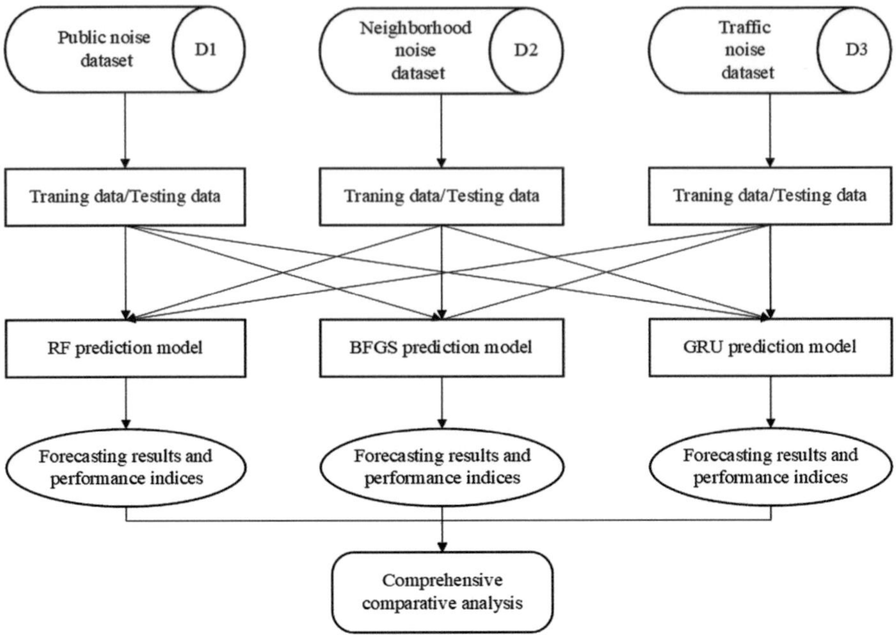

Fig. 10.1 The general framework of the chapter

performance of different models for different types of noise. This part is the summary and sublimation of the prediction models.

10.2 Original Urban Environmental Noise Series

10.2.1 Original Sequence for Modeling

In the chapter, three types of noise data are selected for prediction experiments. They are public noise data, neighborhood noise data, and traffic noise data. Figures 10.2, 10.3, and 10.4 show the data distribution of the three datasets. From Figs. 10.2, 10.3, and 10.4, it can be seen that public noise fluctuates violently and has strong non-stationarity. Compared with public noise, neighborhood noise changes more smoothly and may have longer periodicity. Among the three kinds of noise, the amplitude of traffic noise waveform is the largest and the periodicity of change is the strongest. For convince, public noise data is named D1, neighborhood noise data is named D2, and traffic noise data is D3.

In order to make a more comprehensive analysis of the selected noise datasets, Table 10.1 shows the respective statistical index values of the three datasets, which contain minimum, maximum, mean value, standard deviation, skewness, and kurtosis. Among those, skewness is the third-order ratio of central moment and standard

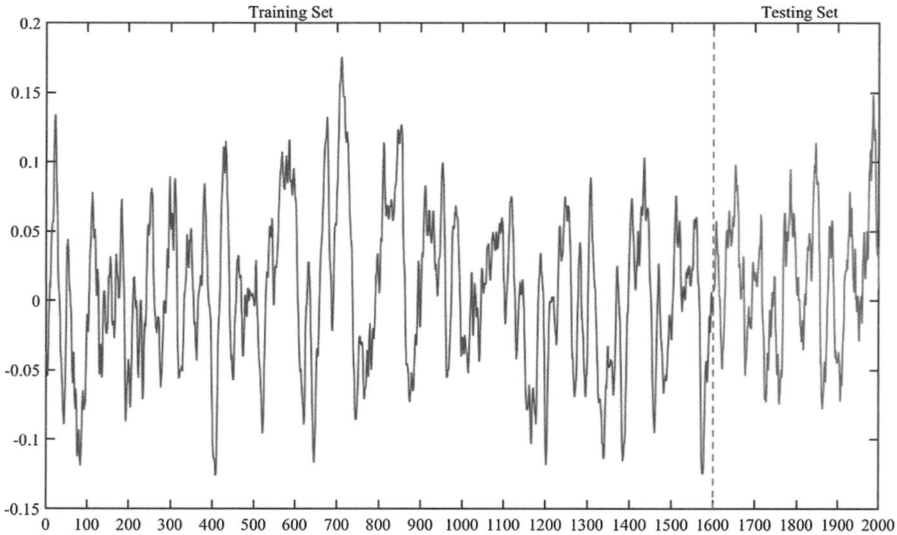

Fig. 10.2 Original public noise series

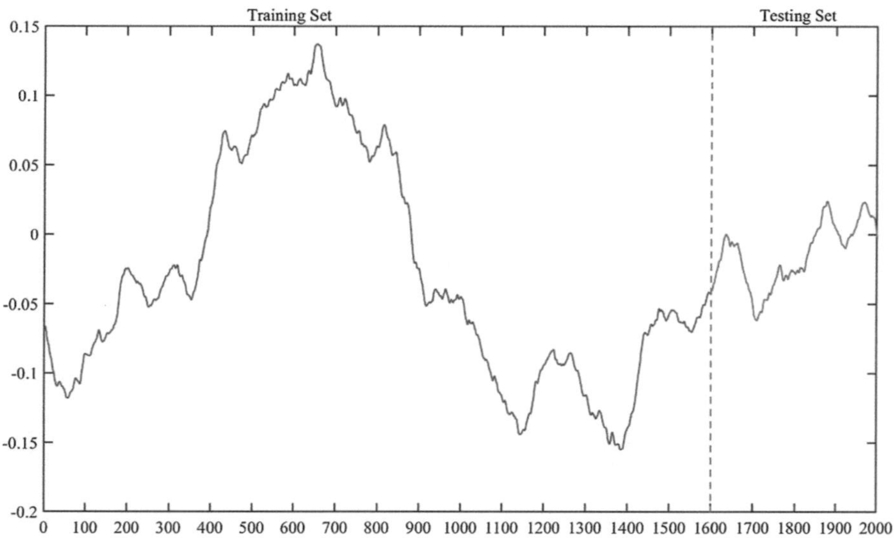

Fig. 10.3 Original neighborhood noise series

deviation of time series, which reflects the deviation of time-series data distribution relative to symmetric distribution. Skewness value is negative, its distribution is skewed to the left, when skewness value is positive, its distribution is skewed to the right, and when skewness value is 0, its distribution is symmetrical. Kurtosis is the ratio of the fourth order central moment of time series to the fourth power of standard deviation, which reflects the outliers of data. The kurtosis value adopted is the

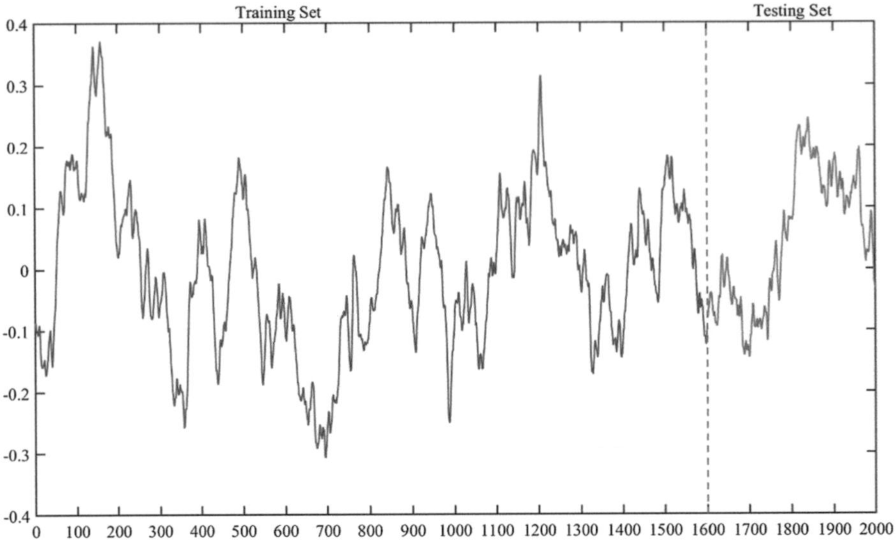

Fig. 10.4 Original traffic noise series

Table 10.1 Statistical characteristics of the original series

Original series	Minimum	Maximum	Mean value	Standard deviation	Skewness	Kurtosis
D1	−0.1258	0.1757	0.0072	0.0543	0.0637	2.6937
D2	−0.1549	0.1375	−0.0261	0.0727	0.4160	2.2995
D3	−0.3073	0.3715	0.0051	0.1259	0.0880	2.7081

kurtosis value relative to the standard normal distribution. When the kurtosis value is positive, it indicates that the dispersion degree of the distribution is greater than the standard normal distribution.

10.2.2 Separation of Sample

In order to construct the model and evaluate the models' performance, the original noise series for simulation need to be grouped. The three groups of time series in this chapter all contain 2000 sample points and divide the data into training dataset and testing dataset. The 1st to 1600th sample points are used as training samples to train the prediction model, and the 1601st to 2000th sample points are used as testing samples to test the obtained model after the model training is completed, so as to obtain the prediction output of the model, calculate the prediction error of the model, and evaluate the prediction performance of the model. The results of the separation of sample are shown in Figs. 10.2, 10.3, and 10.4.

10.3 The RF Prediction Model for Urban Environmental Noise

10.3.1 The Theoretical Basis of the RF

Random forest is a flexible and widely machine learning algorithm, which can get good results in most cases even without hyperparametric tuning. It is also one of the most commonly used algorithms because it is simple and has a good performance on classification and regression (Liaw and Wiener 2002).

The RF belongs to supervised learning method. The constructed forest is an integration of decision trees, which is mostly trained by the bagging method (Javaherian et al. 2018). The bagging method uses random selection of training data to construct the classifier, and finally combine them (Yuchi et al. 2019).

Random forest is a special bagging method that uses decision tree as the model in bagging. First, the number of *m* training sets is generated by the bootstrap method. Then, a decision tree is trained for each training set. When nodes search for features are split, not all features can be found to maximize the index. Instead, it randomly extracts some features from the features, finds the optimal solution among them, and applies it to node splitting. The method of random forest is equivalent to sampling samples and features, so overfitting can be avoided (Ishwaran and Lu 2019).

10.3.2 Steps of Modeling

The steps of using the RF model to predict noise data are as follows (Couronné et al. 2018):

a. The training sample set and test sample set are divided according to the method in Sect. 10.2.2. In the step, the training set and test set of the three kinds of noise data will be obtained respectively;
b. Setting up the necessary training parameters of the random forest model. The number of decision trees (*ntree*) is 500 according to the empirical value. The number of features in a split feature set (*mtry*) is 3. In general, *mtry* takes one-third of the total number of characteristic variables. The RF model here has 10 input characteristic variables, so *mtry* value is 3. These two factors have great influence on the prediction accuracy and generalization ability of stochastic forest regression model;
c. *M* groups of sub-training sample sets are randomly extracted from the original test sample set with the bootstrap method. Since the original training sample set has 10 input variables, 3 random variables from 10 input variables are randomly selected at each node of each decision tree to be divided into subsets, and then optimal branches are selected according to the criterion of goodness.
d. The *M* decision trees generated are formed into the random forest regression model by the bagging method. The training test is carried out with the test sample set.

10.3.3 Forecasting Results

Figures 10.5, 10.6, and 10.7 show the forecasting results of three kinds of noise data using the RF model. From them, it can be concluded that:

a. The RF algorithm model has a good prediction effect on the noise data selected in this chapter. From those figures, it can be seen that all the prediction curves well fit the main trend of the original noise data.
b. In Figs. 10.5 and 10.7, noise datasets D1 and D3 fluctuate violently. Therefore, the errors of the prediction curve are mainly concentrated around the peaks and troughs, and the prediction results have a certain delay. In the practical application, time compensation can be added to obtain the prediction results with higher accuracy.
c. In Fig. 10.6, the curve of dataset D2 is relatively smooth due to the flat change of neighborhood noise. Therefore, the prediction error of the RF prediction model is mainly caused by overfitting. From the forecasting graph, it can be seen that the RF model is seriously overfitting at this time. It can reduce overfitting by reducing the number of decision trees of the RF model, to improve the prediction accuracy of the model.

In order to make a more accurate evaluation of model prediction performance, Table 10.2 gives four evaluation indices for the forecasting results of three datasets. According to Table 10.2, the following conclusions can be drawn:

a. The RF prediction model has the best prediction results for dataset D3. The performance indices of D1 are 0.0136, 167.9111%, 0.0180, and 0.0180, respectively.

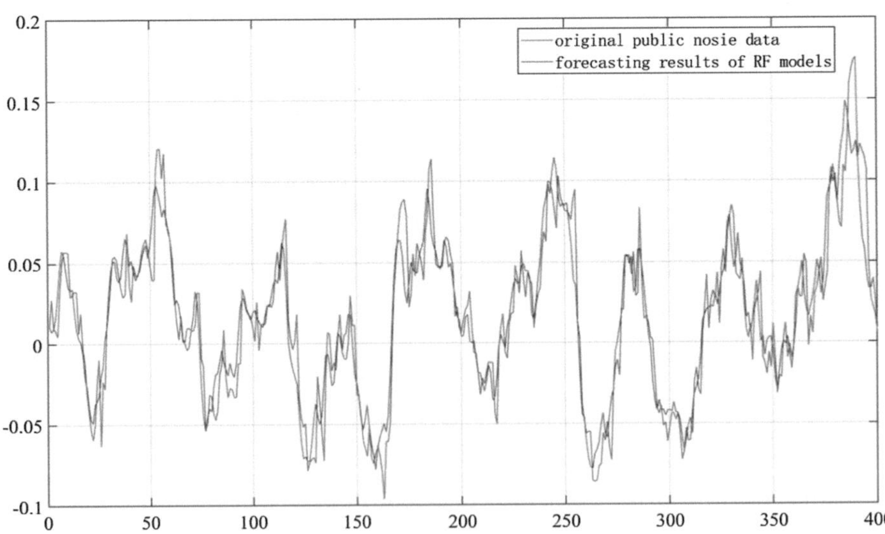

Fig. 10.5 Forecasting results of the RF model for D1

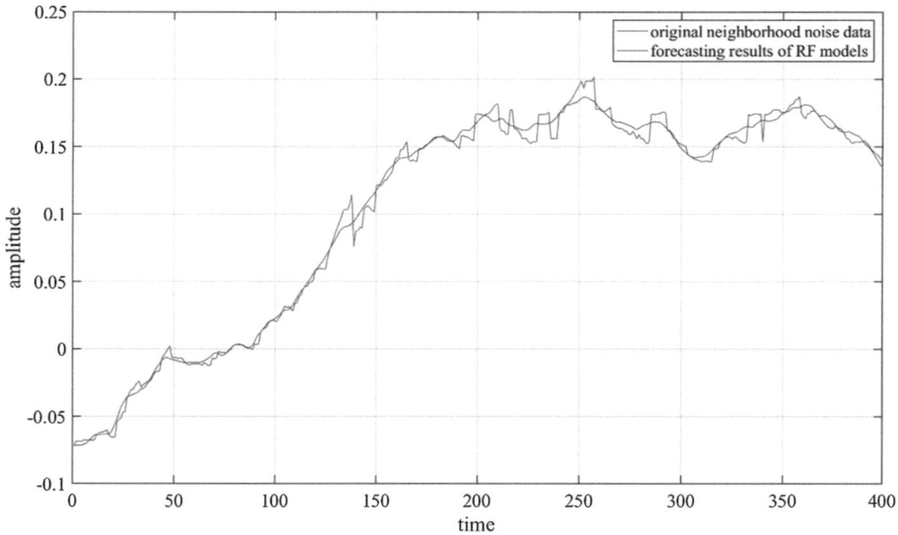

Fig. 10.6 Forecasting results of the RF model for D2

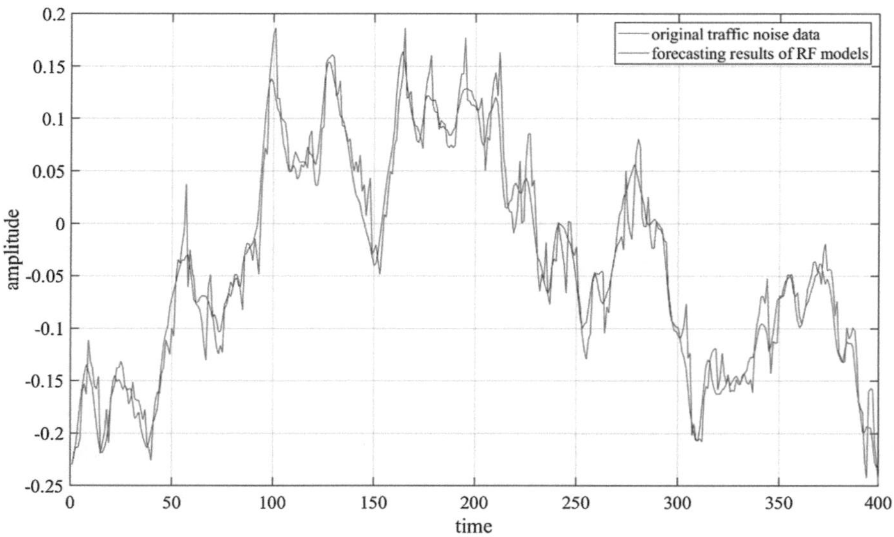

Fig. 10.7 Forecasting results of the RF model for D3

The performance indices of D2 are 0.0181, 64.9091%, 0.0236, and 0.00234, respectively. The performance indices of D3 are 0.0044, 11.5155%, 0.00058, and 0.0057, respectively. All the evaluation indices of dataset D3 are smaller than those of the other two datasets. These statistical indices prove that the RF model has excellent prediction performance for traffic noise data.

Table 10.2 The forecasting performance indices of the RF model

Model	Dataset	MAE	MAPE (%)	RMSE	SDE
RF	D1	0.0136	167.9111	0.0180	0.0180
	D2	0.0181	64.9091	0.0236	0.0234
	D3	0.0044	11.5155	0.0058	0.0057

b. The forecasting accuracy of D1 is poor. Although the MAE, RMSR, and SDE values show that it can predict the overall trend of test data, its MAPE value is relatively high. When the values of the three datasets are in the same order of magnitude, the higher MAPE value indicates that this prediction deviation is large. This is the prediction error and slight time delay of the peak and trough in the figure.
c. The forecasting result of dataset D2 is better than that of dataset D1, but not as good as that of dataset D3. The RMSE value and SDE value of dataset D2 are the highest predicted results among the three datasets. This may be caused by serious overfitting of the prediction results of dataset D2.

10.4 The BFGS Prediction Model for Urban Environmental Noise

10.4.1 The Theoretical Basis of the BFGS

Similar to the BP algorithm, the BFGS algorithm is an iterative optimization algorithm applied in the training iteration process of neural network, so it can be regarded as a kind of neural network model (Yuan 1991). As a quasi-Newton method, the BFGS is often used to solve nonlinear optimization problems. For the unconstrained minimization problem $\min_{a} f(a)$, where $a = (a_1, a_2, \ldots, a_N)^T \in R^N$. The BFGS algorithm constructs the following iteration formula (Badem et al. 2018):

$$a_{k+1} = a_k + \lambda_k d_k \tag{10.1}$$

where a_0 is initial value, a_k is current value, λ_k is step size in search, d_k is search direction. They can be calculated by the following formulation:

$$\lambda_k = \arg\min \ f(a_k + \lambda d_k) \tag{10.2}$$

$$d_k = -B_k^{-1} r_k \tag{10.3}$$

where B_k is a $N*N$ matrix, $r_k = \nabla f(a_k)$, the key step of the BFGS algorithm is to determine B_k, which is usually required to be a symmetric positive definite matrix and satisfies the following equation (Yang and Wang 2018):

$$B_k s_k = c_k \tag{10.4}$$

$$s_k = a_{k+1} - a_k \tag{10.5}$$

$$c_k = r_{k+1} - r_k \tag{10.6}$$

B_k can be calculated by the following iterative formula:

$$B_{k+1} = B_k + \frac{c_k c_k^T}{s_k^T c_k} - \frac{B_k s_k s_k^T B_k}{s_k^T B_k s_k} \tag{10.7}$$

10.4.2 Steps of Modeling

According to the method in Sect. 10.4.1, the steps of using the BFGS model to predict noise data are as follows (Liu et al. 2018):

a. The training sample set and test sample set are divided according to the method in Sect. 10.2.2. In the step, the training set and test set of the three kinds of noise data will be obtained, respectively;
b. Initializing the value of the variables, such as initial value a_0, precision threshold ξ, and iterations $n = 0$. Initialize the Hyson approximate matrix and inverse approximate matrix. Then initial search direction is obtained by Eq. (10.3);
c. When all training parameters are set, the model training calculation is carried out. Search for optimal step length λ_k according to damped Newton method. Then c_k is obtained by Eq. (10.6) (Ge et al. 2018);
d. Updating $a_{n+1} = a_n + \lambda_n d_n$ and $c_{n+1} = r_{n+1} - r_n$;
e. Hyson positive definite matrix B_{n+1} is calculated by Eq. (10.7);
f. Determining whether the modulus of the search direction is less than the set precision threshold. If so, complete the BFGS model training at the end of the iteration. If not, update $n = n + 1$, return to step (c) for the next round of calculation.

10.4.3 Forecasting Results

The test sample set is input into the trained BFGS neural network to get the predicted results. Figures 10.8, 10.9, and 10.10 show the prediction result curves of the BFGS neural network model for three test sample sets. From the figures it can draw the following conclusions:

a. The BFGS neural network model has good prediction effect on three test sets. The fitting degree of both the main trend component and the peak and trough of the wave is close to the real value of the test set. It can be concluded that the BFGS neural network has good effect on noise prediction.

b. The prediction accuracy of neighborhood noise and traffic noise is higher than that of public noise. From the figures, it can be seen that the prediction error of the public noise prediction curve near the peak and trough is larger than that of the other two kinds of noise. This may be because the sound composition of public noise is more complex, which leads to more violent fluctuations of public noise waveform, resulting in decreased prediction accuracy.
c. From the figures, it can be concluded that the BFGS neural network does not have a high prediction accuracy for the situation where the data changes dramatically before and after. This is also the reason why the BFGS neural network has the highest prediction accuracy to test set D2 and the lowest to test set D1 in the experiment. The waveform of test set D2 changes gently and the curve is smooth. The fluctuation of dataset D1 is severe, and the data jump before and after is large.

In order to better analyze the prediction performance of the BFGS neural network for three kinds of noise datasets, Table 10.3 gives some statistical indices of the forecasting results of the BFGS neural network for three kinds of datasets. According to Table 10.3, the following conclusions can be drawn:

a. The BFGS neural network has the best prediction effect for neighborhood noise dataset. The MAE, MAPE, RMSE, and SDE of forecasting results for D1 are 73.922, 0.0087%, 0.0109, and 0.0109, respectively. The MAE, MAPE, RMSE, and SDE of forecasting results for D2 are 6.0050, 0.0006%, 0.0008, and 0.0020, respectively. The MAE, MAPE, RMSE, and SDE of forecasting results for D3 are 24.5034, 0.0053%, 0.0227, and 0.0223, respectively. The statistical indices of forecasting result of dataset D2 by the BFGS neural network are far smaller than that of other two datasets, which has the best prediction effect. This is

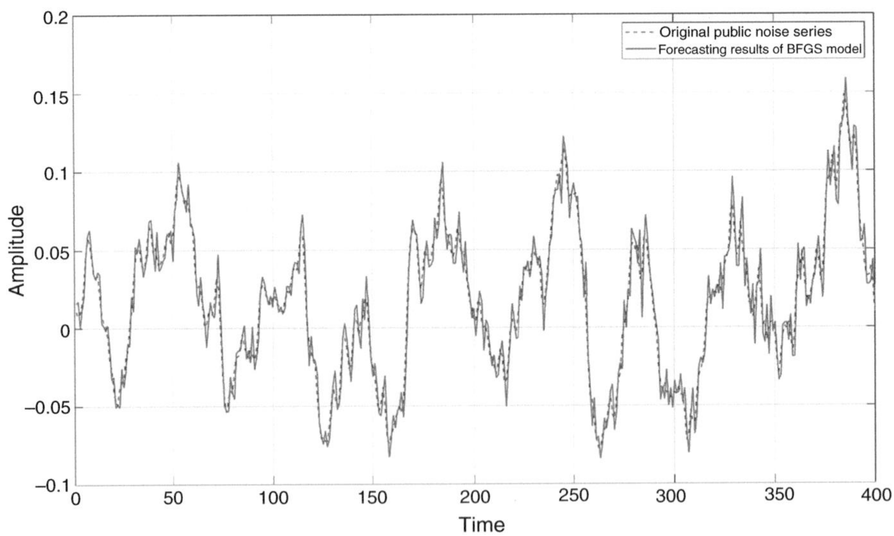

Fig. 10.8 Forecasting results of the BFGS model for D1

consistent with the conclusion got from the figures, which reflects the rigor of the experiment.

b. From the data in the tables, it can be seen that the BFGS neural network has a better prediction effect on the test dataset D3 than on the test dataset D1.This is because dataset D3 compared to dataset D1, its data fluctuations are smoother, so that the prediction accuracy of the BFGS neural network is higher.

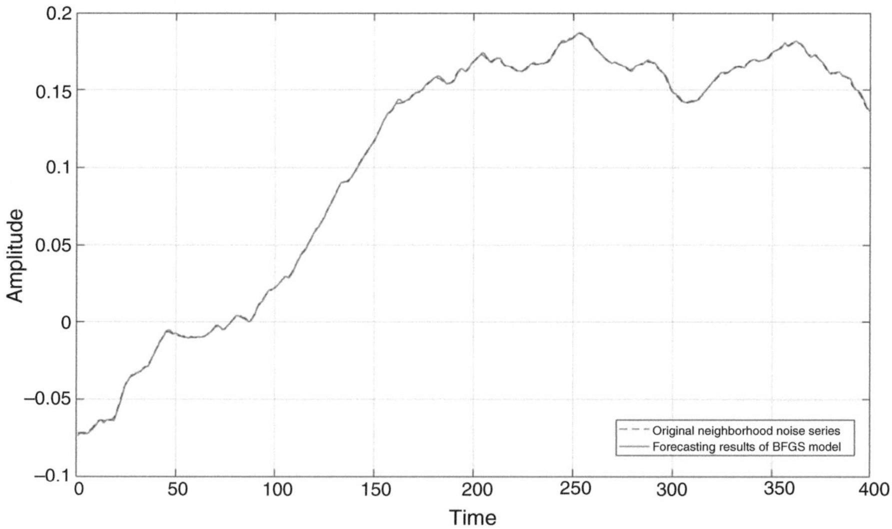

Fig. 10.9 Forecasting results of the BFGS model for D2

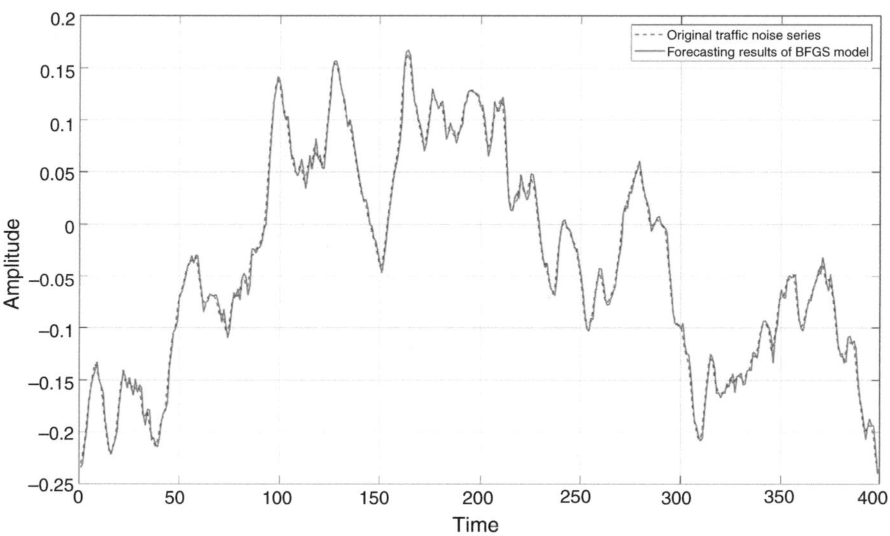

Fig. 10.10 Forecasting results of the BFGS model for D3

Table 10.3 The forecasting performance indices of the BFGS model

Model	Dataset	MAE	MAPE (%)	RMSE	SDE
BFGS	D1	73.922	0.0087	0.0109	0.0109
	D2	6.0050	0.0006	0.0008	0.0020
	D3	24.5034	0.0053	0.0227	0.0223

10.5 The GRU Prediction Model for Urban Environmental Noise

10.5.1 The Theoretical Basis of the GRU

The GRU is a variant of the LSTM (Borovykh et al. 2019). However, compared with the LSTM, the GRU has only two gates and no cell state, which simplifies the structure of the LSTM. Figure 10.11 is the structure of the GRU network model used in this chapter. It substitutes the single update gate for the forgetting gate and the input gate, which determines how much information is transferred to the current hidden state in the previous hidden state. There is also a reset gate in the GRU, whose calculation operation is similar to that of the update gate, except that the weight matrix is different. Its purpose is to determine how much information to forget about the previous hidden state. The GRU also mixes the cellular state with the hidden state, among other changes that allow the GRU to have fewer parameters and lower computational costs to run (Zahid et al. 2019).

In Fig. 10.12, x_t and h_t represent the input and output of the GRU, respectively. h_t can be obtained by the following formulations (Aimal et al. 2019):

$$z_t = \sigma\left(W_{hz} h_{t-1} + W_{xz} x_t + b_z\right) \tag{10.8}$$

$$r_t = \sigma\left(W_{hr} h_{t-1} + W_{xr} x_t + b_r\right) \tag{10.9}$$

$$\tilde{h}_t = \tanh\left(r_t * W_{hh} h_{t-1} + W_{xh} x_t + b_h\right) \tag{10.10}$$

$$h_t = \left(1 - z_t\right) * h_{t-1} + z_t * \tilde{h}_t \tag{10.11}$$

where z_t represents the output of the update gate. r_t represents the output of reset gate. W_{hz} is weight matrix between output at $t - 1$ time and update gate. W_{xz} is weight matrix between input at t time and update gate. W_{hr} is weight matrix between output at $t - 1$ time and reset gate. W_{xr} is weight matrix between input at t time and reset gate. $\tilde{h}_t$ and h_t are candidate hidden state and hidden state, respectively. They control how much information is forgotten and how much information is saved from the previous hidden state. W_{hh} is weight matrix between output at $t - 1$ time and candidate hidden state. W_{xh} is weight matrix between input at t time and candidate hidden state. b_z, b_r, and b_h are bias of update gate, reset gate, and candidate hidden state, respectively.

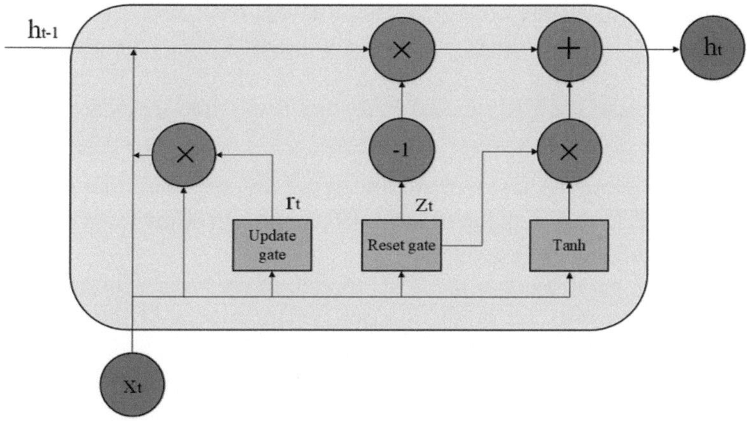

Fig. 10.11 The structure of the GRU

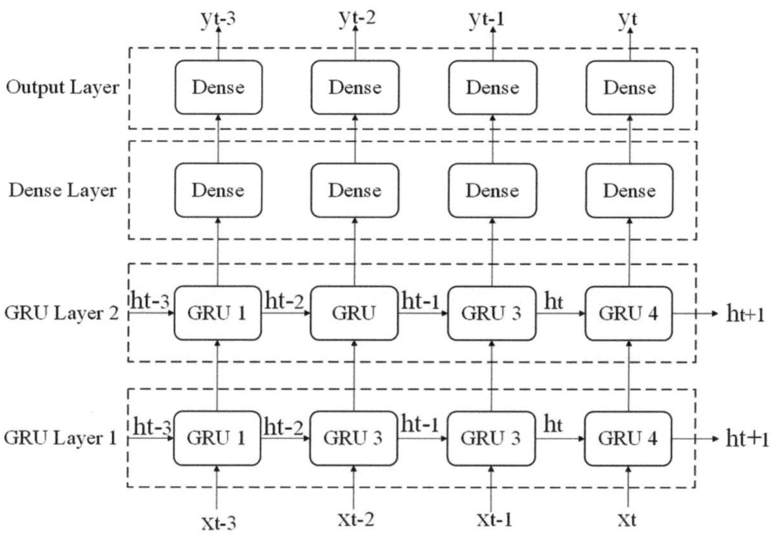

Fig. 10.12 The framework of the GRU in the section

10.5.2 Steps of Modeling

The GRU can be stacked to form multiple layers. In this chapter, the two-layer GRU network is used for noise prediction. The network structure consists of two GRU layers and one dense layer. At *t* time, the input of the model is the training sample dataset established above, and the output is the noise data value at future time.

More specifically, the first GRU layer takes historical data as input and outputs as hidden states. The output of a GRU layer above the second GRU layer is the input, and the output is the hidden state of the second GRU layer. The second GRU layer is connected to a fully connected dense layer. The dense layer remains connected to all the neurons in the upper layer and finally outputs the predicted results of the model.

The steps of using the GRU model to predict noise time series are described as follows:

a. The training sample set and test sample set are divided according to the method in Sect. 10.2.2. In this step, the training set and test set of the three kinds of noise data will be obtained, respectively;
b. Initializing the parameters of the GRU model. In the process of model training, the loss function uses the prediction performance index of the MAE, and the RMSprop algorithm is used for optimization in the process of model training. The model parameters include the training times, learning rate, hidden layer size, and regularization parameters;
c. Training the GRU model. The final training of the GRU model needs to go through several iterations. Each iteration randomly selects training samples to avoid sample order interfering with prediction results;
d. Testing the trained GRU model. The divided test set is used to conduct prediction experiments on the trained GRU model. The MAE, MAPE, RMSE, and SDE are used to compare and analyze the prediction results of different noise datasets.

10.5.3 Forecasting Results

The test sample set is input into the trained GRU to get the predicted results. Figures 10.13, 10.14, and 10.15 show the prediction result curves of the GRU model for three test sample sets. From the figures the following conclusions can be drawn:

a. The GRU neural network model has good prediction effect on three test sets. The fitting degree of both the main trend component and the peak and trough of the wave is close to the real value of the test set. It can be concluded that the GRU neural network has good effect on noise prediction.
b. The prediction accuracy of neighborhood noise and traffic noise is higher than that of public noise. From the figures, it can be seen that the prediction error of the public noise prediction curve near the peak and trough is larger than that of the other two kinds of noise. This may be because the sound composition of public noise is more complex, which leads to more violent fluctuations of public noise waveform, resulting in decreased prediction accuracy.
c. From the figures, it can be concluded that the GRU neural network does not have a high prediction accuracy for the situation where the data changes dramatically before and after. The waveform of test set D2 changes gently and the curve is smooth. The fluctuation of dataset D1 is severe, and the data jump before and after is large.

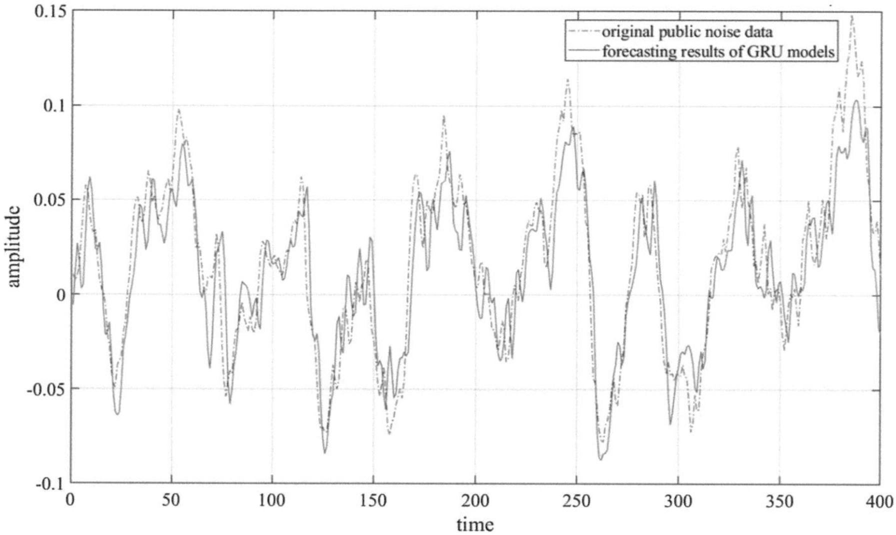

Fig. 10.13 Forecasting results of the GRU model for D1

In order to better analyze the prediction performance of the GRU model for three kinds of noise datasets, Table 10.4 gives some statistical indices of the forecasting results of the GRU neural network for three kinds of datasets. According to Table 10.4, the following conclusions can be drawn:

a. The GRU model has the best prediction effect for neighborhood noise dataset. The MAE, MAPE, RMSE, and SDE of forecasting results for D1 are 73.922, 0.0087%, 0.0109, and 0.0109, respectively. The MAE, MAPE, RMSE, and SDE of forecasting results for D2 are 6.0050, 0.0006%, 0.0008, and 0.0020, respectively. The MAE, MAPE, RMSE, and SDE of forecasting results for D3 are 24.5034, 0.0053%, 0.0227, and 0.0223, respectively. The statistical indices of forecasting results of dataset D2 by the GRU are far smaller than that of other two datasets, which has the best prediction effect. This is consistent with the conclusion got from the figures, which reflects the rigor of the experiment.
b. From the data in the tables, it can be seen that the GRU has a better prediction effect on the test dataset D3 than that on the test dataset D1. This is because dataset D3 compared to dataset D1, its data fluctuations are smoother, so that the prediction accuracy of the GRU model is higher.

10.6 Big Data Prediction Architecture of Urban Environmental Noise

With the rapid pace of smart city construction, the level of urban informatization has been greatly improved. The deployment of a large number of noise sensors can obtain a large amount of original noise data. These data cover a wide range of noise

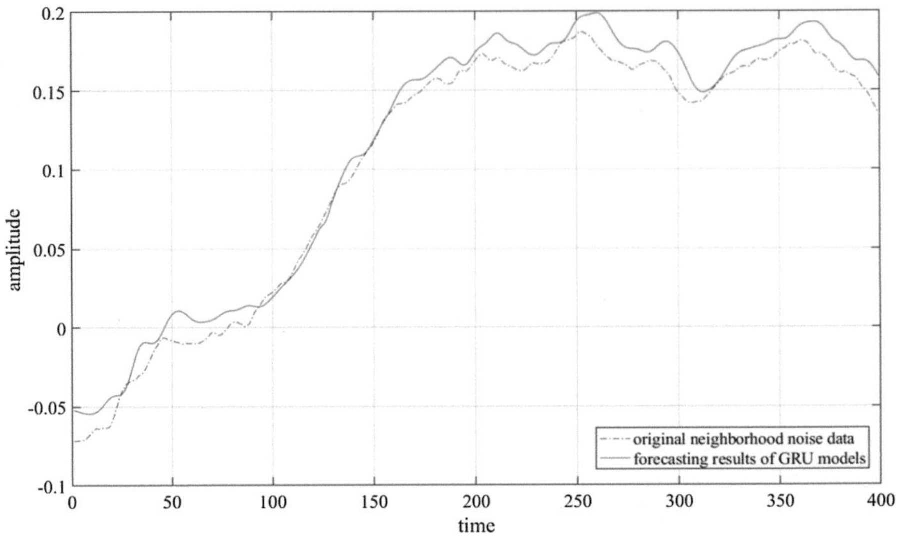

Fig. 10.14 Forecasting results of the GRU model for D2

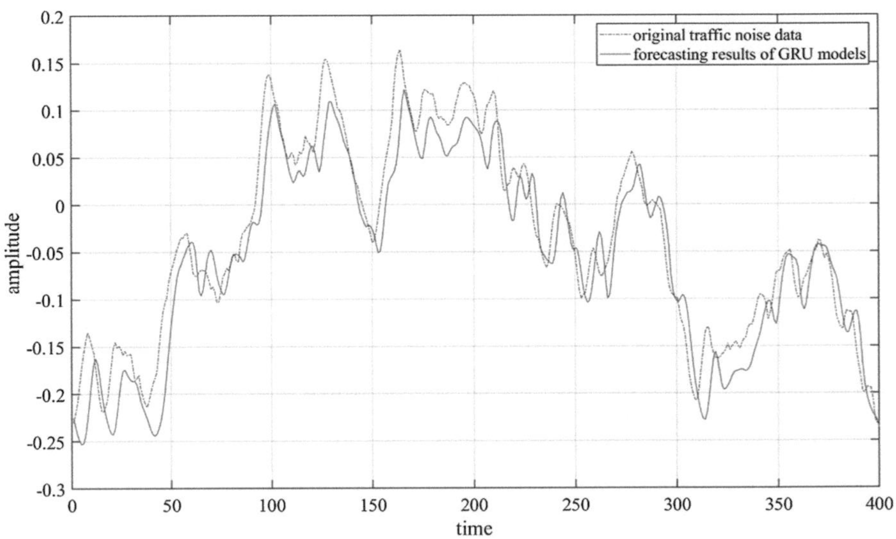

Fig. 10.15 Forecasting results of the GRU model for D3

data types. By means of data mining and statistical analysis, a lot of useful information can be mined from the original data. This information can help people better control urban noise pollution and improve the ecological construction capacity of smart cities. Therefore, data-driven management models and management ideas have been more and more recognized and practiced.

Table 10.4 The forecasting performance indices of the GRU model

Model	Dataset	MAE	MAPE (%)	RMSE	SDE
GRU	D1	0.0202	329.6954	0.0249	0.0232
	D2	0.0129	70.1137	0.0146	0.0110
	D3	0.0258	110.8410	0.0322	0.0319

However, the increasing amount of data and the variety of data also bring many new challenges. The first is the storage of a large amount of noise data. There is noise interference in the environment where people live all the time, so the noise of the variety of data is rich, the amount of data is huge. How to integrate and store these massive multi-dimensional data requires a breakthrough in technology. In addition, noise data continue to be generated. Therefore, when building a noise processing platform oriented to big data, it is necessary to consider not only the capacity of the system to handle large-scale data but also the scalability and dynamic scalability to adapt to the fast and changeable market demands. Second, although large-scale multi-dimensional data can provide richer dimensional information and more abundant training samples for data-driven models, it will also bring great difficulties to model training. Therefore, the noise information platform for data processing should not only have good reliability but also meet the reasonable accuracy and real-time requirements.

As a brand new technology concept and means, big data technology has shown its powerful data processing ability and has gradually become a new way of data governance and thinking. The combination of big data technology and smart city can not only solve a large number of practical problems but also provide an effective way for the construction of smart city.

10.6.1 Big Data Framework for Urban Environmental Noise Prediction

In this chapter, the Spark big data platform is selected to build the RF noise prediction model of smart cities. Spark cloud computing platform is the most widely used distributed computing platform. Spark is particularly suited to handle complex tasks such as machine learning with large amounts of data that require iterative computation, and its MLlib library is developed to solve various machine learning problems (Meng et al. 2016). In addition, Spark supports interactive query, streaming processing, graph computing, and other computing scenarios. Spark Streaming is a streaming technology based on the Spark platform that allows the system to respond in seconds and continuously process online.

The RF model has many advantages that make it suitable for Spark platform prediction. Based on bagging and random subspace, the random forest algorithm establishes multiple decision trees to predict the results, which reduces the complexity of calculation and improves the accuracy of prediction in terms of data volume, calculation volume, and other aspects. It is precisely due to the application of these two ideas that the overfitting problems in the decision tree, such as data complexity or noise, can be effectively avoided in the random forest. Figure 10.16 shows the framework of the Spark-RF model.

10.6.2 Big Data Storage for Urban Environmental Noise Prediction

HDFS (Hadoop Distributed File System) is widely deployed on cheap PC to store super large files in the mode of streaming data access and run on commercial hardware clusters. HDFS is a file system that is stored across multiple computers in the management network, and it is a distributed file system with high fault tolerance. The use of HDFS greatly improves the data throughput of the system and the processing speed of data. Therefore, HDFS is suitable for the application scenarios that need to process massive datasets but require little time delay. HDFS is not suitable for applications that require low latency data access, store large amounts of small files, write to multiple users, and modify files arbitrarily.

HDFS provides Spark with a highly fault-tolerant, robust, reliable, and convenient file system. First, the data is ensured to exist on multiple data nodes. When one node fails, data can still be retrieved through other nodes. Second, HDFS checks the status of nodes through a periodic heartbeat mechanism. When the name node does not receive heartbeat information from the data node, it is flagged as a failed node and resumes the task performed by the node.

10.6.3 Big Data Processing of Urban Environmental Noise Prediction

a. Spark's operation process

Spark programs are running in a cluster. First it needs to create a SparkContext object, which is used to connect with the cluster (including Standalone, Mesos, YARN, etc.). And then the Spark in the cluster manager assigns resources, respectively, in each node to start the executors, and sends the code to executors. Finally SparkContext will send executors tasks to run.

b. Spark's scheduling process

The DAG scheduler calculates a directed acyclic graph for each job submitted to the cluster and finds the best execution path. The specific execution process is to submit the task set under the stage to the TaskScheduler object, which will submit to the cluster to request resources and finish execution.

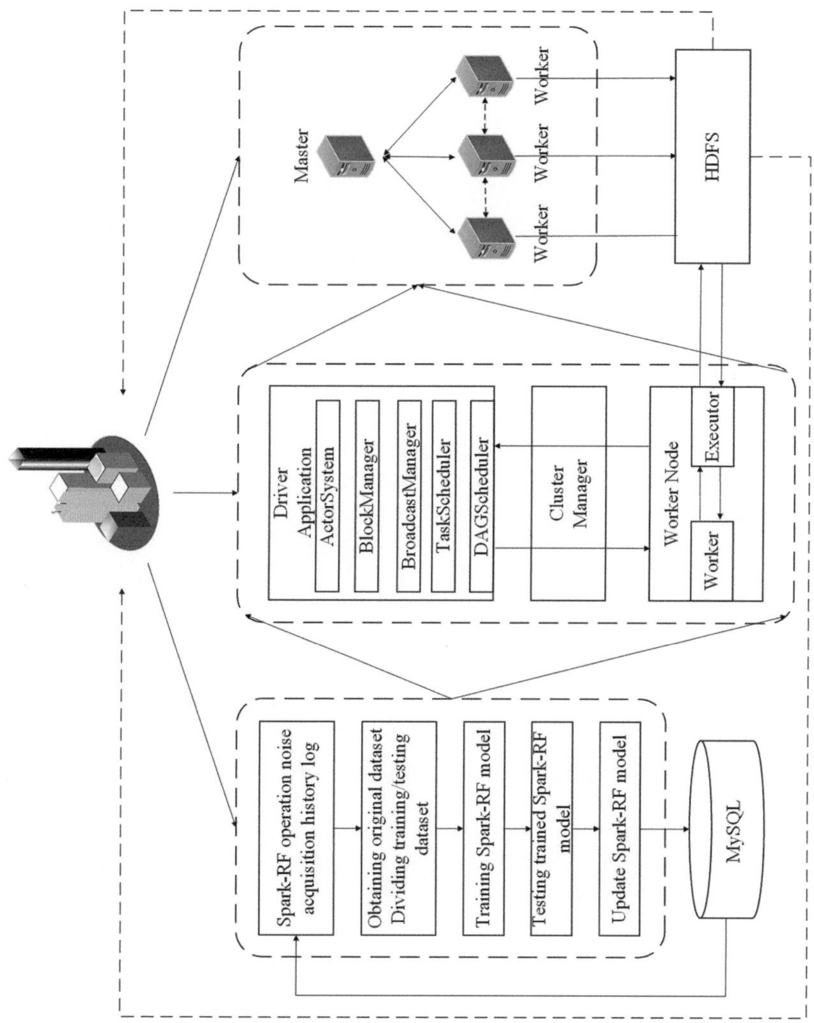

Fig. 10.16 The framework of Spark-RF model

10.7 Comparative Analysis of Forecasting Performance

The following conclusions can be drawn from Figs. 10.17, 10.18, and 10.19 and Table 10.5.

a. For the prediction of noise series, the BFGS model has the highest prediction accuracy, followed by the RF model, and the prediction accuracy of GRU models decreases. For example, the MAE, MAPE, RMSE, and SDE of the BFGS model on D1 are 73.9222, 0.0087%, 0.0109, and 0.0109, respectively; the MAE, MAPE, RMSE, and SDE of the RF model on D1 are 0.0136, 167.9111%, 0.0180, and 0.0180, respectively; the MAE, MAPE, RMSE, and SDE of the GRU model on D1 are 0.0202, 329.6954%, 0.0249, and 0.0232, respectively. According to the four evaluation indices, BFGS model has the highest degree of longitude predicted on noise series in this experiment.
b. From the prediction results of the three datasets, the prediction fitting effect of dataset D2 is the best. It can also be seen from the data in Table 10.5 that the three prediction models have the smallest evaluation index for the prediction results of dataset D2. This may be due to the characteristics of dataset D2 itself. The neighborhood noise itself changes gently, and the data contain less characteristic information. In the process of model training and prediction, simple data relationship often leads to good model performance. Taking the results of BFGS as example, the MAE, MAPE, RMSE, and SDE of the BFGS model on D1 are 73.9222, 0.0087%, 0.0109, and 0.0109, respectively; the MAE, MAPE, RMSE, and SDE of the BFGS model on D2 are 6.0050, 0.0006%, 0.0008, and 0.0020, respectively; the MAE, MAPE, RMSE, and SDE of the BFGS model on D3 are 24.5034, 0.0053%, 0.0227, and 0.02223, respectively.
c. The BFGS model performs best in combination with the predicted results of three datasets. It can be seen from the figures that the BFGS model has a good prediction effect on the overall trend of the three datasets. The main disadvantage of it is that the prediction effect of wave peak and trough is not good. Taking dataset D1 as an example, it can be seen from the data in Table 10.5 that the MAE value of BFGS model's prediction results on dataset D1 is large, but the other three indices are relatively small. Combined with Fig. 10.17, it can be seen that the predicted value of BFGS at this time has a poor prediction effect on wave peaks and troughs, but has a good prediction effect on the overall trend.
d. The prediction error of the RF model is mainly caused by overfitting. As can be seen from the data in Table 10.5, MAPE value is relatively large in the evaluation index of the RF model's prediction results, while the other three indices are relatively small. Combined with the figures, it can be seen that it is due to the overfitting phenomenon caused by the RF model. Taking dataset D2 as an example, the prediction results of the RF model fluctuate around the original data, and the overfitting phenomenon is serious. The phenomenon of overfitting may be caused by the high parameter setting of RF model.
e. The GRU model has the lowest prediction accuracy among the three prediction models. It can be seen from the figures that the GRU model has a poor prediction

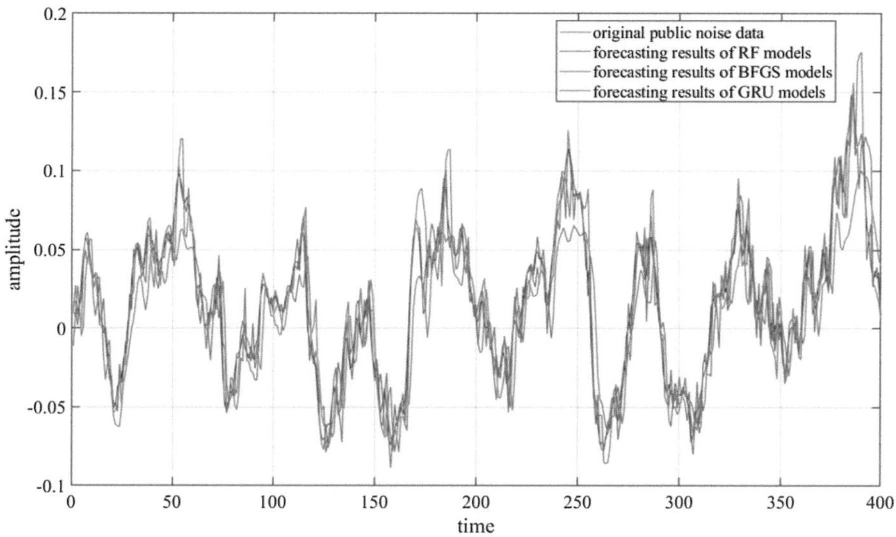

Fig. 10.17 Forecasting results of proposed models for D1

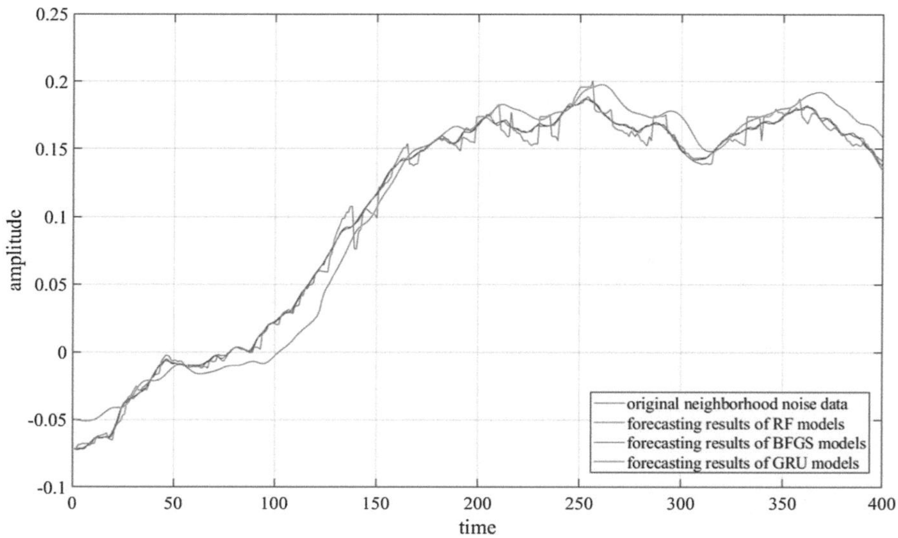

Fig. 10.18 Forecasting results of proposed models for D2

effect on the peaks and troughs of the three datasets. Moreover, the GRU model has serious phase lag effect. This conclusion is also reflected in the data in Table 10.5. The reason may be that the GRU model has too few neurons. The optimization algorithm can be used to improve the parameter setting of the GRU model to improve the prediction accuracy.

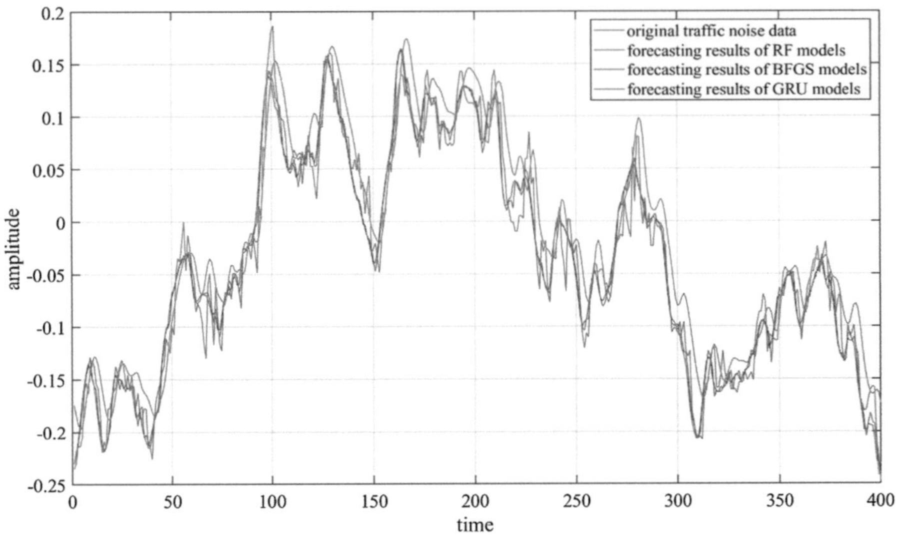

Fig. 10.19 Forecasting results of proposed models for D3

Table 10.5 The comprehensive forecasting performance indices of proposed models

Model	Step	MAE	MAPE (%)	RMSE	SDE
RF	D1	0.0136	167.9111	0.0180	0.0180
	D2	0.0181	64.9091	0.0236	0.0234
	D3	0.0044	11.5155	0.0058	0.0057
BFGS	D1	73.922	0.0087	0.0109	0.0109
	D2	6.0050	0.0006	0.0008	0.0020
	D3	24.5034	0.0053	0.0227	0.0223
GRU	D1	0.0202	329.6954	0.0249	0.0232
	D2	0.0129	70.1137	0.0146	0.0110
	D3	0.0258	110.8410	0.0322	0.0319

10.8 Conclusion

In this chapter, three kinds of algorithm models are adopted to conduct prediction experiments on different types of noise, so as to analyze the sensitivity of different algorithm predictors to different types of noise and the feasibility of predicting noise. The three algorithm models studied in this chapter are the RF model, BFGS model, and GRU model. The noise data types selected in this chapter include public noise, neighborhood noise, and traffic noise. In the prediction experiment, historical noise data are used as input and future noise data as output to carry out model training and testing. The following conclusions can be drawn:

a. The BFGS model is the optimal prediction model with the highest prediction accuracy and the best prediction timeliness.

b. The prediction results of the RF model have overfitting phenomenon. The prediction results of GRU model have serious phase lag effects. These conditions lead to the low prediction accuracy of these two models for noise data. However, the accuracy of these two models may be improved by properly setting the model parameters.
c. The three prediction models studied in this chapter have the best effect on the prediction of neighborhood noise. This may be because the change of neighborhood noise data is relatively gentle and the internal characteristic information of data is relatively simple.

References

Acharya UR, Oh SL, Hagiwara Y, Tan JH, Adeli H (2018) Deep convolutional neural network for the automated detection and diagnosis of seizure using EEG signals. Computers in Biology 100:270–278

Aimal, S., Javaid, N., Islam, T., Khan, W. Z., Aalsalem, M. Y., & Sajjad, H. (2019). An efficient CNN and KNN data analytics for electricity load forecasting in the smart grid. In *Workshops of the International Conference on Advanced Information Networking and Applications* (pp. 592–603). Springer.

Anthopoulos, L., Janssen, M., & Weerakkody, V. (2019). A unified smart city model (USCM) for smart city conceptualization and benchmarking. in *Smart cities and smart spaces: Concepts, methodologies, tools, and applications* (pp. 247–264). IGI Global.

Auger N, Duplaix M, Bilodeau-Bertrand M, Lo E, Smargiassi A (2018) Environmental noise pollution and risk of preeclampsia. Environmental Pollution 239:599–606

Badem H, Basturk A, Caliskan A, Yuksel ME (2018) A new hybrid optimization method combining artificial bee colony and limited-memory BFGS algorithms for efficient numerical optimization. Applied Soft Computing 70:826–844

Boggs PT, Byrd RH (2019) Adaptive, limited-memory BFGS algorithms for unconstrained optimization. SIAM Journal on Optimization 29(2):1282–1299

Borovykh A, Bohte S, Oosterlee CW (2019) Dilated convolutional neural networks for time series forecasting. Journal of Computational Finance, 22(4):73–101

Chang D, Sun S, Zhang C (2019) An accelerated linearly convergent stochastic L-BFGS algorithm. IEEE transactions on neural networks and learning systems, 30(11):3338–3346

Couronné R, Probst P, Boulesteix A-L (2018) Random forest versus logistic regression: A large-scale benchmark experiment. BMC Bioinformatics 19(1):270

Dewey RS, Hall DA, Guest H, Prendergast G, Plack CJ, Francis ST (2018) The physiological bases of hidden noise-induced hearing loss: Protocol for a functional neuroimaging study. JMIR Research Protocols 7(3):e79

Ge F, Ju Y, Qi Z, Lin Y (2018) Parameter estimation of a gaussian mixture model for wind power forecast error by Riemann l-bfgs optimization. IEEE Access 6:38892–38899

Grange SK, Carslaw DC, Lewis AC, Boleti E, Hueglin C (2018) Random forest meteorological normalisation models for Swiss PM 10 trend analysis. Atmospheric Chemistry Physics 18(9): 6223–6239

Ishwaran H, Lu M (2019) Standard errors and confidence intervals for variable importance in random forest regression, classification, and survival. Statistics in Medicine 38(4):558–582

Javaherian M, Abedi A, Khoeini F, Abedini Y, Asadi A, Ghanjkhanloo EJG (2018) Survey of noise pollution in Zanjan, and comparing them with standards. Journal of Applied Science 1(1): 01–08

Khan J, Ketzel M, Kakosimos K, Sørensen M, Jensen SS (2018) Road traffic air and noise pollution exposure assessment–A review of tools and techniques. Science of the Total Environment 634:661–676

Krizhevsky A, Sutskever I, Hinton GE (2012) Imagenet classification with deep convolutional neural networks. Advances in Neural Information Processing Systems 2:1097–1105

Kumar A, Kumar P, Mishra RK, Shukla A (2018) Study of air and noise pollution in mega cities of india. In: Environmental pollution. Springer, New York, pp 77–84

Liaw A, Wiener MJRN (2002) Classification and regression by randomForest. R News 2(3):18–22

Liu H, Duan Z, Han F-Z, Li Y-F (2018) Big multi-step wind speed forecasting model based on secondary decomposition, ensemble method and error correction algorithm. Energy Conversion Management 156:525–541

Meng X, Bradley J, Yavuz B, Sparks E, Venkataraman S, Liu D et al (2016) Mllib: Machine Learning in Apache Spark 17(1):1235–1241

Purwaningsih NMS, Alli MSA, Shams OU, Ghani JM, Ayyaturai S, Sailan AT et al (2018) Analysis of noise pollution: A case study of Malaysia's university. Journal of International Dental 11(1):330–333

Rastegari M, Ordonez V, Redmon J, Farhadi A (2016) Xnor-net: Imagenet classification using binary convolutional neural networks. In: European conference on computer vision. Springer, Cham, pp 525–542

Sengers F, Späth P, Raven R (2018) Smart city construction: Towards an analytical framework for smart urban living labs. In: Urban living labs. Routledge, New York, pp 74–88

Singh D, Kumari N, Sharma P (2018) A review of adverse effects of road traffic noise on human health. Fluctuation 17(1):1830001

Yang Z, Wang J (2018) A hybrid forecasting approach applied in wind speed forecasting based on a data processing strategy and an optimized artificial intelligence algorithm. Energy 160:87–100

Yuan Y-x JI (1991) A modified BFGS algorithm for unconstrained optimization. Numerical Analysis 11(3):325–332

Yuchi W, Gombojav E, Boldbaatar B, Galsuren J, Enkhmaa S, Beejin B et al (2019) Evaluation of random forest regression and multiple linear regression for predicting indoor fine particulate matter concentrations in a highly polluted city. Environmental Pollution 245:746–753

Zahid M, Ahmed F, Javaid N, Abbasi RA, Kazmi Z, Syeda H et al (2019) Electricity price and load forecasting using enhanced convolutional neural network and enhanced support vector regression in smart grids. Electronics 8(2):122

Zhu X, Du X, Kerich M, Lohoff FW, Momenan R (2018) Random forest based classification of alcohol dependence patients and healthy controls using resting state MRI. Neuroscience Letters 676:27–33

MIX
Papier aus verantwortungsvollen Quellen
Paper from responsible sources
FSC® C105338

If you have any concerns about our products,
you can contact us on
ProductSafety@springernature.com

In case Publisher is established outside the EU,
the EU authorized representative is:
Springer Nature Customer Service Center GmbH
Europaplatz 3, 69115 Heidelberg, Germany

Printed by Libri Plureos GmbH
in Hamburg, Germany